Fueling Culture

Fueling Culture

101 Words for Energy and Environment

Imre Szeman, Jennifer Wenzel, and Patricia Yaeger
Editors

FORDHAM UNIVERSITY PRESS
NEW YORK 2017

Fordham University Press also publishes its books in a variety of electronic formats. Some content that appears in print may not be available in electronic books.

Visit us online at www.fordhampress.com.

Library of Congress Cataloging-in-Publication Data available online at http://catalog.loc.gov.

Printed in the United States of America

19 18 17 5 4 3 2 1

First edition

for Patsy
for Helen

CONTENTS

HOW TO USE THIS BOOK

Our compendium of keywords is organized alphabetically. In addition to listing the keywords in the Table of Contents, we offer some further aid to navigation with two systems of cross-reference. When one of our contributors uses another *Fueling Culture* keyword in his or her discussion, we indicate the link by formatting the term in SMALL CAPS. Also, at the end of each entry, you will find a "See also" section with a list of keywords that intersect or overlap with it in less obvious ways—sometimes ironically and, in a few cases, humorously. In the e-book version, the keywords are hyperlinked to their destinations for ease of navigation through the book.

Both Jennifer Wenzel's Introduction and Imre Szeman's Afterword offer maps of the terrain that our contributors stake out in *Fueling Culture*. In so doing, they draw out some conceptual threads that you might use to guide your reading. These include:

- important geographical sites and paradigmatic spaces of energy production, consumption, and conflict;
- the many substances and forces that humans have used to produce energy;
- several technological developments associated with energy and its infrastructure;
- the myriad unwanted side effects and unintended consequences of energy extraction and use;
- several forms of cultural production; and
- a slew of abstract nouns, many with the Latinate suffix *-tion*, which touch upon various affects, social formations, and political predicaments of energy.

This book will offer insight no matter how you use it. Choose your own path. Plug in. Let the sparks fly.

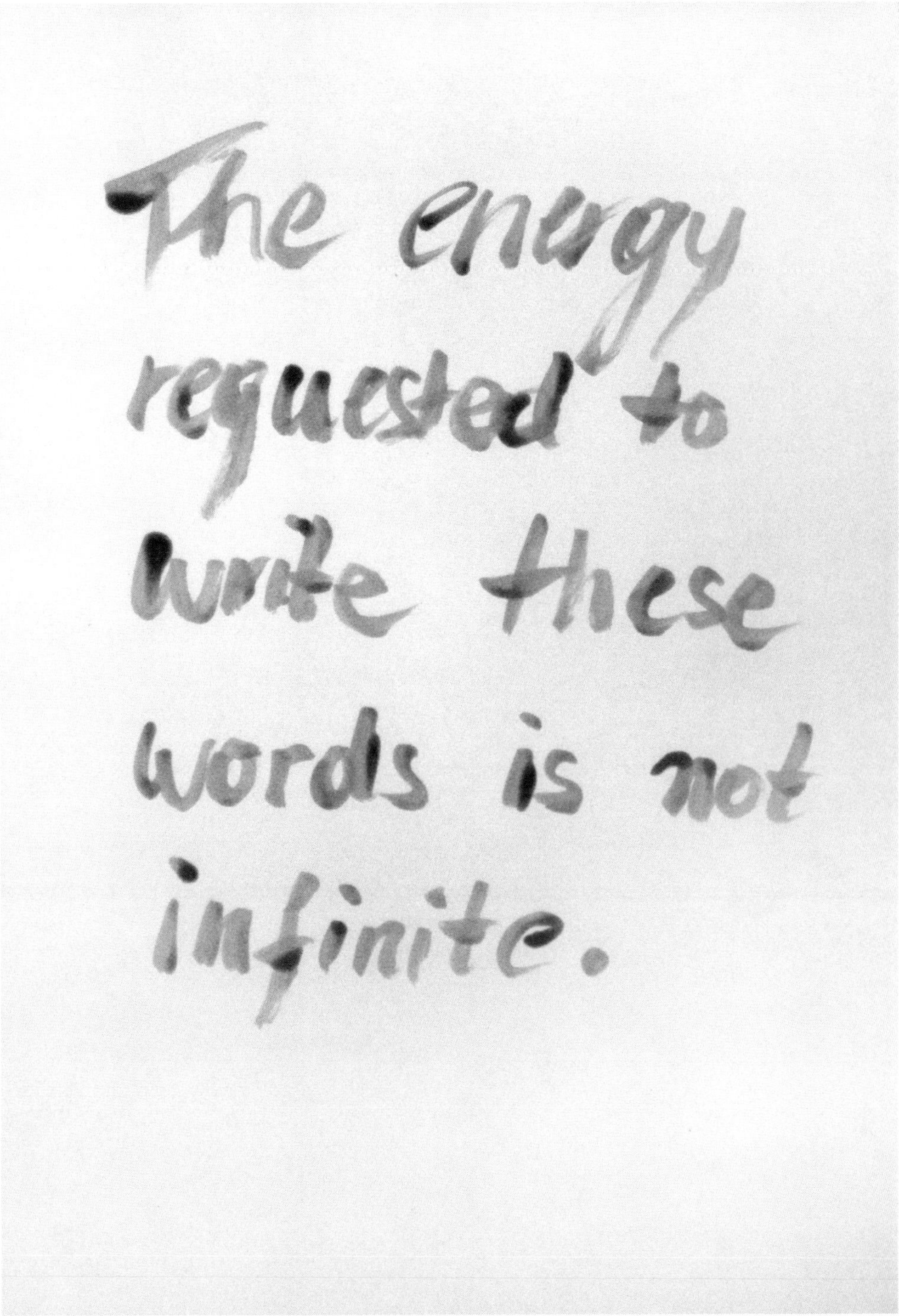

The energy
requested to
write these
words is not
infinite.

Figure 1. "Infinite." (Ernst Logar)

Introduction

Jennifer Wenzel

One of the first acts of armed struggle undertaken by Umkhonto we Sizwe (MK), the military wing of the African National Congress (ANC) that launched on December 16, 1961, was the dynamiting of an electrical pylon in Durban and a power station near Port Elizabeth.[1] Launched with a series of coordinated attacks, the sabotage campaign of 1961–63—whose targets included major infrastructure as well as pass offices and other government buildings—resulted in the arrest of Nelson Mandela and other ANC leaders, who were charged with sabotage and treason and sentenced to life in prison. When the anti-apartheid struggle revved up again in the wake of the 1976 Soweto Uprising, repeated targets of sabotage included electrical pylons, oil refineries, PIPELINES, and fuel depots; coal and petrol rail tankers (as well as the railways and buses more broadly); the electrical grid's power stations, transformers, and power lines; and even NUCLEAR power plants. When comrades and historians describe these targets as "economic," they generally intend to distinguish them from government targets—police stations, South African Defence Force installations, and pass offices—or to insist on the movement's rejection of human

1. December 16 is a date of overdetermined significance in the history of South Africa. It has loomed large in Afrikaner nationalism: celebrated first as Dingane's Day, and later as the Day of the Vow or Day of the Covenant, it was the date of a major victory of Voortrekkers over Zulu warriors at the Battle of Blood River in 1838. But precisely because of this significance in Afrikaner nationalism, activists have used December 16 to launch resistance to segregation and apartheid, dating back as far as 1910. In post-apartheid South Africa, December 16 is the Day of Reconciliation.

targets and its commitment to avoiding injury and loss of life. But this tactic—described in the 1980s as a campaign to make the black townships, and South Africa more broadly, "ungovernable"—is perhaps better understood as an attack on infrastructure and, more specifically still, as an attack on the infrastructure of energy.

Part of what fueled this tactic was the racialization of access to energy in apartheid South Africa. Most black South Africans lived, perforce and by force, off-grid. Although the legendary jazz culture of Sophiatown in the fabulous 1950s might be described as "electric," it was, in fact, illuminated mostly by KEROSENE lanterns and candles, as is evident in the shebeen scene of Lionel Rogosin's film *Come Back, Africa* (1959), where Sophiatown intellectuals Can Themba and Lewis Nkosi debate the virtues and limits of liberalism until Miriam Makeba arrives to light up the room with a song. Rogosin shot *Come Back, Africa* illicitly, and the remarkable footage early in the film, from deep underground in a mine shaft, inadvertently reveals the stark contrast between ELECTRICITY haves and have-nots in apartheid's energy regime. Within South Africa's minerals-energy complex (Fine and Rustomjee 1996), the mines and industrial smelters were the beneficiaries of the world's cheapest electricity (much of it generated with particularly inferior and dirty coal, subsidized by the World Bank), while electrification was either denied to the black masses as they were forcibly moved around the country or, after the mid-1980s, sold to some of them at exorbitant rates (Bond and Ngwane 2010, 204).[2]

From the slantwise angle of a downed pylon, it is possible to understand the anti-apartheid struggle as a war waged on—and for—ENERGY. Such a counter-history would recognize that, in addition to dignity, equality, democratic citizenship, a redistribution of resources, and the right to live, work, travel, play, love, and marry freely, where or with whom one might choose, the struggle was also for light at the flick of a switch, water at the twist of a tap, and a proper roof over one's head.[3] To live, as Niger Delta poet Ogaga Ifowodo writes memorably from another site of militant struggle over energy and its infrastructure, within petromodernity's "chain of ease" (Ifowodo 2005, 5). In the early 2000s, Soweto once again became a site of struggle over energy, this time with protests at the ANC government's failure to deliver to the politically liberated masses the basic services of electricity, water, and shelter. With cheap, subsidized electricity still flowing to the mines, and more than 60 percent of residential customers in Soweto having their service cut off for nonpayment of electric bills at rates several times what the mines paid, the Soweto Electricity Crisis Committee spearheaded a new campaign of defiance—reprising the Defiance Campaign of the 1950s, this time against undemocratic, neoliberal "cost recovery" measures rather than apartheid's unjust laws (Fiil-Flynn 2001, 17). Nonprofessional electricians began re-

2. In Soweto, for example, widespread electrification came only in the 1980s and was then drawn into boycotts on paying for municipal services (Bond and Ngwane 2010, 199).

3. Among the desiderata outlined in the 1955 Freedom Charter: "Slums shall be demolished, and new suburbs built where all have transport, roads, lighting, playing fields, crèches and social centres."

attaching dwellings to the grid, thereby restoring service that had been cut off (Bond and Ngwane 2010, 200–1). The national utility Eskom found itself without the infrastructural capacity to meet the energy needs of the new and democratic South Africa, as opposed to the privileged few. As in the 1980s, when comrades burst out of the townships and took the struggle to city streets to make South Africa ungovernable—and to demonstrate that apartheid was untenable, we might even say unsustainable—Eskom's lack of capacity became apparent to the nation as a whole in early 2008, with unscheduled load shedding (in American parlance, rolling blackouts) darkening entire cities at once. Once again, but in a very different vein, the struggle-era call and response slogan—*Amandla! Awethu!* Power! To the People!—was in the air.

Since the moment in 1911 that Winston Churchill decided that the British navy should run on oil rather than coal, we have become accustomed to thinking of geopolitics as another name for struggles over the control of energy, and particularly oil. But the Durban pylon sabotaged by MK, and the subsequent attacks on, and fight for, the INFRASTRUCTURE of energy in South Africa, make clear the stakes of more local struggles involving varied energy sources. Such struggles are deeply embedded in the texture of everyday life, even the nearly bare lives of millions living in a state of emergency, when something so seemingly neutral and apolitical as an electrical grid was riven by race and disrupted by militant politics. We offer this account of the importance of energy in South Africa, and its legibility in *Come Back, Africa*, in order to set in motion the major terms that animate *Fueling Culture*: energy, culture, history, and politics. The reprise of the "Amandla! Awethu!" slogan in the post-apartheid era is a suggestive example of our aim to develop the critical and imaginative capacity to think, at a range of scales, between the two senses of *power*.

Fueling Culture grew out of an editor's column in *PMLA*, where Patricia Yaeger invited several scholars of literary and cultural studies to speculate about the significance of energy in their period or field of specialization. Here we continue and expand this inquiry in several ways. We have brought together more than one hundred scholars, practitioners, and activists, from around the world and across the humanities and social science disciplines, who offer brief meditations on keywords related to energy. Our contributors explore the significance of energy across the social, political, and cultural spectrums (considering its role in the world-historical processes of industrialization, decolonization, modernization, globalization, and digitization) and at a variety of scales, from the global to the intimacies of the body, both human and nonhuman. The volume attends to place-specific concerns as well as the more far-flung spatial relations entailed in different forms of energy use, and to the relationships among energy, accumulation, modes of production, and inequality. Our contributors consider how the humanities and social sciences can rethink the relation of energy to specific places (e.g., CANADA, NIGERIA, RUSSIA, TEXAS), to particular historical periods (e.g., early modern, Enlightenment, ANTHROPOCENE), and to disciplinary protocols of interpretation (i.e., how to "read" for energy and how energy makes reading, as part of a broader sociology of culture, possible).

Fueling Culture's compendium of keywords offers something more than a catalog or encyclopedia of existing knowledge. We asked our contributors to be suggestive and exploratory rather than purely informative or summative: to stretch our thinking by telling us what we do not quite know about energy as the source and limit of culture. Mindful of the contemporary predicament that Imre Szeman describes as the impasse of "know[ing] where we stand with respect to energy" and environment but being unable to act (or to take action at a scale and scope adequate to the challenges we face) (Yaeger et al. 2011, 324), *Fueling Culture* aims to chart and venture beyond the LIMITS of current discourse—much of which focuses on the irresolvable contradictions of dependence upon unsustainable energy forms and is often articulated in the key of CATASTROPHE. Resisting the not-so-implicit imperative to find "solutions" in the face of crisis, we intend the form of this volume—brief, risk-taking think pieces aimed to open up further thought—to bring a collective, constellated intelligence to bear on the intersections among energy, environment, political economy, and cultural imagining. Our contributors consider *culture* both in the aesthetic/humanities-based sense of literary, visual, and other forms of poesis, as well as the social science sense of the broader forms of collective human experience. Energy fuels culture in both of these senses of the term.

Taken together, the arguments and methodological experiments in *Fueling Culture* set an agenda for an emergent academic field of energy studies. We aim to spark new insights into the social circulation of energy and the importance of energy for critical investigations and interpretations of history and culture today: thinking about the intersections between ENERGY REGIMES and cultural production allows us to understand both in new ways. "ENERGY SYSTEMS are shot through with largely unexamined cultural values, with ethical and ecological consequences," Stephanie LeMenager writes in *Living Oil* (2014, 4). We sought out contributors like LeMenager who have begun to examine and articulate these (and other) consequences of energy, but we also challenged interesting thinkers in a range of disciplines to explore what difference energy might make to the kinds of questions they tend to ask. Put to work in this collection are insights from political economy, political ecology, environmental history, literary and cultural studies, media studies, postcolonial theory, globalization studies, and materialisms old and new, including "thing theory" and actor network theory. This constellation of voices—including those of artists and activists whose perspectives range beyond the conventional academic disciplines—constitutes a significant rethinking of what it means to do interdisciplinary intellectual work.

Our guiding assumption has been that a more complete and complex understanding of energy pasts is indispensible in confronting the energy challenges of the present and near FUTURE. While all of our contributors inhabit the petromodern present, *Fueling Culture* is not (only) a book about oil or climate change, not merely concerned with the current state of affairs and its uncertain future. After all, resource depletion is not a new phenomenon, nor is energy anxiety—the paralyzing knowledge that the particular forms of energy on which lives and livelihoods depend may be harmful, unsustainable, inaccessible, exhaustible, or, more likely, some combination of these. The end of oil, whenever and however it arrives (if, in fact, it does arrive), will not be the first transition to a new energy regime.[4] We aim to bring insights about past energy crises and energy transitions to bear on our current challenges, no matter how unprecedented they may be, and thereby to defamiliarize the anxious present. How has our relation to energy changed over time? What differences do specific energy sources make to human values and politics? How have changes in energy resources transformed culture? What insights do earlier energy transitions—like that from WOOD to COAL—offer for our current situation? How do questions of energy become legible in moments of CRISIS?

In other words, our collective inquiry is deeply historical in at least three ways: first, we are broadly curious about the myriad substances and forces from which humans have produced energy, including dung, tallow, plant oils, wood, charcoal, water, steam, coal, whale oil, kerosene, petroleum, natural gas, nuclear, biofuels, solar, wind, wave, and our own labor power. Second, we are interested in what happens to previous understandings of how history works when questions of energy become central. What is the role of energy availability, scarcity, and profligacy in historical change? Does a focus on energy lead us to revise received historical narratives or merely entrench their power (as, say, might be argued about the reframing of the European Enlightenment as the onset of the Anthropocene geological epoch)? Where does energy fit into the Marxian trinity of land, labor, and capital?[5] How do we understand energy's role in the relations among base, superstructure, and ideology? How is access to energy (or *energy poverty* as a lack thereof) or vulnerability to the harms of its extraction, production, and consumption a mode of social difference and inequality that we might consider alongside those of race, class, and GENDER? And, perhaps most important, how can a more explicit focus on the difference that energy makes in our understanding of history transform previous ways of knowing rather than merely adding to them?

A crucial example of energy's capacity to transform what and how we know is its relationship to notions of historical periodization, the third way in which our inquiry is deeply historical. Climate change asks us to understand the history of modernity anew, as a history of rapidly increasing emissions of CO_2 and other greenhouse gases, which fueled directly, as historians are beginning to argue, such developments as urban settlement, the Enlightenment, industrialization, organized labor, and democracy itself. In *Carbon Democracy* (2011), Timothy Mitchell compares wood, coal, and oil in terms of the forms of social and political organization they entail. In a series of influential essays, Dipesh Chakrabarty (2009, 2012, 2014) reframes the European Enlightenment's "age of freedom" as the onset

4. Technological developments in the past decade—like hydraulic fracturing, horizontal drilling, and steam or acid injection—have made previously inaccessible petroleum deposits newly accessible (even if expensive and risky) to extract, thus prompting significant recalculations of the peak oil calendar. Given the environmental consequences of extraction and combustion for the earth's oceans and atmosphere, it is possible to think that the most pressing problem with oil is that there is not too little but rather too much left to burn. For a discussion of the shift from peak oil to "Tough Oil," see LeMenager (2014, 3).

5. For the beginning of an answer, see Diamanti (2015).

of an energy-intensive (initially steam-powered) economy that inaugurated what many see as a new geological epoch, the Anthropocene. Kenneth Pomeranz (2000) identifies access to coal as a factor that explains the "great divergence" between rates of economic growth in Europe and China in the nineteenth century. Consider also Michael Pollan's (2006) account of the shift from the sun to fossil fuels as energy inputs in the twentieth-century industrialization of food. As Imre Szeman asks incisively, "What if we were to think about the history of capital not exclusively in geopolitical terms, but in terms of the forms of energy available to it at any given historical moment?" (2007, 806). As a counterpart to a Marxian narrative organized around modes of production, what would a modes of combustion narrative look like? What difference do various social, material, technological, and spatial relations in the production of energy make in particular periods? How is an energy transition—the shift from one dominant mode of energy production to another—shaped by and also shaping other social, ecological, and economic pressures and other kinds of transition that have been more central to familiar historical narratives? And how do we make historiographic (and political) sense of the fact that the aftermath and afterlife of this carbon history will remain with us and those who come after us, in the not-yet realized atmospheric effects of a few centuries of burning fuels that it took millions of years to fossilize?

The multiscalar approach of *Fueling Culture* juxtaposes these new grand narratives of a succession of energy regimes with varied reports from more local contexts, which

FIGURE 2. Men cut wood from the Mangrove trees in the swamps to use for drying the fish in Bonny Island, Nigeria on June 5, 2006. With the Mobil Exxon Gas Plant across the water, fishermen arrive to drop their catches of bonga fish, gold fish, silver fish and crayfish. Because of pollution caused by the oil companies, the catches have decreased in quality and quantity. This has caused major problems of unemployment for the local young men. This scene is in the fishing village of Finima, which is a newly relocated community caused by the rapid growth of the Nigerian Liquefied Natural Gas plant on Bonny Island. None of the locals are given work within any of the gas and oil facilities on Bonny Island, which has caused widespread resentment and frustration. (Ed Kashi/VII)

demonstrate the untidiness and unevenness inherent to a history according to energy. Periodization is no simple matter when we consider that different modes of energy use persist simultaneously (the oil era is also the coal era and, for millions around the globe, also the era of dung, wood, and charcoal) not merely between different sites across the world, but within them. One of Ed Kashi's most revealing photographs of the Niger Delta (collected in Watts and Kashi's *Curse of the Black Gold* [2008]) depicts men chopping and stacking wood for local use as fuel in the shadow of a massive oil storage tank—a striking sight in this epicenter of petroleum extraction (see Figure 2).

To think beyond the paradox of this IMAGE, the concept of underdevelopment (the simultaneous generation of wealth for some and poverty for others) can help us understand how a petro-state like Nigeria is often plagued by fuel shortages. More broadly, ideas like energy simultaneity and energy poverty offer new ways to complicate Eurocentric progress narratives: they can help generate a better sense of the complexity and unevenness of the present and invite reflection on the fraught constituency of the "we" invoked here. It is not only possible but indeed necessary to begin to understand energy in terms of simultaneous scarcity and surplus. Even in this era of hydrocarbon modernity, too many people have too little access to energy, and the exorbitant consumption of a relative few will shape the future of all for millennia to come.

In addition to periodization, issues of scale are central to thinking about energy. In her entry on TALLOW, Laurie Shannon considers the moment when "small-scale, premodern practices of animal slaughter that were local and integrated into daily life" gave way to WHALING, the first globalizing energy industry. This shift from the "energy intimacy" of the household to antipodal oceans as the sites of energy production entails a radical broadening of energy's social, spatial, and ecological relations.[6] Similarly, Ken Hiltner shows how a scarcity of wood led to the "skyrocketing" (and sky-blackening) use of coal in England in the decades around the turn of the seventeenth century, so that air pollution increased substantially with the intensifying spatial concentration of London. This environmental phenomenon, Hiltner observes, is legible in the differences among literary depictions of Hell: Dante's *Inferno* lacks sulfurous imagery, while Milton's *Paradise Lost* (like London's air) is choked with it (Hiltner quoted in Yaeger et al. 2011, 316–17). These analyses of the early modern period provoke a question of urgent concern to the present: Are all forms of energy "dirty" (or otherwise problematic) when scaled up to meet demand?

A more exuberant strain in the history of energy formations tracks the dizzying equivalences and exponential multiplications associated with the use of fossil fuels, which are, after all, mineral deposits of intensely compacted organic matter and sunlight. Indeed, fossil fuels are the quintessential dead metaphor, because we tend to forget that they are made of stuff that was once alive. Timothy Mitchell offers some helpful (and remarkable) ratios: "a single litre of petrol used today needed about twenty-five metric tons of ancient marine life as precursor material, . . . [and] organic matter the equivalent to all of the plant and animal life produced over the entire earth for four hundred years [which then fossilized over millions of years] was required to produce the fossil fuels we burn today in

6. For a theorization of "energy intimacy," see Warren Cariou's ABORIGINAL entry.

a single year" (2011, 15). If using fossil fuels is, in some real sense, burning or harvesting compressed time, it also has the effect of creating and expanding space. By the end of the nineteenth century, the transition from wood to coal in Great Britain involved both massive increases in energy consumption and significant reductions in the amount of land (as woodlots) devoted to energy production: the energy derived from coal would have required forests at least eight times the size of the entire country (ibid.). This startling ratio gives a subterranean dimension to British industrial and imperial expansion in this period, when the British were also busy making use of other people's forests in colonial India, Kenya, Mauritius, and elsewhere. These ratios involving time and space entail not time-space compression so much as expansion to a planetary scale and a duration that feels infinite. Yet note the difference between them. Within a drop of petrol, Mitchell asks us to recognize the actuality of something like the temporal eternity that English poets Andrew Marvell and William Blake imagine in a drop of dew or a grain of sand. What coal enabled for nineteenth-century Britain, however, was something like a counterfactual version of the "spatial fix" that David Harvey (2001) identifies as one of capitalism's favorite ways of overcoming its contradictions by opening up new territories (e.g., colonies): coal allowed the country to run as if it had eight more Englands without actually having to have them. (Although, of course, actual spatial fixes through actual landgrabs and the founding of new little Englands and Scotlands were also afoot.)

Similar ratios attend industrialization's shift from a muscular economy to a mechanical economy in the nineteenth and twentieth centuries. By 1955, the inputs of animal and human work to US industry were dwarfed by those of fossil fuels: 0.7 and 0.9 percent, compared to 90.8 percent for coal and oil (Diamanti 2014)![7] In *Men and Machines* (1929), popular economist Stuart Chase describes this transition to mineral energy in terms that echo the counterfactual multiplication of British landmass described above: it was as if "'a billion wild horses'" had been harnessed and put to work (quoted in Bob Johnson 2014, 21). Chase's analogy is not mere wild imagining, of course; in the shift from muscles to minerals as the source of industrial and locomotive energy, the horsepower became (and remains) an important unit of measure. Since the early twentieth century, attempts to quantify that mineral energy in terms of the human capacity for work have been persistent, statistically problematic, and ideologically troubling.[8] "'Three billion hard working slaves'"—or "'the service equivalent of thirty servants'" for every inhabitant of the United States—was the calculation that Smithsonian minerals specialists Chester G. Gilbert and Joseph Ezekiel Pogue arrived at in *Power: Its Significance and Needs* (quoted in Bob Johnson 2014, 41). Gilbert and Pogue published their museum bulletin in 1918, barely half a century after the Emancipation Proclamation had freed more than three million enslaved humans in the United States from the "'wear and tear and hopelessness of a servile life,'" of which, Gilbert and Pogue wrote, fossil-fueled machines as a source of captive labor "'knew noth-

7. Diamanti cites Frederick Dewhurst, *America's Needs and Resources* (1955).
8. See Bob Johnson (2014, 190n2) for a detailed explanation of the statistical difficulties of converting from horsepower to energy slaves, particularly after the onset of the automobile age.

ing'" (ibid.). Buckminster Fuller was more enthusiastic in his "World Energy" map that graced the cover of *Fortune* in February 1940: "Mechanization, the harnessing of energy, is man's answer to slavery," Fuller declared as he compared the earth's human population of two billion with its quantity of "inanimate energy slaves," nearly thirty-seven billion. As the master of more than half of these slaves, "an army of 20,000,000,000," the United States was leading the way.

Liters to oceans; years to centuries and eons; mines and wells to nonexistent woodlots, imaginary horses, and armies of inanimate slaves: at some point, the equivalences calculated in this arithmetic give way to an alchemy that turns dirty energy to gleaming gold. They enable an economy and infrastructure of the as-if, where one reaps the benefits of resources that one does not actually have. These magical equations help to explain the emergence of what I theorize elsewhere (drawing on the observations of Ryszard Kapuściński on Iran, Fernando Coronil on Venezuela, and Michael Watts on Nigeria) as petro-magic, which promises wealth without work, progress without the passage of time: all surplus! all the time! (see Wenzel 2006 and 2014b). And now that we have entered what Cecily Devereux (2014) wittily calls the "rearview mirror stage" of late petromodernity, which forces us to confront the necessity and difficulty of an energy transition toward some as-yet undetermined alternative fuel source (call it UNOBTAINIUM), a new abolitionist impulse casts an anxious eye at the costs, complicities, and constraints entailed in being human masters of so many energy slaves.[9]

This is the rub evoked in poet Ogaga Ifowodo's suggestive image, "the chain of ease." The speaker of the poem describes a Niger Delta scene with

> petrol and paraffin piped away
> from rotting dugouts and thatched huts
> to float ships and fly planes,
> to feed factories and the chain of ease. (2005, 4)

Note the multivalence of this metaphor, which entails both freedom and constraint, desire and complicity, the gear of the motor and the shackle of the slave. At the dawn of the twentieth century, anxiety about the displacement of human labor by machines in the United States was palpable in Gilbert and Pogue's *Power*; "fossil fuels had destabilized on a basic somatic level both Americans' access to work and the modern body's relationship to its material world," argues Bob Johnson in *Carbon Nation* (2014, 41–42). At the dawn of the current century, by contrast, the notion of "energy slaves" evokes anxiety about the putative masters' dependency on an unsustainable, unethical system in which it is the earth itself, rather than the individual laborer, that each day shows more signs of what Gilbert and Pogue called "wear and tear." Writing from a region that has seen successive waves of energy commodities being extracted and taken elsewhere—slaves, palm oil, petroleum—Nigerian historian G. Ugo Nwokeji notes that the task of the abolitionist movement was to transform slave labor from a seeming economic necessity to a moral scourge; "a time may

9. See, for example, Nikiforuk (2012), Jancovici (2013), and Mouhout (2011).

come," he imagines, "when oil will be viewed in a manner not unlike eighteenth-century slavery, the greenhouse gases emitted from hydrocarbons perhaps akin to slave-produced sugar, and free labor as a parable for renewable energy" (2008, 65).[10] The eight imaginary Englands that coal opened up in the nineteenth century have a chastening contemporary analogue in another somatic metaphor, the ecological footprint (precursor of the carbon footprint). This metric of biocapacity calculates specific forms of human demand placed on land, water, and atmosphere in terms of how many earths would be required to support all of humanity with environmental services at that level of demand: the typical US lifestyle, if adopted by all humans, would require more than four earths (Wackernagel and Rees 1996). For a time, coal allowed the British economy to grow as if it had additional Englands, but we now confront the limits to such growth: there is only one earth.

The counterfactual excesses and alchemic imaginings in the history and ideology of energy formations can seem stranger than FICTION, more evocative than poetry, more vivid than film: in more ways than one, energy fuels the imagination and powers cultural production. As literary and cultural critics, the three of us are particularly fascinated with the import of energy for understanding literary history and aesthetic movements and moments. As Patricia Yaeger asks in her *PMLA* editor's column, "Instead of divvying up literary works into hundred-year intervals (or elastic variants like the long eighteenth or twentieth century) or categories harnessing the history of ideas (Romanticism, Enlightenment), what happens if we sort texts according to the energy sources that made them possible?" (2011, 305). Is it possible to periodize cultural production according to dominant modes of energy use or moments of energy transition?

The proliferation of *petro-* as a prefix—petrofiction, petroculture, PETROREALISM, petro-magic-realism—is implicitly a periodizing move, with the caveat mentioned above, that the "petromodern," denoting the period characterized by fossil fuel use, involves thinking simultaneously the disjunctive timescales and discrepant speeds of gradual sedimentation and fossilization in the prehistoric past, near-instant combustion and the fetish of acceleration in the hypermodern present, and environmental effects persisting into the distant future. This temporal layering poses a challenge to literary representation and to narrative's working out of cause and effect. In a seminal essay (actually a book review of the English translation of *Cities of Salt*, the first volume of Abdelrahman Munif's quintet of novels about the oil industry in a fictionalized Saudi Arabia), Amitav Ghosh wonders why the "oil encounter" has "proved so imaginatively sterile"; in other words, why the arrival of multinational oil companies (and their attendant expatriate personnel, cutting-edge technologies, and Western ideologies) in remote sites of extraction has not produced a body of novels in the same way that the spice trade inspired Luís de Camões's *Os Lusíadas*, among

10. By "free labor," Nwokeji presumably means labor that is self-proprietary and emancipated from slavery rather than the Marxian sense of labor alienated by capital from the means of production. The resonances between renewable energy and each of these models of free labor invite further thought.

other great literary works in the early modern period (1992b, 29–30). Ghosh's answer to his own question looks to the qualities of the substance, which allegorize the social and spatial relations that surround it: "Oil smells bad. It reeks of unavoidable foreign entanglements." "The history of oil is a matter of embarrassment verging on unspeakable, the pornographic" (ibid.). To attempt to write about oil, Ghosh argues, requires confronting its "slipperiness . . . , the ways in which it tends to trip fiction into incoherence" (30). The oil industry is distant and secretive, multilingual and multispatial—qualities Ghosh sees as inimical to the novel: "The truth is we do not yet possess the form that can give the Oil Encounter a literary expression" (31).

One response to Ghosh's provocation has been to compile lists of novels and other texts about oil that challenge or update his claim (made more than two decades ago now) about the paucity of petrofiction (a term he introduced). As literary and cultural critics begin to read for energy, there is always the hook of the thematic: texts and other cultural objects about energy or where energy regimes become unmistakably manifest. As with most "turns" in interpretive studies, once you start looking for it, you see it everywhere. Certain subgenres lend themselves to this kind of analysis: think of the road novel (and movie) in the US postwar era of cheap gas and new highways or the strike novel in the era of coal. Constructing an archive or enumerating a canon of what Graeme Macdonald calls "energy classics" is a necessary first step (see FICTION), beyond which remain other paths of inquiry with more potential to disrupt our understanding of cultural production and interpretation writ large.

Our more significant methodological curiosity is in identifying protocols of reading and modes of inquiry that can perceive the pressure that energy exerts on culture, even and especially when energy is not-said: invisible, erased, elided, so "slippery" (as in Ghosh's account of oil) and ubiquitous as to elude representation and critical attention. Invoking Pierre Macherey's notion of the *non-dit*, where "'what is important in the work is what it does not say,'" Yaeger wonders whether "energy invisibilities may constitute different kinds of erasures" and offers the idea of an "energy unconscious" (along the lines of Fredric Jameson's political unconscious) as a way of probing the presence and absence of energy within a given text or generic form (2011, 306, 309). Indeed, energy has been the great not-said (or, in terms of reception, not-seen, unread) in cultural production during the unprecedented and unrepeatable moment of abundant cheap energy in the past century or more. This is the great paradox of fossil fuel imaginaries: in literature as in life, oil in particular is at once everywhere and nowhere, indispensable yet largely unapprehended, not so much invisible as unseen.[11] This silence in our literary fictions has been complicit in maintaining what Szeman calls an ideological "fiction of surplus": "not only the belief

11. In marking the absence of energy as a topic of critical discussion, we are mindful of the limits of visibility/invisibility as an explanatory framework, as if the work of the critic is merely to make the previously invisible visible and thus amenable to change. Like many objects of interventionist critical attention, energy often hides in plain sight or is spectacular yet remains politically unapprehended. In other words, the work of making things visible can remain trapped within the impasse between knowledge and action that Szeman (2011) describes.

that there will always be plenty of energy to go around" but also the failure to reckon with how nearly every aspect of what passes for modern life is premised upon access to cheap and easy energy. "It is not just energy that constitutes a limit" but also our understanding of its importance (Yaeger et al. 2011, 324).

To move beyond such limits would require an approach to the intersections of cultural production and energy that understands "petrofiction" (or "petroculture" more broadly) in terms of the fundamental parameters of a period or chronotope rather than a subgenre or theme: this expansive view would take the "oil novel" as what happens to the novel writ large in the era of oil rather than (as Ghosh would have it) a subset of texts expressly concerned with the oil industry.[12] From this perspective, one characteristic aspect of petroculture—in the maximalist sense of culture in the era of fossil fuels—is its constitutive failure to reckon with the indispensability of mineral energy, the bedrock of life as we have known it.

The everyday tedium of filling the gas tank, or in earlier times the drudgery of feeding the coal stove, brings us into contact with peculiar forms of matter to which we tend to give hardly a thought. The ratios I previously cited will blow your mind if you let them, but we mostly do not let them: that is part of the secret of petro-magic's conjuring trick, the nonobvious obvious, the staggering facts that are easy to forget as we go about our oil-soaked lives. The failure to recognize or reckon with these transformations in the everyday lives of those who benefit from them and take them for granted amounts to a massive failure of the imagination, a normalization of the make-believe world of the as-if economy that ignores what is actually happening. In this sense, the characteristic mode of thinking about energy (for those with secure access to it) is actually not having to think about it: the unseen privilege of taking energy for granted. Although this perspective reframes Ghosh's notion of oil's "unspeakability" in terms of structure and ineffability rather than embarrassment or salaciousness, there is actually something scandalous about a society that so fails to apprehend the enabling conditions of its existence. (Marx's account of the commodity fetish and the necessity of its demystification is a helpful—but incomplete—model for overcoming this "massive avoidance," as Edward Said wrote of imperialism [1993, 60]: to be sure, energy and its sources are commodities, with all the mist and mystery they entail for Marx, but our hypothesis in *Fueling Culture* is that they are also something more.) Thus, our reading of energy and culture aims to understand their relations reflexively and dialectically, to gain critical purchase not only on the pressure that energy exerts on culture but also the pressure that culture exerts (or, in the case of the fiction of surplus, fails to exert) on energy regimes. How do we get beyond pointing to energy as a structuring impasse, an absent center? How might fictions of surplus give way to a reckoning with limits?

12. Note the parallel between this systemic, periodizing (rather than thematic) approach to petroculture (and, by extension, energy more broadly) and Timothy Mitchell's argument that we consider as petro-states not only countries whose political economy and political ecology have been primarily and notoriously shaped by their status as oil producers (e.g., Nigeria, Venezuela, Saudi Arabia) but also the industrial democracies that have been major oil consumers: "Without the energy they derive from oil their current forms of political and economic life would not exist." To what extent is democracy itself "carbon-based," Mitchell asks (2011, 5–6).

Beyond questions of theme, how else do genre and other matters of form look different when we consider energy? What work do particular cultural forms do in making our relation to energy visible or obscuring it from view?

This line of analysis demands new ways of thinking that bring together questions that have often been quarantined from one another: questions about aesthetic form, on the one hand, and about the material aspects of cultural production and circulation, on the other. We might understand this double vision—and construe *matters of form* in a rather different way—by revising Roland Barthes's distinction between *text* and *work* (and, unlike Barthes, refusing to privilege one over the other). That is, by attending to formal strategies, generic conventions, and intertextual resonances that shape the presence or absence of energy on the text's hermeneutic "surface," while also being cognizant of the materiality of the page itself, in the manufacture of paper and ink or screen and processor, and in the broader infrastructural and energy regimes in which cultural objects circulate, fueled by substances extracted from the earth's geological depths. Taking energy as a central analytic requires a new kind of materialist analysis, attuned to carbon as well as capital and class. How do we read for energy in relation to both the sociology and the materiality of cultural production and distribution, as well as our own intellectual practice? Among the most groundbreaking aspects of LeMenager's *Living Oil* is the appendix, a "Life Cycle Assessment of a Conventional Academic Print-Book" prepared by environmental engineer Sougandhica Hoysal, that estimates the energy inputs (excluding paper production) in the book's conception, production, and transport. Environmentally concerned readers may be relieved to learn that, even if book publishing is the fourth largest industrial source of greenhouse gas emissions, a book is a far less profligate object than a cheeseburger.

This kind of reflexive materialism can begin to grapple with the double bind of energy in relation to the interpretive disciplines: these disciplines, on the one hand, are well-suited to generate the kinds of critical insight that might help transform entrenched cultural narratives like the fiction of surplus, and yet, on the other hand, they remain deeply embedded within the built environments, infrastructures, and energy regimes that, increasingly, are themselves the objects of critique. This double bind is perhaps most palpable in the classroom and in civic debate about energy policy and proposed energy infrastructure projects. Learning to read for energy can evoke in students (at least North American ones) an entirely new kind of paralyzing liberal guilt. Charges of hypocrisy (whether directed at oneself or others) are often a depoliticizing gesture that works in favor of Big Energy and the status quo. Ever since Al Gore flew on Air Force Two to rescue the Kyoto climate treaty negotiations in 1997, pointing out that one uses significant quantities of dirty energy while advocating for a cleaner future has served to delegitimize critique by deflecting attention away from the structural predicament of inhabiting an energy regime and toward the atomizing inertia of individual "choice."[13] (In OFF-GRID, Michael Truscello shows how one remains imbricated within the grid even after deciding to unplug.)

13. The hypocrisy problem obviously precedes Al Gore; see, for example, American conservationist Aldo Leopold's famous meditation on complicity from 1932: "But have we not already

Most if not all of you reading this book are oil subjects—subjects of oil in an era when oil is often equated with life itself, as is evident in running-out-of-gas narratives like Italo Calvino's 1970s oil-shock short story "The Petrol Pump," in which an empty gas tank is conflated with the end of oil, which looks an awful lot like the end of the world. The challenge is how to grapple with the contradiction between the excesses of the fossil-fueled imagination and the imaginative failure to reckon with what it means to be a subject of oil. Confronting the ways in which energy regimes shape subjectivity and intersubjectivity can generate a critical awareness of one's embeddedness within a broader system (an ontology, even) and, indeed, of how one's own EMBODIMENT is a function of oil or other modes of combustion. Far from being a depoliticizing gesture, such an energy inventory (to adapt an idea that Edward Said borrows from Antonio Gramsci) can generate a self-knowledge of one's relation to energy that recognizes accusations of hypocrisy or impulses of disavowal for the depoliticizing distractions they are.[14]

After all, energy in the petromodern present is not merely an unpleasant ADDICTION or unfortunate NECESSITY but instead bound up just as much with pleasure and desire as with GUILT and complicity: the thrill of acceleration; the smell of gasoline, which for Marcel Proust evoked the future just as powerfully as the taste of a madeleine evoked the past (Neuman 2012). LeMenager writes incisively about the aspects of "living oil" that can only be described as "loving oil" (2014, 102).[15] These attachments to energy, which LeMenager aptly calls "ultradeep" (2014, 4), are only partially explained as bad love or "cruel optimism" (Berlant 2011). Our sense is that critical ambivalence (with some grief in the mix)[16] is a politically more productive energy affect to cultivate than naïve asceticism or abstemiousness.

The nod to politics here indicates our sense of responsibility to the perennial question "What is to be done?" voiced by Vladimir Lenin more than a century ago and reprised by our students every semester. We intend *Fueling Culture* not so much to provide answers as to expand and complicate our understanding of what the questions might be. We see this project as a contribution to the emergent interdisciplinary formations of the environmental humanities and, more recently, the energy humanities, both of which seek to bring the critical intelligence of the interpretive and imaginative disciplines to bear on some

---

compromised ourselves? . . . When I submit these thoughts to a printing press, I am helping to cut down the woods. . . . When I go birding or hunting in my Ford, I am devastating an oil field, and re-electing an imperialist to get me rubber." Voicing the perennial question of "what to do?" Leopold dismisses the option of living in the wilderness ("if there is any wilderness left") and advocates instead for creating "new cogs" within the "economic Juggernaut" to reactivate the "residual love of NATURE" (1992, 165–66).

14. In the introduction to *Orientalism*, Said quotes Gramsci: "'The starting point of critical elaboration is the consciousness of what one really is, and is "knowing thyself" as a product of the historical process to date, which has deposited in you an infinity of traces, without leaving an inventory'" (1978, 25). The language of sedimentation is fortuitous here.

15. For an exploration of the pedagogical import of such love, see Wenzel (2014a).

16. On environmental grief, see LeMenager (2014), Sandilands (2010), Morton (2007).

of the most urgent challenges of our time. Announcing the energy humanities, Dominic Boyer and Imre Szeman (2014) write: "What we energy humanists contend is that today's energy and environmental dilemmas are fundamentally problems of ethics, habits, values, institutions, beliefs and power—all traditional areas of expertise of the humanities and humanistic social sciences." (These new humanities initiatives constellated around energy and the environment are complementary but not identical: our aim in *Fueling Culture* has been to keep distinct and in tension the concepts of energy, ecology, and economy rather than collapsing any of them into the others.) With the rise of market-friendly SUSTAINABILITY discourse as a response to crisis, these interdisciplines urge dwelling critically and reflexively within problems and questions rather than rushing toward "solutions." Crisis talk about the humanities is just as rife as on energy or the environment; although scholars in these emergent formations are mindful of the precarious institutional situation of the humanities in higher education and public life, they are also attuned to, and advocates for, the humanities as a discipline, method, and habit of mind that is singularly incisive on these burning questions. Indeed, there is a dawning and not entirely unproblematic sense that since science and economics have failed us so badly, perhaps the humanities can save us[17]—witness, for example, the turn to wild, unruly imagining and unreason on the part of such seemingly mild-mannered activists as Al Gore ("it's time to out-crazy the crazy" of the hydrocarbon status quo [2009]), Bill McKibben ("this is fucked up" [2010]), or Warren Cariou (his "Tarhands" manifesto on Canadian tar sands, which mobilizes irrationality after the failure of reasoned discourse in the public sphere [2012]).

With renewed confidence and urgency, these interdisciplinary formations dedicate themselves to the work of making unforeseen connections between matter and metaphor, cultivating the capacity to read images and narratives creatively and critically, and training a supple imagination to apprehend the possible but as-yet unimagined. These reconfigured and remobilized interdisciplines can help us recognize and think through the profound ideological work that cultural objects do all the time, whether we want them to or not, while also resisting the instrumentalization of the aesthetic as a useful, pliable, and predictable tool for consciousness raising.

In this sense, we find a kind of model in Edward Said's last book, *Humanism and Democratic Criticism* (2004), in its affirmation of the possibilities of humanist self-knowledge and self-critique and its argument for the agency of the literary and of intellectuals in public life. For Said, humanism is equally a practice of reading and a practice of citizenship, a model well-suited to the commitment of the environmental and energy humanities to speak to contemporary crises and injustices at multiple scales. Our contributors to *Fueling Culture* demonstrate an attunement both to the capacity of metaphor to open up the imagination to alternative possibility and to the revelatory power of understanding how things (whether light bulbs or legislatures) actually work.

Neither climate change nor the various economic and environmental challenges surrounding energy is merely an engineering problem: these are political problems, narrative

17. See, for example, Sörlin (2012).

problems, and, ultimately, problems of the imagination. This is the unremarked contradiction in Dipesh Chakrabarty's influential analysis of the Anthropocene: for him, humans are at once so powerful as to have changed the basic physical processes of the planet and yet so puny that their traditions and conventions of political deliberation are woefully inadequate to address this predicament. We share the sense that our politics are inadequate, that we need new narratives and new imaginative capacities that can leap (like electrons?) among enormous technical challenges, engrained habits of thinking, and the textures of everyday life. The multilayered materialism that our contributors perform here can grasp the ways in which the future is already here—for some, already or still tenuous; for others, flickering with possibility—and begin to give substance to the idea that things might be, must be, otherwise.

Banff and Ypsilanti, Summer 2014

Aboriginal

Warren Cariou

When we think of Aboriginal people in the context of fuel, one common image is of Aboriginal protestors voicing resistance to energy projects like PIPELINES, DAMS, and uranium mines. One reason for the prevalence and the visibility of such protests is that Aboriginal people are disproportionately affected by energy megaprojects—in fact, it is difficult to find a megaproject that has *not* displaced indigenous people or in some way threatened their cultural and physical survival. This situation can hardly be accidental. Low population densities, ongoing histories of colonial disempowerment, and the existence of alternate (noncapitalist) value systems within Aboriginal communities make them particularly attractive targets for the incursions of large energy developments. While such projects generally create some economic activity and jobs in the affected regions, they also enact what Rob Nixon (2011) calls "slow violence": contamination of land, water, and air; health and safety crises; disruptions of the social fabric and family structures; and, perhaps most devastating, erosion of the people's connection to their land. Thus, it is not surprising that indigenous people are often in the vanguard of resistance movements that aim to stop or disrupt the fuel regimes of modernity. However, this resistance is motivated by something far more profound than NIMBYism or a simple reaction to the perceived negative consequences of development. It is instead deeply rooted in the philosophical and SPIRITUAL contexts of specific indigenous nations and their particular territories. This short entry cannot do justice to the immense variety of Aboriginal cultures and the panoply of energy-related challenges they face, but it is an attempt to map out some of the recurrent patterns in their philosophies and pragmatics of energy use, with the goal of indigenizing contemporary global practices of energy production and consumption.

Traditional Aboriginal energy-use practices are characterized by what might be called *energy intimacy*, in which every community member necessarily has direct and personal relationships with the sources of their energy. In hunter-gatherer societies, it is a matter of survival to be able to locate, process, and utilize energy sources for oneself, whether these sources are derived from WOOD, ANIMAL fat, food, or other fuels. This has philosophical and spiritual implications. Energy in such a context is based primarily upon the relationship between the people and their land, and in Aboriginal cultures this relationship is not one of mastery or objectification but rather of kinship, respect, and responsibility. Laguna writer Leslie Marmon Silko describes one aspect of this relationship when she writes, "Pueblo and Apache alike relied upon the terrain, the very earth itself, to give them protection and aid" (1998, 12). This may seem like a description of an idyllic life in NATURE, but Silko makes an important qualification when she adds, "Human activities or needs were maneuvered to fit the existing surroundings and conditions" (ibid.). Thus, while the people might receive "protection and aid" from the land, they are still expected to adapt themselves to nature rather than impose human will upon it to transform it toward their own ends. The land is depicted here not as a reservoir of resources to be exploited but as a source of gifts that humans must accept with gratitude. The reciprocity of the gift relationship results in a conception of energy fundamentally different from today's prevailing Western ideologies of energy extraction, commodification, and ownership.

In most Aboriginal cultures there is little interest in generalized concepts of energy or fuel as they are understood in Western cultures, but instead there are teachings about the vitality of all beings, including the earth itself. Therefore, energy in Aboriginal contexts is almost always about relationships and inevitably about ethics. Two of the most important teachings in many Aboriginal traditions are that no one should make demands upon nature and no one should ever waste resources by using more than immediate needs dictate. Omushkego Cree Elder Louis Bird explains his people's ethical obligation toward the environment when he says, "there were rules about respecting nature and the environment—the animals and the birds. If one of these were broken by a member of the family . . . the punishment was a retraction of the benefits from nature" (2007, 75–76). Bird goes on to tell the story of a caribou that appears to a community and says, "because I have been abused by this or that person I will not give up my life for you" (ibid., 77). Here, the primary factor in the success of the hunt is not the hunter's skill or technical knowledge but rather the community's respectful treatment of the animal, which then of its own free will decides whether or not to provide the energy-giving sustenance of its body for the use of the community. Hunting then—understood as the gathering of energy-yielding resources—is not an activity of taking but one of respectful receiving. While Bird's examples focus on animals, the code of ethics implicit in this teaching can also be generalized to a broader picture of humans' relationship to the environment. He discusses the concept of *pastahowin*, the "sin against nature" (ibid.) involved in any action that shows disrespect to the natural world, such as wasting resources or failing to give thanks for the gifts received from the land. As in the story of the caribou, the punishment for an act of pastahowin is that nature withholds further gifts from the transgressor and his or her community. Re-establishing an ethical relationship with the natural world then becomes a matter of immediate survival.

In addition to the teachings about the ethics of hunting and gathering, there are also stories in several Aboriginal traditions about people who are able to harness the energy of natural phenomena through medicine power. Louis Bird writes about Cree *mitews* or shamans who are able to shape fire "into a human form to understand it, and communicate with it, and have a command over it" (2007, 93). The emphasis here is on forming an intimate relationship with fire in order to gain the ability to use it for human ends. Similarly, Dene Elder George Blondin (2006) writes about the medicine man Teleway, who was able to make the wind blow behind his boat with such intensity that he and his fellow voyagers traveled 250 miles in one day. Blondin explains that Teleway was gifted with medicine power from the Creator, but he also had to develop his power through careful observation and intimate knowledge of wind: "He understood how the wind was created and understood the power of the wind. He knew exactly how the wind worked. He had all the necessary medicine power to take action with the wind" (2006, 108).

As Bird and Blondin both show, the spiritual medicine powers of shamans are not really supernatural but are in fact derived from particularly detailed and personal engagement with nature itself and from the cultivation of a favorable relationship with it. People with medicine power have especially close relationships with the energies of the natural world. Thus, these teachings about medicine power can be seen as further elaborations of the concept of energy intimacy.

A question that remains is whether the ethics of energy intimacy can be applied to contemporary globalized energy practices and whether modern energy regimes can be indigenized. I believe that we do not necessarily have to adopt hunter-gatherer economies in order to apply teachings that have informed the traditional practices of Aboriginal people. The industrial infinite-growth model of energy in Western cultures is clearly not viable over the long term, so it is vital that we seek alternatives wherever they can be found. Energy intimacy means that energy is always contextualized, always specific to a particular place with which the energy user must establish an intimately familiar connection. This is the opposite of the contemporary Western corporate logic of energy extraction, which by its very definition is about taking energy out of its context, turning it into a commodity that can be circulated in a global economy wherein its value is guaranteed by virtue of its sameness, its uniformity. When energy becomes decontextualized and commodified, it no longer seems to be in a relationship to us, and therefore we cease to feel responsibilities in regard to it.

This situation is where Aboriginal theories of energy might be most useful in providing a way forward. If we are able to indigenize energy practices on a large scale, it will involve becoming more connected on an intimate bodily level with the sources of our energy—understanding where it comes from and how that source location is affected when the energy is extracted, processed, and delivered to us. We will need to become like locavores, sensitive to the environmental costs and contexts of our energy. This will also require a fundamental CHANGE in the way nature is conceived and a move toward understanding energy as gift rather than as commodity.

See also: ANIMAL, COMMUNITY, EMBODIMENT, SPIRITUAL, TALLOW.

Accumulation

Daniel Gustav Anderson

For those of us who wish to carry on living—as bodies and as "us," a totality of lived relations—ENERGY is needed. How do we get enough calories to get up and do it again or to create something? Collaboration is necessary, but on what terms? For now, the terms by which people must engage with one another to meet their needs for energy are capitalist ones. "The capitalist production process," our best theorist on the matter explains, is "a process of accumulation" (K. Marx 1991, 324). Accumulation of what, by whom, and with what consequences? These questions of calorie- and capital-accumulation, production of value, and reproduction of life and violent social relations intersect.

The familiar narrative has some truth to it. For centuries, social groups collaborate to stay warm and reproduce their lives, and in doing so learn how to survive in particular contexts—this source of fuel, that source of calories, in this place. As such, repertoires are refined, societies become rich in practical knowledge—and conflicts. The accumulation of these tactics is called *culture*. Giambattista Vico's *New Science* (1999) describes how our species' need to eat is exploited strategically by those with the means to do so—whole mythoi are elaborated and enforced to maintain a strict division between those who must struggle to feed themselves and those who struggle to continue to be fed by the labors of the former and of the earth. This is not merely Enlightenment-era speculation on antiquity; today, the function Vico attributes to mythos in ancient times is performed by what the Situationist International, Raoul Vaneigem most lucidly among them, describes as the spectacle: the relentless repetition of lies made desirable in the service of capital in everyday life. Vico frames the problem correctly as a conflict between those with the means to determine how they and others feed themselves (those who accumulate the crystallized labor-energy known as capital) on one side and everyone else (those whose means to maintenance is determined by others and whose lived energy is absorbed by others as calories or as capital) on the other side. Hence, if culture is a body of knowledge for reproducing life under given and variable conditions, then culture is the stuff of conflict *now*, and this conflict can be named.

The social regime of capital accumulation produces situations in which it asserts and reproduces itself now but with little regard for FUTURE consequences and crises, such as the accumulation of greenhouse gases and the depletion of resources, the billions of bodies made to seem disposable, and hence the diminution of conditions for survival of the totality, much less the possibility for living. Capital accumulation is the engine of this EXHAUSTION. Persons working collaboratively to ensure their own survival develop tactics (culture) for getting along in this milieu, and these also accumulate. At this moment, the neoliberal moment, the planet is a theater of war. Violence is done against all those who wish for breathing to remain possible and lack the technological means (and energy) to mitigate the RISK.

Thinking conflict dialectically, Carl von Clausewitz observes that the lived realities of particular engagements do not correspond to commonplace make-believe about war—a lacuna in evidence now between neoliberal realities and neoliberal ideology. Clausewitz calls this gap *friction* (appropriately enough for an inquiry on energy) and describes it in environmental and social terms: the unpredictability of the weather and terrain; the unpredictable motivations of individual persons under command, motivations that may not align with their commander's (1976, 119–21). The accumulated friction of the present miasma is also discernable socially and ecologically—ecologically in that the "easy" oil, minerals, and timber were taken decades ago, and hence extractive industries must apply more and more capital (and use more and more energy) to drill deeper and deeper on the promise of taking less and less. The emergence of such friction was beyond the horizon of the early Marxian vision. Apart from human labor, which he had conceptualized as energy by 1846 (K. Marx 1977, 46)—note that the *Kraft* of such central Marxian concepts as *Produktivkraft* connotes energy, power, force—Marx accounts for energy in production (and, in passing, the production of energy) in terms of the cost to acquire, transport, or store fossil fuels (K. Marx 1992, 219–20), the benefits of energy conservation both for the capitalist (K. Marx 1991, 191–95) and, in utopian terms, for labor to liberate the "energy" of labor-power from appropriation by capital toward the "development of human powers as an end in itself" (K. Marx 1991, 959). Engels had thought much of this through as early as 1845 in observing working life in Manchester. However, Marx and Engels do not anticipate the problem of the scarcity of energy (fuel for machines, food for labor-power) produced by production *on production* in evidence now.

James O'Connor (1998) describes this predicament as a second kind of CRISIS of overproduction: capital's overproduction of air, water, and land pollution, coupled with its overextraction of finite and increasingly scarce resources, threatens its very capacity to reproduce itself; even if the accountants determine that a profit might appear possible, poisoned and malnourished labor cannot be made to feed nonexistent fuel to broken machines. And socially this "second" crisis of capital, coupled with the tendency of the rate of profit to fall (K. Marx 1991, 317–75) has made it necessary for capital to implement the series of violent policy interventions known as neoliberalism, as well as increasingly implausible financial innovations to reproduce the spectacle of quantifiable capital accumulation and, by this means, the lived reality of re-entrenched class divisions globally. Altogether, these crises produce the extraordinarily uneven access to means-to-living that characterizes the class nature of globalization, from which Clausewitzian friction writ large emerges, and of the precariousness that capital has engineered with regard to itself and its own (and "our") conditions of possibility. In much of the world, the absence of petrochemical fertilizers means famine for an inflated population; rising sea levels mean already-squalid homes disappear; arms and resentments freely circulate. Little imagination is needed to envision what may emerge from such a convergence. It follows that serious strategic attention must be directed to the question of "What next?" before the sooner-or-later exhaustion of the petrochemical sub-base finally undermines the capital-accumulation base and superstructure. The task is to imagine intentional alternatives around which a strategy can be pursued in opposition to this regime that is premised upon the accumulation of both capital and carbon, under the diminished, resource-scarce, and overpopulated conditions said regime has produced (see D. G. Anderson 2012a and 2012b).

Practical knowledge is necessary to such a strategy. It begins with seeing through the "metaphysical subtleties and theological niceties" of the commodity as a crystallization of the lived energies of contemporary labor and millennia of natural history (K. Marx 1990, 163–65) inclusive of fossil fuels and ancient forests, swept up gratis (K. Marx 1991, 751; D. G. Anderson 2012c, 38–42). Marx emphasizes the visual capacity as an immediate means to knowledge in *Capital*, like recognizing the true history behind an everyday lie—analogous perhaps to waking up from a dream-image as explored in Walter Benjamin's *The Arcades Project* or, for Clausewitz, establishing "a concrete recognition" of the facts of engagement when "the reports turn out to be lies, exaggerations, errors" (1976, 117). This post-Hegelian dialectic of seeing is a strategic practice of recognition with transformational capacity: refusing the given, the "commonsense," or false consciousness is, as Herbert Marcuse (1960) presents it, a political intervention. The point of coming to grips with the present moment is to contest the means to the future, not the accumulated DETRITUS of the present order (Guattari and Negri 1990, 13).

"Commonsense" tactics accept the terms of engagement—right now, the "environmental type of intervention" that is neoliberalism, as an explicitly Clausewitzian Foucault envisions it (2008, 260)—as given, and hence as desirable, and demand that one play by the given rules; think of helicopter parents hectoring their children's teachers with the knowledge their kids must get a leg up on the neighbors' to maintain their position or at least stay fed. This strategy of accommodation is informed by the false consciousness of the spectacle; it is entirely lacking in vision. Far from negating the lie, this strategy realizes the lie in practice. By contrast, tactics serving a conscious or critical strategy effect a partial or complete withdrawal from—or, better, strive for a transformation of—these social relations, such that the totality of earth-bound life can meet its needs and develop its capacities. The fact of these critical tactics shows that, perhaps, the accumulation of practice-tactics called culture is not only a site to be contested but also a means of contestation (Bloch 1986, 674–86; Trotsky 1991, 277–84).

A world without conflict of any kind may not be possible, but because of its own consequences, a world without the accumulation of capital is inevitable. What then?

See also: CHANGE, ENERGY REGIMES, LEBENSKRAFT, METABOLISM.

Addiction

Gerry Canavan

A July 16, 2010, segment on *The Daily Show*, three months after the catastrophic Deepwater Horizon oil spill in the Gulf of Mexico, offers an amusing supercut of every president since 1970 promising to eliminate US oil dependence. Barack Obama in 2010 announces that "now is the moment" to "seize control of our own destiny" back from the oil on which we depend; George W. Bush in 2006 promises to "make our dependence on Middle Eastern oil a thing of the past"; his father in 1992 asserts "there is no security for the United States in further dependence on foreign oil"; and so on, all the way back to Richard Nixon, who swore that by 1980 "the United States will not be dependent on any other country for the energy we need" (Stewart 2010).

The repeated word *dependence* suggests an intriguing medicalization of the material conditions of oil capitalism. The fourth edition of the Diagnostic and Statistical Manual of Mental Disorders (DSM-IV), which offers standard definitions of psychological disorders and pathologies for use by clinicians and psychiatrists, defines "substance dependence" as "a maladaptive pattern of substance use, leading to clinically significant impairment or distress" (American Psychiatric Association 2000, 197). From the beginning of the petroleum age, literary works have associated oil prosperity with precisely this risk of ecstatic, drunken excess; in the delirium of abundance made possible by oil there is always the possibility that one might consume too much. Among other alcoholics in his 1927 novel *Oil!*, Upton Sinclair describes "Old Chief Leatherneck, of the Shawnees," whose oil money has purchased him "a different colored automobile for each day of the week, and he figures to get drunk three times every day" (Sinclair 1927, 323). James Dean's Jett from George Stevens's *Giant* (1956)—much like his 2007 counterpart Daniel Plainview from Paul Thomas Anderson's *There Will Be Blood*—moves quickly from oil entrepreneur to oil casualty, ending the film an embarrassing wreck, drunk and disgraced at a party he has thrown to honor himself. Even the perception-altering "spice" that makes interstellar navigation possible in Frank Herbert's *Dune* (1965)—found only on the desert planet of Arrakis, which is populated by religious fundamentalists who ultimately unleash an intergalactic "jihad"—carries with it the risk of a hallucinogenic "spice trance," an overdose.

The commonplace metaphorical connection between petroleum and alcohol reimagines oil capitalism itself as a kind of abiding, society-wide moral weakness, particularly insofar as our collective level of "dependence" has continued to degenerate over decades without any material progress toward a solution. Now, as each year brings oil that is both harder to find and harder to extract than what had been available before, each year necessarily brings us closer to the end of this century-long civilizational bender. And the only thing that will be worse than our current orgiastic excess, the thinking suggests, will be the hangover we face the morning after the oil finally runs out. "Dear future generations," writes Kurt Vonnegut in a late work to the people of a future world devastated by resource scarcity and climate change, "Please accept our apologies. We were roaring drunk on petroleum" (quoted in Blais 2007).

The DSM-IV offers up several familiar criteria to define this maladaptive pattern of substance dependence that are easily applied to oil capitalism. There is *tolerance*, the need for ever-growing amounts of the substance in order to achieve the same effect, namely the smooth functioning of the economy; and of course there is *withdrawal*, the emergence of physical symptoms—in oil's case, catastrophic economic downturns—when insufficient amounts of substance are procured. And on we go, through each of the seven criteria:

3. The substance is often taken in larger amounts or over a longer period than intended.
4. There is a persistent desire or unsuccessful efforts to cut down or control substance use.
5. A great deal of time is spent in activities necessary to obtain the substance, use the substance, or recover from its effects . . . .
6. Important social, occupational, or recreational activities are given up or reduced because of substance use.

Most crucial of all, perhaps, is number 7: "The substance use is continued despite knowledge of having a persistent physical or psychological problem that is likely to have been caused or exacerbated by the substance" (American Psychiatric Association 2000, 197). Even knowing full well that the carbon release from fossil fuels is aggressively destabilizing the climate—even after all the smog, all the spills, all the recessions, all the wars—even though the oil is bound to run out anyway, whether or not we choose to adapt in time—the only thing we can think to do is drill harder.

"Dependence," then, seems like an unusually perspicacious euphemism for describing the relationship of technological modernity to oil. But the revised and updated DSM-V, released in May 2013, suggests that even the category "dependence" does not go quite far enough. In the DSM-V, the language around substance abuse has been altered significantly; the word "dependence" is abandoned in favor of a return to the once-eschewed language of "addiction." Charles O'Brien, who headed the American Psychiatric Association's Substance-Related Disorders Work Group for the DSM-V, explained the change at the 21st Annual Meeting of the APA in December 2010 (Fox 2010). The DSM-IV category of "substance abuse and dependence" had unhappily blurred the distinction between functional and dysfunctional relationships with drugs; a person prescribed antidepressants (for instance) may develop many of the markers of substance dependence, such as tolerance and withdrawal, but this relationship is not necessarily pathological. "The term dependence is misleading, because people confuse it with addiction, when in fact the tolerance and withdrawal patients experience are very normal responses to prescribed medications that affect the central nervous system," summarizes O'Brien in a later press release. "On the other hand, addiction is compulsive drug-seeking behavior"—uncontrolled, erratic, destructive, even suicidal self-harm in the name of procuring more of the drug at any cost (Curley 2010). "The adaptations associated with drug withdrawal," O'Brien writes with Nora Volkow and T. K. Li in 2006 in support of this revision, "are distinct from the adaptations that result in addiction, which refers to the loss of control over the intense urges to take the drug even at the expense of adverse consequences" (O'Brien et al. 2006, 764–65).

In the surreal, years-later coda of 2007's *There Will Be Blood*, a miniature history of the oil age, we find a metaphorical vision of the inevitable endpoint of this addictive downward spiral. The film's oilman protagonist, Daniel Plainview, is rediscovered living in decadent opulence in Hollywood in a mansion complete with a private bowling alley. He is roused from an alcoholic blackout by his nemesis, the young preacher Eli, with a terrible shout: "Wake up, Daniel Plainview! The house is on fire!" Some crisis is at hand, but Plainview cannot be moved. Ruminating sullenly on the death of God, having already exhausted through slant drilling the very oil field Eli has come to sell him, Daniel soon erupts into a sudden paroxysm of violence, ultimately beating Eli to death with a bowling pin before collapsing in defeat and shrugging, "I'm finished." Cut to black. This is one dream of how the oil age ends. No future, no hope, no possible alternatives; we are just finished.

In the DSM-V's terms, then, American presidents should clearly have spoken explicitly of addiction rather than dependence. The wars in the MIDDLE EAST, the obsessive exploitation of tar sands and dangerous deep-sea deposits, hydrofracking in the suburbs, ongoing government handouts to fossil fuel companies, the continued refusal to invest in clean energy and GREEN INFRASTRUCTURE in any serious way, the absolute denialist refusal to even discuss the transition to post-carbon economics at any point during the endless 2012 presidential campaign—this is compulsive drug-seeking behavior, this is addiction. And in this logic, perhaps, the "long emergency" of peak oil—imminent, if not already here—becomes the dreaded but longed-for moment called "hitting rock bottom": the inevitable, epiphanic final crash, which simultaneously offers up to the addict his only two possible futures: death in the gutter or healing and redemption, but only after you have admitted you have a problem.

See also: FICTION, LIMITS, PETROREALISM.

Affect

Matthew Schneider-Mayerson

We must internalize the externalities. By this I mean not that we must reform neoliberal capitalism so that global markets account for pesky "externalities" like a functional biosphere but that we must internalize and embody the consequences of our heretofore disastrous ENERGY choices. Not as theater or exercise but as a step toward action.

Reading Elizabeth Kolbert's (2014) recent book on mass extinctions, I was reminded that upon the detonation of the first atomic bomb in New Mexico in 1945, J. Robert Oppenheimer claimed that a line from the *Bhagavad Gita* sprung to mind: "Now I am become death, the destroyer of worlds." With only mild exaggeration, I would posit this sentiment as apt for our uniquely energetic age, the ANTHROPOCENE. In our energy choices (yes, choices) and daily actions, in our steady but nonetheless substantial contributions, we are responsible for the destruction of worlds big and small, near and far, human and nonhuman, existing and still to come. This is a fact. For most of us, however—for reasons that sociologists, social psychologists, and nonfiction authors have documented for over a decade now—this fact enters our minds and bodies only fleetingly: the newspaper headline on climate change or ecosystem collapse that elicits a sinking sadness, a muscular contraction that is quickly but only temporarily relieved by turning to the sports or Sunday style section. As environmental humanists have begun to explore (e.g., LeMenager 2014), this awareness is not merely intellectual but corporeal, lodged in our nerves and tendons, and it presents us with two options: to turn away or to act.

This is, then, a call for the development or cultivation of what might be termed an Anthropocene affect, "affect" referring (as Gregory Seigworth and Melissa Gregg put it) to those "visceral forces beneath, alongside, or generally other than conscious knowing, vital forces insisting beyond emotion—that can serve to drive us toward movement" (2010, 1).

To this end, we can draw lessons from the past. For most of history and all of prehistory, human beings had an intimate awareness of connections to the world around them that did not require lessons in energy consumption or ecology but was necessary for daily life. Until the eighteenth or nineteenth century (depending on geographical location and class), it was almost impossible for people to be unaware of the resources and energy (measured in human and ANIMAL work) that went into the manufacture of all the artifacts they came into contact with, be they flint arrowheads, animal-hide clothing, or toothbrushes. With the explosion of trade from distant regions and the harnessing of ancient sunlight in the form of fossil fuels, such awareness has become distant, theoretical. By the twentieth century, living within an ecology of cheap oil, Americans, citizens of industrialized nations, and, increasingly, elites and others around the world have been able to live as if energy and resources simply did not matter. This too shall pass, for better or worse.

We can also draw lessons from the present. In my work (2015) on the subculture and politics of American hard-core peak oil believers in the mid- to late 2000s, I show how they developed something resembling an Anthropocene affect. They based their life decisions on the threats of oil depletion and climate change and adopted a dissident ideology in which daily practices constructed an alternative affective landscape: driving more slowly, refusing to fly, retrofitting their homes, etc. They formed an insurgent emotional habitus, which, as Deborah Gould writes, "contains an emotional pedagogy, a template for what and how to feel, in part by conferring on some feelings and modes of expression an axiomatic, natural quality" (2012, 97). These (currently) "outlaw emotions," to borrow a phrase from Alison Jaggar (1989), take root in particular social conditions. A number of older "peakists" first developed these tendencies during the oil crises of the 1970s, when Americans were encouraged to be cognizant of energy consumption, both as citizens and as consumers. Such a historical perspective is a useful reference for our anticipation of FUTURE developments; questions about the plausibility—or, indeed, the inevitability—of future affective shifts might be answered not only with quotes from Spinoza ("no one has yet determined what the body can do" [1959, 87]) but also with the work of historians of emotion, who have shown that affective dispositions are not nearly as natural, timeless, and universal as they seem.[1]

As a result of the copious work on crowd psychology and the influence of scholars such as Teresa Brennan (2004), we tend to think of the transmission of affect via face-to-face interactions, but it should go without saying that digital NETWORKS offer ample possibilities for such exchange. In my book on the peak oil subculture, I critique the metaphor of the crowd as a representation of the typical virtual experience, but the collective construction of outlaw emotions goes a long way toward answering one of my unanswered questions: Why did so many peak oil believers in the mid-2000s radically alter their lives in preparation for the impending collapse, despite having never met another peak oil believer

1. See the work of Peter Stearns (1989), Barbara Rosenwein (2006), and William Reddy (2001).

in person (i.e., in "the real world")? One explanation would look to the isolated, rational individual gathering information and making decisions, but a focus on the spread of dissident affective networks, even across fiber-optic network cables, is perhaps a more compelling picture. How can we lubricate the transmission of such an affect?

We can also draw lessons from possible futures. For example: in Paolo Bacigalupi's twenty-third-century, post-petroleum, climate-chaos novel *The Windup Girl*, human beings have primarily returned to a somatic energy regime, once again dependent on human muscle, "the joules of men," and geneticists have resurrected fifteen-foot Pleistocene proboscideans to wind "kink-springs" that serve as energy storage units (2009, 8). Once the seas have subjugated entire regions and wait patiently beyond the dikes like invading armies, the environmental consequences of carbon consumption become palpable. When one character comes upon a room of working computers, "the amount of power burning through them makes" her "weak in the knees. She can almost see the ocean rising in response. It's a horrifying thing to stand beside" (ibid., 215). The fear and trembling of anthropogenic climate change in this passage is not just imagined but embodied, which Bacigalupi emphasizes by highlighting his character's physical proximity to the machines. These connections are reinforced by religious practices—the Environment Ministry has its own shrines around Bangkok (one of the few extant megalopolises), for example, and one of the most popular religious figures is a "biodiversity martyr," recalling the prayers of the eco-cult God's Gardeners in Margaret Atwood's MaddAddam trilogy. This radical shift in awareness of energy and environment—as this passage shows, an understanding that is affective and not merely intellectual, corporeal as well as rational—could certainly be seen as a harbinger of things to come, and perhaps works of FICTION such as *The Windup Girl* can even encourage such connections.

How solid the present feels—its political orders, its infrastructures, its ideologies, but so too its affective predispositions. If we are to avoid the worst of the dystopic forecasts suggested by climate scientists, disaster movies, and cli-fi novels alike, change will begin—has begun?—not in the voting booth (policy) or the market (consumption) but in our bodies, our selves. The last ten years of environmental political inaction have shown that the information deficit model is flawed: few people will take proportionate action until they *feel* the consequences of their (our, my, your) choices. The economists are right, for once: we must internalize the externalities. Let's begin that work.

See also: ABORIGINAL, EMBODIMENT, MEDIA.

America

Donald Pease

The discovery of oil did not create American capitalist society, but the form of capitalism responsible for the United States' rise to global dominance can trace its origins to the day in 1859 that Colonel Edwin Drake struck rock oil near Titusville, Pennsylvania. The "American way of life" would be unimaginable without an oil economy marked by surplus production, militarism, AUTOMOBILITY, unregulated markets, and mass consumption.

Oil capitalism shaped significant turns in US national history. President Monroe imagined the Americas as a national protectorate, but Big Oil installed the transportation, commercial, military, and geopolitical networks that guaranteed US seigniorage over the hemisphere. Following its discovery in California, Oklahoma, and TEXAS in the 1890s, oil facilitated the transportation NETWORKS interconnecting the West and East Coasts. The oil economy underwrote the United States' spectacular military and economic growth throughout the "American century." An exceptional relationship to oil supplied power for the key weapons systems—warships, bombers, submarines, oil tankers, aircraft carriers, tanks, explosives, and a large portion of sea and land transport—that fortified victories in the First and Second World Wars.

After World War II, the United States structured its global hegemony on a complex of military and commercial processes that regulated the uneven production and distribution of global oil. Five of the seven oil conglomerates that dominated the international oil industry from the 1920s to the 1970s were American companies (Sampson 1975). By 1940, the United States accounted for more than two-thirds of world oil production. During the

Cold War, an archipelago of military bases allowed the United States to project its power into almost every region of the world and thereby control oil reserves (Leffler 2005).

Oil also triggered seemingly irresolvable crises in the American capitalist order and structured socioeconomic programs capitalizing on them. The surplus production of oil contributed to the precipitous fall in its price and mass unemployment during the Great Depression. When Franklin Delano Roosevelt took office, he faced the possible collapse of the capitalist order and working class revolution. Roosevelt's New Deal introduced housing, highway, and INFRASTRUCTURE policies that created irreversible changes in the distribution and consumption of ENERGY. American elites had purchased cars and suburban houses in the 1920s. By redefining suburbia as a "middle-class" way of life that he rendered affordable to millions of workers, Roosevelt's New Deal policies supervised a mass exodus to the suburbs (see Huber 2013).

The modern American state form emerged with the administration of New Deal policies. By undercutting support for public transportation, the Roosevelt administration encouraged the use of cars and taxis in cities throughout the United States. The number of registered automobiles increased from 3.4 million in 1916 to 23.1 million by the end of the 1920s. Oil accounted for one-fifth of US energy consumption in 1925; that figure rose to one-third at the outset of World War II. On VJ Day, Americans, who composed 6 percent of the world's population, owned 75 percent of its cars (Yergin 1991, 208). The 1949 Housing Act facilitated low-cost financing of suburban tract development, and the interstate highway system made the automobile-accessible world more than forty thousand miles larger. From 1945 to 1973, car registration in the United States quadrupled from 25 million to over 100 million. The imagined correlation of freedom with automobility transformed travel between home, WORK, and leisure into Americans' normative cultural geography (LeMenager 2014, 81–89).

By 1972, oil was responsible for 45.6 percent of US energy consumption, but that was the final year of the United States' reign as the world's leading oil producer (Darmstadter and Landsberg 1976, 21). During the 1970s, the United States vied for control of world oil with Saudi Arabia and other Middle Eastern countries that had become the world's top oil suppliers. In responding to the energy CRISIS caused by OPEC's 1973 oil embargo, the Pentagon devised a coalition between the nation's financial petro dollar and military weapon dollar strategies, linking the capacity to issue debt without redemption to the threat of unlimited military force. In the aftermath of the 1970s energy crises, US military strategists formulated a security policy correlating energy, financial, ecological, and military crises within an integrated grand strategy of imperial power (M. Cooper 2010).

Most American citizens do not live next door to an oil extraction site; many wish to stay ignorant of the nation's relation to the political economy of global oil. Oil remains nonetheless indispensable to the production and reproduction of America as an imagined national COMMUNITY. The ritual of filling car tanks with gasoline perseveres as the primary everyday practice through which Americans live and imagine national IDENTITY. Consequently, a world without open highways, electrified homes, and cheap and abundant gasoline, would, for the vast majority of Americans, be unimaginable.

As a substance that was once live matter, and that bursts into visibility with a force expressive of a vital form of life, oil invites mystification. Fears and aspirations stick to it. The innumerable derivatives of this mineral resource can destabilize countries and regions. Oil's extraction, distribution, and exchange require painstaking work, but the vividly spontaneous manifestation of a "gusher" appears to offer a liberation from labor. After its 1859 discovery, American abolitionists considered oil a replacement for slave labor (Nikiforuk 2012, 22–26). Oil created the tycoon, the financier, and the cowboy millionaire, and oversaw their spectacular rise from ignominy to fame. According to Frederick Buell, "oil replaced coal's back-breaking labor and widening of social caste with an energy infrastructure that encouraged individual enterprise" (2012, 282).

The imbrication of fossil fuels in all forms of American social and economic endeavors accounts for petroleum's foundational place in American life. American petroculture comprises a veritable cornucopia of synthetic materials and objects. Aspirin, antibiotics, toothpaste, artificial colors and flavors, petroleum jelly, insecticide, lipstick, perfume, nail polish, chewing gum, and paint are made of petroleum derivatives. So are adhesives, carpets, clothing, diesel fuel, gasoline, pillows, disposable plastic bags, and most other PLASTICS. Oil capitalism's infrastructure—a vast layout of refineries, water pipes, storm drains, sewers, electric cables, and natural gas lines—is present everywhere. Yet, as Imre Szeman and Maria Whiteman observe, it is "hidden in plain sight" (2012, 55).

This ENERGY SYSTEM is saturated with catastrophic economic and ecological consequences. Oil capitalism's processes of resource extraction have destroyed wildlife habitats and natural vegetation, contaminated water supplies, sapped topsoil from agricultural fields, depleted prairie and forested areas, damaged archaeological sites, and contributed to disastrous atmospheric conditions that threaten the survival of the human species. Ecological thinkers hoped to CHANGE prevailing values when they advocated a shift in attitudes toward oil, from its affirmation as the nation's lifeblood, indispensable to the American way of life, to the recognition that oil is a toxic substance to which Americans' continued ADDICTION will bring about the destruction of planetary life. Yet immersion in the oil economy has rendered Americans unwilling or unable to disentangle themselves from this deadly attachment. The lack of will demonstrates how petro-capitalism has engendered structures of denial and disavowal as pervasive cultural attitudes. Apropos of this predisposition, Fredric Jameson has famously observed, "it seems to be easier for us today to imagine the thoroughgoing deterioration of the earth and of NATURE than the breakdown of late capitalism; perhaps that is due to some weakness in the imagination" (Jameson 1994, xii). But why is it so difficult for Americans to imagine an end to a twentieth-century energy infrastructure so detrimental to planetary life in the twenty-first century?

One reason inheres in core mythological figures embedded in the national imaginary —American Adam, Virgin Land, Errand into the Wilderness—out of which Americans struggle to shape a consensus about their collective tasks. When evoked to address issues related to oil capitalism, these figures assign the agents, actions, and outcomes of the serial acts of ACCUMULATION-through-dispossession of the continental landmass with a quasi-theological calling. Working in tandem with scientific and political explanations, these images are sedimented within a network of meaning-making narratives that influence public perceptions of ecological crises and legitimate state policies supportive of oil capitalism. When cited within presidential addresses and governmental policy statements, these images buttress pseudo-analytic processes of invention, decontextualization, and erasure that remove the visible signs of oil's mode of production from the geographic landscapes the oil economy has disfigured.[1] Individually and collectively, these mythemes form a relay of connected beliefs: that reclamation can return the land to its predevelopment state; that petro-capitalism's putative enhancement of social power can remunerate its theft of political agency; that jobs, fines, and taxes adequately compensate environmental pollution; and that surplus production and mass consumption of oil can continue without any permanent cost.

President George W. Bush used the nation's composite myth to solicit consent to his blood-for-oil policies in Iraq and Afghanistan when he declared, "America is a nation with a mission . . . we understand our special calling. This great republic will lead the cause of freedom" (Bush 2004). After the Bush and Obama administrations described "foreign oil" as the source of the nation's addictive dependency, they enjoined Americans to forget about ecological catastrophe and take up the mission of reclaiming Americans' "right" to global dominance through the FRACKING, horizontal drilling, and strategic "Saudization" of the national geography.

Rather than desiring to be freed of such destructive attachments, Americans continue twentieth-century habits through competition over scarce resources in the twenty-first. Before Americans can uncouple the American way of life from the petroleum infrastructure to which they have become habituated, they must first supplant the assemblage of images with which they envision change with an alternative lens through which to recognize the real consequences of the oil economy and to imagine and aspire to make real a heretofore unimaginable post-carbon FUTURE.

See also: AUTOMOBILE, ENERGOPOLITICS, LIMITS, MIDDLE EAST, PETRO-VIOLENCE, ROADS.

1. Perry Miller demonstrates this strategy of occultation in his account of the epiphany spurring his vision of the national destiny "on the edge of a jungle of central Africa . . . while supervising . . . the unloading of drums of case oil flowering out of the inexhaustible American wilderness" (1956, 3).

Animal

Melissa Haynes

Why are animals absent from so many histories of ENERGY development? Given the significance of animal energy to the growth of human civilization, the oversight is puzzling. Perhaps the animal is too close to us to be recognized: animals may be "good to think" (Levi-Strauss 1991, 89), but the immediacy of their energy seems to make them difficult to think *about*. Reading animals into the history of energy might require us to begin at a distance, asking first when and how we have used animal energy, before we can see what it is and what it means to us.

Temporality

Humans have relied on animals from the beginning. Conceptually, "human" only acquires its legibility in contrast to all other sentient lives: "we" become humans when "they" become animals, objects for our use. Materially, preindustrial societies grew in step with their reliance on animals. Scavenging carcasses left by predators gave way to hunting and farming; domesticating cattle produced draft power, transportation, and meat (Smil 1999). Cattle were such a valuable form of energy that in many European languages "cattle" and "capital" were synonymous (Rifkin 1993, 28). The domestication of horses was likewise transformative: through the harnessing of horsepower, forest became fields and timber, quarried stone became cathedrals, and people and goods became armies and cities

(Smil 1999, 116). Industrial WHALING provided oil for illumination, lubrication, and consumption (and a range of consumer goods) long before fossil fuel became the dominant energy source (Dorsey 2013). Animals not only saturate history—they also animate it through the multiple forms of energy they have provided human beings.

Animal energy also pervades discourses of futurity, as the measure and means of re-established human control over planetary futures. Global warming discourse insistently displays cute animals (conveniently lacking nationality, race, or religion) to represent the threatened life that needs our protection. In contrast to those anxious appeals, optimistic reports about genomic and biomimetic innovations build fantasies that animal-derived technology will generate sustainable energy and remediate pollution. Such hopes that "microbial electrosynthesis" or "aquatic zooremediation" might be sufficient means to resolve ecological CRISIS constitute forms of "[bio]technological utopianism": the belief that scientific entrepreneurialism and invention may forestall the need for substantive political action in addressing energy problems (Szeman 2007).

The use of animal energy is far from obsolete in the present, even if overshadowed by the scale of current petroleum use. Considering how we use animal energy may offer insight into why these uses are hidden away, so deeply concealed from view.

Technique

The most intimate and obvious use of animal energy is the oldest: we eat the concentrated energy of animals in muscle energy and fat. Homeopathic logics equating the consumption of muscle with the achievement of strength led to an association between meat and masculinity (C. Adams 2010); more generally, muscle is imagined as the manifestation of agency. The word *muscle* functions as a synonym for power, influence, and (particularly coercive) force. Fat is stored energy and, like muscle, bears substantial linguistic significance. As a metaphor, *fat* stands for accumulated potential and wealth, as when figurative "fat cats acquire fat bank rolls off of the fat of the land" (while obesity is imagined as excessive ACCUMULATION and, therefore, wasted energy). In addition to their use as food, animals have been a protean source for material goods, providing leather, bone, wool, gelatin, pharmaceuticals, and countless other products. Even vegans cannot escape deriving value from animals; every part of the animal is a resource for capitalist society. Without the manufacture of these additional products from animal bodies, slaughtering animals for meat may not be sufficiently profitable. In the current system, "even the squeal return[s] as capital" (Shukin 2009, 95).

The disavowal of animal energy exemplifies the willed ignorance that enables other modes of energy extraction. Numerous physical and conceptual architectures distance eaters from the lives that make meat. Just as oil extraction sites stand at a distance from gas stations, concentrated animal feeding operations (CAFOs) and meatpacking plants are kept at a remove from urban centers, supermarkets have largely replaced butchers, and bloodless packages of beef and pork are hard to imagine as ever having been cows and pigs. The reduction of animals to meat is secondary to an earlier disavowal: the bodies described as animal were already marked as beings that can be killed but not murdered (Derrida 1991). The word *animal* functions like the CAFO, circumscribing and concealing our ethical decision to flourish at the expense of others. The capacity to make—and disavow—this decision is a power on which we draw in all our energy uses: it enables us to put gas in our cars while decrying the violence of its production and the consequences of its combustion.

The animal motor power that was crucial to the development of civilization continues today, albeit to a greater extent in less-industrialized economies in the Global South than in the developed North.[1] And even if other forms of energy have largely superseded the mechanical work of animals, the latter still serves as the measure of other forms of energy. James Watt's term *horsepower* (33,000 foot-pounds of work per minute), invented to demonstrate the economic superiority of steam engines over animals, continues as the common unit of power (Tarr and McShane 2008). Animals have also been the measure of ELECTRICITY since Luigi Galvani first demonstrated its stimulating effect on animal muscles, and both electroconvulsive therapy and execution by electric chair were developed and demonstrated on animal bodies (Burt 2001). Thomas Edison famously electrocuted an elephant on film to publicize the danger of alternating current (AC) electricity as well as "the ability of Edison's moving picture camera to mimetically capture reality [and] the painless immediacy of both electrical and cinematic affect" (Shukin 2009, 150).

Animals also measure the safety of energy use. Their presence, for example, is offered as proof of oil-field remediations—if animals are romping through what were once zones of oil extraction, everything must be okay. More often, though, we measure RISK in animal deaths. In the popular imagination, each energy source has an animal harbinger of the CATASTROPHE of its presence: oil drowns penguins, wind turbines cut down migratory birds, hydroelectric DAMS trap fish, NUCLEAR power mutates frogs, tailings ponds kill ducks. Such images pressure the public to demand or resist changes in ENERGY REGIMES. In this light, mass extinction not only gauges the speed of ecological collapse but also eradicates our ability to measure the crisis. We find ourselves caught in a feedback loop in which every animal death is also the erasure of an epistemological sign of DISASTER: each represents a lost marker of our capacity to catalog loss.

Target

Considering not only food but also WORK complicates the definition of animal energy: people are animals, and thus human labor is always already animal energy. This observation is not meant to dehumanize those who "work like a dog" but to shed light on the varying degrees of suffering, agency, and recompense applied to bodies. Some animals are

1. In modern homes in the Global North, perhaps a more familiar form of animal work is the emotional energy people derive from their pets.

affective laborers with fine homes; some humans are enslaved into work that extracts their most intimate energies without reward.

Wildlife, too, goes unregistered as a type of animal energy. And yet, the deaths of undomesticated animals register the ongoing energy crisis; in many respects, their deaths *are* the crisis. Though we only indirectly instrumentalize the energy of wild animals, we grasp their significance in the form of NATURE or ecology. No profit or production, or any form of everyday life, would be possible if not for the work of animals. An essential element of the larger system that we have come to call "nature," animals pollinate and provide reproductive assistance to plants, engage in the work of decomposing waste into humus, and so on. Resource exhaustion describes declining biodiversity as much as waning petrocarbon reserves; both are dwindling vital sources of energy.

Following from the difficulty of enumerating animals' energetic inputs, philosophy tends to see the animal as an abstraction of energy itself. John Berger suggests that animal energy inaugurated symbolic thought; for Berger, the "similar/dissimilar lives" of animals led humans to develop figurative language (Berger 1991, 7). Akira Mizuta Lippit (2000) argues that the animal can "never entirely vanish" (1), because as life unmediated by consciousness, it persists as an "animetaphor" (165) that animates *logos*. Nicole Shukin reads the animal as the fetish that sutures the material and symbolic; it is "moving . . . in both the physiological and affective senses" (Shukin 2009, 42). Gilles Deleuze and Félix Guattari (1987) describe "becoming-animal" as a kind of relating that transforms beings' capacity for CHANGE. Perhaps this understanding of the animal *as* energy explains our persistent, unsubstantiated faith in biotechnological salvation. Equating animal life and the abstract quality of liveliness or animacy (Chen 2012) reassures us in our belief that life goes on; it allows us to write energy narratives that ignore animals and the catastrophe of extinction. However, this metaphor finds its truth in the fear that animal extinction heralds the end of the world. We sense that animals are as indestructible as energy itself, but our human experience of energy—that is, our very existence—can persist only so long as animals do.

See also: ABORIGINAL, KEROSENE, MEDIA, METABOLISM, TALLOW.

Anthropocene 1

Dipesh Chakrabarty

The International Union of Geological Sciences (IUGS) names the current epoch the Holocene ("entirely recent"), which began about 11,700 years ago, after the last major ice age (Stromberg 2013). Many students of the Earth's climate argue that, in view of human effects on the biosphere, this name is no longer adequate. They suggest that we may have entered a new geological epoch when humanity acts on the planet as a geophysical force: the Anthropocene. The first statement in this regard was made jointly by Paul J. Crutzen, a Nobel Prize–winning chemist from the Max Planck Institute, and Eugene F. Stoermer, a former biologist. In a note published by the International Geosphere-Biosphere Programme (IGBP) in 2000, they argued that the "impact of human activities on earth and atmosphere . . . at all, including global, scales" made it "more than appropriate to emphasize the central role of mankind in geology and ecology" (Crutzen and Stoermer 2000, 17). In an essay in *Nature*, Crutzen restated the argument that "anthropogenic emission of carbon dioxide" could make the global climate "depart significantly from natural behaviour for many millennia to come" and proposed *Anthropocene* to name "the present, in many ways human-dominated, geological epoch" (2002, 23).

However, Crutzen and Stoermer were not the first to make this kind of argument; the "idea of an epoch of the natural history of the Earth, driven by humankind" has a longer history (Steffen et al. 2011, 843–45). The Italian Catholic priest and geologist Antonio Stoppani (1824–1891) wrote of the *anthropozoic era*, a term the American environmentalist George Perkins Marsh borrowed in *Earth as Modified by Human Action* (1874). The term *noosphere* (world of thought) became popular in Paris after the Great War, when the Russian geologist Vladimir I. Vernadsky, the French Jesuit and geologist Pierre Teilhard de Chardin, and the mathematician and philosopher Édouard Le Roy jointly coined it in 1924 to mark "the growing role played by mankind's brainpower and technological talents in shaping its own future and environment" (Crutzen and Stoermer 2000, 17; Steffen et al. 2011, 843–46). In *Global Warming* (1992), the journalist Andrew C. Revkin coined *Anthrocene* to describe a "geological age of our own making" (55).

Crutzen and his colleagues date the beginning of the Anthropocene to the "latter part of the eighteenth century," the period to which recent analyses of air trapped in polar ice date the increase in global concentrations of carbon dioxide and methane (Crutzen and Stoermer 2000, 17; Zalasiewicz, Crutzen, and Steffen 2012, 1036). They further identify a "Great Acceleration" after the Second World War, when human population, consumption, and greenhouse gas emissions all exploded (Steffen, Crutzen, and McNeill 2007). Other scholars date the onset of the Anthropocene far earlier, linking it to the invention of agriculture (Ruddiman 2003, 2005, 2013; Ellis 2011).

In 2008, the Stratigraphy Commission of the Geological Society of London created an Anthropocene Working Group, charged with submitting a report to the Subcommission on Quaternary Stratigraphy (a division of the International Commission on Stratigraphy answerable to the International Union of Geological Sciences). A decision is expected in 2016. The geological judgment regarding the Anthropocene will depend on the available stratal data and other kinds of evidence by which geologists read the past (Zalasiewicz et al. 2008; Zalasiewicz et al. 2011). Bryan Lovell, then President of the Geological Society of London, acknowledged that "if, by our own hand, we create our own extreme warming event," then "the time in which we now live would . . . sadly and justly, surely become known as the 'Anthropocene'" (2010, 196).

Although the Anthropocene has yet to attain the official status of a geological epoch, an increasing scientific recognition that anthropogenic emission of greenhouse gases causes planetary climate change has popularized the term among concerned scientists and large sections of the general readership. The notion that humans have become geological agents with the capacity to determine the future of the planet has inspired among scholars in the humanities and the interpretive social sciences numerous vibrant discussions regarding climate injustice, human agency, the (collapsing) distinction between natural and human histories, interspecies and intergenerational ethics, consumerist cultures, Anthropocene affects, the (post)human condition, and the difficulties of representing the Anthropocene in film, art, and performance (see Chakrabarty 2009, 2012; House of World Cultures 2013; T. Morton 2013; Braidotti 2013; Di Leo 2013). These debates will only become more vigorous as the climate CRISIS unfolds.

As a historian of recorded human history, I am interested in the Anthropocene's implications for how we tell the human story. We do not yet know whether the term will be formalized by geologists. Giving an official name to something that has implications for policy is always a political process, and I imagine that the deliberation will be subject to various pressures, formal and informal, scientific and nonscientific. But the anthropogenic nature of the climate crisis poses interesting challenges to several metanarratives of human history.

First, take the ideas of freedom and justice that saturate most humanist narratives of history. It is deeply ironic that what enabled humans to curtail the use of massive slave or bonded labor in the construction of massive structures such as the pyramids or the Taj Mahal was the discovery of cheap ENERGY in the form of fossil fuels, since profligate use of those fuels is now understood to threaten human futures. Even if we assume that all will be well in the end and that humans will make a smooth transition to renewables, energy is likely to be dearer than at present. So if there is a close connection between consumption-expenditure of energy and the exercise of "freedoms," then our freedoms are going to be more expensive and therefore relatively more scarce. A just distribution of freedom as a scarce resource will demand significant reordering of social hierarchies: society and freedom will need to be reimagined.

Second, consider the arguments that blame the climate crisis on the capitalist mode of production and the unstoppable tendency of capitalism (loosely speaking, for "capitalism" is not the same as "the capitalist mode of production") to accumulate wealth. But if we accept this popular position, we stretch the analytics of "capital" to include information foreign to all received procedures of political-economic analysis. That the logic of capitalist production leads to further ACCUMULATION of capital—and thus to the pursuit of never-ending growth—is not in doubt. But climate change is not a problem that could have been recognized or named within the traditional procedures of political economy. To know what climate change is and to be able to measure it, you need—much more than the theories of the left—geological and paleoclimatological knowledge of this and other planets (for geophysicists study global warming on Earth as a subset of the more general phenomenon of planetary warming seen on other planets as well), the knowledge of climate modelers, instruments to measure trace gases and temperature, including those needed for the extraction of ice core samples, and so on. In other words, while some may see in the dynamics of capitalist production the causes of human-induced climate change, it still remains a problem that—unlike many other crises of capitalist accumulation, including some environmental issues—could not have been predicted from within the frameworks of political-economic analyses alone.

What does that mean? It means that political-economic knowledge about capital alone does not equip one to understand the relationship between the capitalist mode of production and global warming. One has to get beyond the historical life of capital—both backward and forward—to understand that the Earth has seen planetary warming long before there were human beings and that the logic of capitalist accumulation may have interfered with longer-term processes in the history of the earth system and the role of life in that history. In other words, the present crisis reveals the available analytics of capital to be necessary but insufficient. We have to think the history of capital (spanning a few hundred years) and much longer histories (of the earth system and life on it) at the same time.

And, finally, take the question of climate justice. There is surely a case for justice among nations and classes, as only some ten or twelve nations (India and CHINA included) and about one-fifth of humanity account for most greenhouse gas emissions to date. One could legitimately argue that the crisis must be met in a way that addresses this uneven responsibility. Yet consider the problem we face. The calendar of justice among nations, groups, and individuals is an open one. One does not know when the world will be just. But the Intergovernmental Panel on Climate Change (IPCC) calendar for global action is short and finite. According to the fifth report of the IPCC, our budgeted emission of greenhouse gases should come to an end around 2040 in order to avoid "dangerous" warming beyond a 2°C increase. Now suppose that for an indefinite period into the FUTURE, nations do not find a way to come together and some parties hold out—with good reasons—on issues of justice. What would be the result? Given that climate change, while affecting all, will affect the poor of the world more than the rich, those feeling unjustly treated now will most likely find themselves in a world that is even less just, for climate change will have made their situation worse. This is not an argument for not fighting for justice but instead a suggestion that global unity on fighting climate change actually contributes to justice. Here the expression "the Anthropocene"—in its general sense of "the age of humans"—reflects an interesting problem of nomenclature.

Many are suspicious of the category "humanity" and resist notions of "human-induced" or "anthropogenic" climate change. The term *human*, they argue, hides the different responsibilities the rich and the poor bear for the current crisis. Yet climate change, whoever bears responsibility for it, is everybody's problem, for we all share the planet's climate. "Common but differentiated responsibility" is how the Kyoto Protocol puts it. The word *differentiated* acknowledges that we are not all equally responsible for climate change. The developed countries bear greater historical responsibility. But why is the responsibility also described as "common"? In what way could we all be responsible? Is there any name for this horizon of commonality? Is it *humanity*, that much-despised term, at some other level? Climate change thus raises the issue of a shared or common history, but we do not yet have a name for the subject of that history, a name that would not be mired in the ideological trappings of the term *humanity*.

See also: ACCUMULATION, GUILT, LIMITS, STATISTICS.

Anthropocene 2

Rob Nixon

For a growing chorus of scientists, the Holocene is history. Through our collective actions we have jolted the planet into a new, unprecedented epoch, the Anthropocene, which, according to one influential view, dates back to the late-eighteenth-century beginnings of the Industrial Revolution. The ecologist Eugene Stoermer coined the term *Anthropocene* (age of humans) in 2000, and the Nobel Prize–winning atmospheric chemist Paul Crutzen quickly popularized its core assertion that for the first time in Earth's history, a sentient species, *Homo sapiens*, has become not just a biomorphic but a geomorphic force. The grand species narrative that drives the Anthropocene hypothesis is, in both senses of the phrase, epochal: it moves the geological boundary markers while also disturbing conventional assumptions about human agency, IDENTITY, and temporal power. The Anthropocene is a story of massive, lasting anthropogenic changes—to the lithosphere, the atmosphere, the hydrosphere, and the biosphere—that will be legible, in many cases, tens of millennia from now. In other words, over the past two and a half centuries we have been inadvertently laying down in stone a geological archive of human impacts.

This ascendant twenty-first-century grand narrative is unsettling some of our most profound assumptions about what it means to be human—imaginatively, biologically, existentially, ethically, and politically. Since George Perkins Marsh in the 1860s, many thinkers have recognized humanity's capacity to transform the planet, among them Rachel Carson who observed that "only within the moment of time represented by the present century has one species—man—acquired significant power to alter the nature of his world" (1962, 23).

But between Carson's mid-twentieth-century perspective and that of twenty-first-century Anthropocene mandarins, we have witnessed a marked technological and narrative shift. For Anthropocene scientists, the new metrics of accelerating human impacts indicate that we have not only been changing the world environmentally but changing the planet's deep chemistry in ways that demand radically new modes of storytelling.

New metrics demand new metaphors. Taking her cue from paleobiologist Anthony Barnosky, Elizabeth Kolbert argues that "we are the asteroid" (2015). We may be meteoric in our capacity to catalyze mass extinctions and other long-lasting planetary effects, but we are not insensate: the humanity at the center of the Anthropocene is a hurtling hunk of rock that feels. Much Anthropocene theory addresses the complex implications of translating the feeling human species into a species of unfeeling geological agency. But less attention has been paid to the other assumption in this metaphoric equation: if "we are the asteroid," who exactly is this high-impact "we"? What is gained and what sacrificed through this geologic-biologic turn that places, at its center, "we the species"?

Arguably, the central challenge posed by this new version of planetary history is this: How do we tell the story of *Homo sapiens* as an Anthropocene actor in the aggregate, while also insisting that the grand species narrative be disaggregated to reflect the radically divergent impacts that different communities have had on planetary chemistry? To approach the matter in this way is to ask searching questions about the geopolitics of Anthropocene geology's layered assumptions. Historically and in the present century, how have social institutions, cultural practices, and forms of governance affected the way diverse communities have exercised radically different levels of geomorphic and biomorphic power? *Homo sapiens* may constitute a singular actor, but it is not a unitary one. Oxfam reports that in 2013 the combined wealth of the world's richest 85 individuals equaled that of the 3.5 billion people who constitute the poorest half of the planet (Wearden 2014). In 2009, the 1.2 billion inhabitants of low-income countries were responsible for 3 percent of CO_2 emissions, while the 1 billion inhabitants of high-income countries were responsible for 47 percent, an immense difference per capita (World Bank 2009). Moreover, a 2013 study concluded that since 1751—a period that encompasses the entire Anthropocene to date—a mere ninety corporations have been responsible for two-thirds of humanity's greenhouse gas emissions (Goldenberg 2014). That is an extraordinary concentration of earth-altering power.

The advent of the Anthropocene story has profound consequences for how we conceptualize the environmental CRISIS and the inequality crisis, two of the greatest crises of our time, which are joined at the hip, although the join is often invisible. The implications of Anthropocene perspectives for environmentalism have been extensively examined, but there has been little attention to the Anthropocene's implications for how we address—and redress—inequality. In terms of the history of ideas, what does it mean that the Anthropocene has gained credence during the twenty-first century, during a time when, in society after society, we are seeing a widening chasm between the ultrarich and the uberpoor, between resource capture at the top and resource depletion at the bottom (Nixon 2011)? What does it mean that the Anthropocene as a grand explanatory species story has taken hold during a plutocratic age? For "we the species" is being positioned as a planetary actor when, planet-wide, in most societies, what it means to be human is breaking apart economically, exacerbating the distance between extremes of affluence and abandonment. Those extremes are profoundly consequential for the way human impacts are distributed, recorded, and deciphered in earth's geophysical archive.

The story of the Anthropocene links "earthly volatility to bodily vulnerability," as geographer Nigel Clark has noted (2011, xx). Yet the most influential Anthropocene mandarins have marginalized questions of unequal human agency, unequal human impacts, and unequal human vulnerabilities. If, by contrast, we take an environmental justice approach to Anthropocene storytelling, we can better acknowledge the way human actors' geo- and biomorphic powers have involved vast disparities in exposure to RISK and access to resources. In conceptualizing the Anthropocene, then, a critical challenge is how to think simultaneously about geological and social strata. The stratigraphers who are central proponents of the Anthropocene are specialists at reading layers of rock. But in studying the human contribution to those sedimentary layers, we also need to conduct another kind of reading, a reading of social stratification. We should acknowledge that different social strata, historically and increasingly in the twenty-first century, have exerted unequal agency.

Two other terms are pertinent here: the Great Acceleration and the Great Divergence. If most Anthropocene scholars date the new epoch to the late eighteenth century, they also note an exponential increase, beginning circa 1950, in the pace of anthropogenic changes to the carbon cycle, the nitrogen cycle, the water cycle, global trade, resource consumption and habitat clearance, and industrial-scale agriculture. This post-1950 Great Acceleration—which includes the appearance of unprecedented isotopes from atomic bombs—will be registered in the planet's physical systems for tens of millennia to come.

However, the most authoritative accounts of the Great Acceleration fail to position it in relation to the rise of neoliberalism since the late 1970s.[1] By now, more than half of the Great Acceleration has occurred during an era dominated by neoliberal policies: the concentration of wealth through privatization; assaults on the civic and environmental commons; the shredding of social safety nets and public services through structural adjustment and asset stripping; rampant deregulation and union busting; mega-mergers that have created corporations more powerful than the nations they operate in, leading to (often militarized) alliances between unanswerable corporations and unspeakable regimes; and, under the banner of an international free market, an increasing license, on the part of the most powerful, to internalize profits and externalize costs across national boundaries and across generations.

Together these policies have led to what Timothy Noah calls "the Great Divergence" (Noah 2012, 1). Noah's subject is the twenty-first-century economic fracturing

1. Clark (2011), Crutzen and Stoermer (2000), Williams et al. (2011), and Zalasiewiecz (2008 and 2011) are among the influential Anthropocene intellectuals who have either ignored or marginalized the question of Anthropocene inequality.

of AMERICA in the so-called new gilded age. But the Great Divergence is not just an American concern: it scars most twenty-first-century societies, including CHINA, India, Indonesia, South Africa, NIGERIA, Italy, Spain, Ireland, Costa Rica, Mexico, Greece, Jamaica, the United Kingdom, Australia, and Bangladesh. The Great Acceleration is not reducible to neoliberalism's ascent, but any account of human-induced planetary morphology since 1950 needs to keep neoliberalism's durable impacts front and center. Many of these impacts will be legible, long term, in what science writer Peter Brannen calls "the thin glaze of life-supporting chemistry that coats the earth" (2013, 32). A central failure of the dominant mode of Anthropocene storytelling is a failure to articulate the Great Acceleration to the Great Divergence.

Will the Anthropocene proffer, as geographer Nigel Clark believes, "an invitation to relearn what it is to be human" (2011, 17)? We are in the process of finding out. What we do know is that this iconoclastic idea unsettles habitual assumptions about humans as shapers of deep time—our time and the times of other species that share our rapidly changing earth. The Anthropocene debate is moving beyond the zone of interdisciplinary argument and entering a more public sphere, as museums, galleries, and film festivals wrestle with how best to animate this complex, provocative idea in ways that grant it imaginative energy and emotional traction. In so doing, such institutions are contributing to the shaping of environmental publics and the making of environmental policy. The turn toward the public Anthropocene has profound implications for the way humans perceive—and act upon—the planet we have inherited and the planet we will bequeath. But as we engage this public turn, it remains imaginatively, ethically, and politically crucial that we acknowledge the tensions within the Anthropocene, the centripetal and centrifugal tensions within this shared geomorphic story about increasingly unshared resources. We are all in the Anthropocene, but we are not all in it in the same way.

See also: ACCUMULATION, CHARCOAL, ENERGY REGIMES, FUTURE, PLASTIGLOMERATE.

Architecture

Daniel Barber

I

In 1957 the office of Charles and Ray Eames, one of the best known mid-century architecture firms, produced what they called a Solar Do-Nothing Machine. Developed as part of a marketing campaign for the Aluminum Company of America, it consisted of a 24-inch elliptical aluminum platform supporting moving pinwheels and star shapes, all made of brightly colored anodized aluminum. On the side, a freestanding reflector screen of polished aluminum strips captured and reflected sunlight into photovoltaic cells that converted it into ELECTRICITY (Neuhart et al. 1989, 178). As *LIFE* magazine noted in 1958, "the toy has no use and is not for sale, but ALCOA is sending it on tour as an enchanting harbinger of more useful sun machines for the future" ("A Twirling Toy" 1958, 22–24).

The Solar Do-Nothing Machine is evidence, albeit in negative, of mid-twentieth-century architects' intense interest in ENERGY technologies. After World War II, there was a brief but potent panic around how to develop an ENERGY SYSTEM adequate to fuel the economic recovery of the postwar period. As the global NETWORK of oil formed through industry, diplomacy, and economic restructuring, the design fields simultaneously came to be concerned with energy in a broad sense. Proposals for prefabricated and panelized building systems, radiant heating, SOLAR houses, and many other means of designed efficiency consumed large sectors of the international discourse of architecture throughout the 1940s and 1950s (see Barber 2013). The discussion of solar space heating was especially dynamic, as it became clear that the formal, material, and technological disposition of modern architecture was particularly amenable to the needs of solar efficiency. The modern solar house became an important symbol, albeit short-lived, for the possibility of living differently in the future.

Since this fleeting moment, architectural practices, publications, and competitions, and perhaps most especially new ideas and imperatives about engaging knowledge from other fields, have reconfigured architecture as a discursive site to discuss what we now call environmental concerns. If, on the one hand, the Eames's toy indicates the place solar power occupied in the immediate postwar proliferation of energy INFRASTRUCTURE and technology—able to do, if not exactly nothing, then seemingly very little—on the other hand, the Solar Do-Nothing Machine is indicative of a new perspective on architecture and technology and how design strategies in architecture began to focus on the challenges presented by increasing knowledge of global ecological contingencies. Indicative, that is, of a new perspective on the ability of a solar machine, and of ecotechnologies more generally, to *do something*. Architecture became an interdisciplinary frame through which social, technological, and design experimentation encouraged new perspectives and new forms of expertise. These developments, in turn, contributed to the establishment of an intellectual framework, a funding structure, and an ethics of interdisciplinary practice and facilitated a dramatic increase of knowledge about the global environmental system.

II

More evidence: in 2009, the research branch of the Office for Metropolitan Architecture (OMA) submitted a report to the European Climate Foundation called Eneropa. As part of the foundation's Roadmap 2050: A Practical Guide to a Prosperous, Low-Carbon Europe, OMA had been commissioned to provide the graphic narrative that would help communicate the extensive technical, economic, and policy analysis performed by the European Climate Foundation's consulting firms. OMA's contribution redraws the map of Eneropa according to methods of energy generation: Geothermalia in northwestern Europe; Solaria across the Mediterranean south; the Tidal States of the United Kingdom; Biomassburg in the Baltics; North, West, East, and Central Hydropia in the mountain regions. Not only do the names and divisions on this new map toy with geopolitical histories, most potently in the CCSR (Carbon Capture and Storage Republics) written across the former CCCP/USSR. Each region is also fancifully represented with the mechanisms of a new energy technology. A blanket of solar panels, for example, is strewn across the rooftops of Barcelona. As OMA partner Reinier de Graaf notes, by suggesting "the complete integration and synchronization of the EU's energy infrastructure," Eneropa shows how "Europe can take maximum advantage of its geographic diversity towards a complementary system of energy provision ensuring energy security for future generations" (Office for Metropolitan Architecture 2010).

As the work of a leading architectural practitioner, Eneropa suggests that as concerns over energy efficiency have returned, the design fields have again become an important discursive location for debating, understanding, and thinking about environmental complications and about energy in particular. Today, architects are aware of their imbrication in the formal and material conditions of the built environment—global flows of materials, energy, and ideas have been of concern to architects for very pragmatic reasons—and are also aware of a strong disciplinary tradition to focus on projection of a design concept into a FUTURE scenario. Architects have frequently been enlisted, as in Eneropa, to present a vision of future conditions based on research from an array of fields—economics, policy analysis, energy forecasting—and as a means to think creatively about environmental CHANGE.

All the same, if Eneropa is any indication, the disposition of the images and ideas used by designers to represent means of environmental change is very specific and open to criticism. De Graaf's presentation of Eneropa is quite explicit in this regard: he proposed that, despite the redrawn political boundaries, "the most shocking part of [this plan] is how incredibly unshocking it is. Everything that moves is the same and still moves. Only the things that make the things move have all completely changed. It's a situation where everything changes and at the same time nothing changes" (ibid.). Even conceding some ironic self-positioning, de Graaf's words exemplify a broader trend in contemporary architectural strategies toward energy efficiency: through carefully applied technological innovations, this disposition holds, energy systems will be reconstructed in such a way that we almost won't notice the difference. Daily life will stay the same; "nothing changes." Or, more pointedly, the proposal is that in order for the social fabric to remain intact, dramatic changes are needed in technoscientific relations to the natural world. One could look at any number of recent GREEN buildings to see a similar emphasis on technological innovation as the primary means by which our environmental problems can be mitigated, almost without anyone noticing.

III

As we consider the relationship of architecture to energy and culture, we can ask: What other positions are available? As a counterpoint one could explore, as many interested in the history of architectural-environmental issues have done, the experimental practices of the 1970s. Here as well, architectural proposals and projects were a discursive location for explorations of environmental knowledge: in houses made of earth and DETRITUS, in the premise of systems autonomy and self-reliance, and in the interest in participatory design and COMMUNITY development, the design fields served as an important site for debating the prospects and principles of environmental change (see Borasi and Zardini 2007). Distinct from the OMA proposal, many designers of the 1970s saw in technological innovations new impetus for social transformations: for new ways of living and for developing new individual and collective parameters for engaging with the resource base. The countercultural premise was that anticipated energy scarcity could be managed through new forms of social organization (see Boyle and Harper 1976). If Eneropa is based in technological dynamism, the collective work of the 1970s placed its faith in sociocultural dynamism; the solar houses of the 1950s saw themselves somewhere in between, curious as to how technological change could facilitate new social formations. Architecture and energy, in any case, have since been closely, if ambiguously, linked. From this perspective, the history and theory of architecture, and the cultural analysis of energy practices more broadly, is an increasingly important site for the critical analysis of sustainable architectural practices.

See also: CHANGE, GREEN, RENEWABLE, SUSTAINABILITY, URBAN ECOLOGY.

Arctic

Rafico Ruiz

"Arctic" is a cultural figuration that does a lot of work. It marks a nebulous geographic region, an atmospheric condition, and, increasingly, perhaps *the* common place where environmental CHANGE is made manifest. With the proliferation of time-lapse satellite images showing the shrinking polar ice cap, documentary PHOTOGRAPHY and film following the pace of glacial melt, and the prominence of the Northwest Passage as a maritime transportation corridor, the Arctic has a recurring cultural visuality and instrumentality all its own and utterly of the present. Part of the Arctic's figurative work depends on its being perceived as a mappable territory—a place that is not quite a landmass but not exactly an ocean. Rather, it functions as a sea ice barometer and a contested ground, caught between states of "natural" erosion or "human" EVOLUTION. Like a barometer interacting with its surrounding atmosphere, the amorphous boundaries of the Arctic mark and measure the changeable states of environmental being and becoming. For scholars, scientists, indigenous residents and activists, and others invested in the Arctic as sea ice—as a hermeneutic, scientific, and experiential territory both more than and other than open ocean—it is imperative to articulate the various ecological and anthropomorphic conjunctures at which our Arctic has arrived.

If the Arctic is the singular site where the earth's ENERGY effects are made manifest, then it follows that the Arctic is the human species' most collaborative and collective geographic creation. In this reading, energy effects comprise the circulation of past, present, and future CO_2 emissions just as much as the very intentionality of state, corporate, and individual actors in their ways of producing and consuming energy in a broad sense. As Georges Canguilhem has it, "Man, as a historical being, becomes the creator of a geographical configuration; he becomes a geographical factor" (2008, 109). Yet, while the Arctic circulates powerfully as a figuration, its status as a concrete, livable space is undergoing disintegration and drift. The singular Arctic is becoming an innumerable series of Arctics: drifting ice islands that get taken up in the currents of the Beaufort Sea. When these fragmented Arctics start to circulate, drifting at ever-higher speeds due to the increase in ice-free open water, they pose a new a threat to the oil and gas drilling operations that are moving into these far northern environments. Regardless of one's position in natural resource expansion debates, the condition of drift thus becomes one of the defining characteristics of Arctic ecology.

This condition of drift brings to mind Alfred Wegener's famed theory of continental drift and the formation of continents through the movement of tectonic plates (Wegener 1966). Writing in the 1910s, Wegener formulated his theory during his work on climatology in Greenland (McCoy 2006). The difference between these two forms of drift is primarily heuristic, as the Arctic has never been a stable landmass but an ocean, and so its own drift is not geological but climatological. But what of this condition of drift? If the Arctic emblematizes the effects of our energy cultures, then it matters what sorts of figurations it gets shaped into, as those figurations will, in part, determine its FUTURE cultural normativity. This cultural-hermeneutic work falls to those in the humanities and social sciences who turn to the Arctic as a "world object" (following Michel Serres) that is constantly renegotiating its "natural" boundaries (Serres 1995). To think through drift is to assume a degree of theoretical eco-mimicry and become like the aleatory current that is changeable and not entirely knowable, yet—like those Arctics—fast becoming climatological. In a way (at least in the high CO_2 present) this approach to drift makes the disappearing polar ice cap a spur to critical thought that can keep pace with and be tropological in its fields of disciplinarity. This predominantly stochastic mode is one that points to the adaptability of open-ended and variable theoretical premises when it comes to such climatological milieus as the Arctic. For me, to drift is to engage in a form of environmental MEDIA inquiry that is both historical and theoretical. It is to go to that proverbial and problematic "middle of things," where sea ice can be understood as a legitimate and mobile, if unconventional, storage medium that is on a continuum with bound paper and encased data (Sofia 2000). In this sense, the Arctic becomes a tropological "media" problem and as such needs to be thought out from its conditions of environmental mediation in addition to being thought with a host of figures and figurations that foreground, as Gregory Bateson would have it, the relations of relations (Bateson 1972, 154).

Some media scholars contend that a medium "designates a minimal relationality, a minimal openness to alterity, a minimal environmental coupling (in the terminology of contemporary ethological cognitive science), that appears somehow central to our understanding of ourselves as 'essentially' prosthetic beings" (Hansen and Mitchell 2010, xii; see also Hansen 2006). In this sense, *media* constitute our relations to worldly becoming and as such account for an inevitable focus on relationality, emergence, and mediation as important concepts for that process. This understanding of *medium* also suggests that human being is somehow conflated with that always problematically singular anthropogenic *environment*, with or without a human precedent, as originally having been one medial substance in which all life does its living. Such an understanding is in line with much environmental historical scholarship that debunks the myth of a "NATURE" that is neatly divided off from human incursion and influence. However, thinking about the relationship between human beings and the environment as one of ongoing mediation and coemergence can also establish closer ties between supposedly "human" and "natural" systems that can be taken apart and made distinguishable.

For the media historian, drift theory could thus take into account the "relations of geographical milieu to man by saying that doing history came to consist in reading a map, where this map is the figuration of an ensemble of metrical, geodesic, geological, and climatological data, as well as descriptive bio-geographical data" (Canguilhem 2008, 107). For instance, Arctic media historiography in this reading can encompass such an exemplary event as the USSR's establishment of a temporary research station on a drifting ice floe at the geographic North Pole in March 1937 (Fyodorov 1939). Dubbed the "North Pole" Drifting Observatory by the expedition led by Ivan Papanin, it drifted for 274 days, travelling roughly 1600 miles to reach the Greenland Sea (Papanin 1939). "North Pole" obtained some of the first scientific observations from this high latitude, including measurements of ocean depth, water temperature, meteorological conditions, as well as water and bottom-soil samples (Woods Hole Oceanographic Institution 2015). The expedition recorded bottom relief along its drift line; it also ascertained that warm Atlantic water reaches the Arctic Ocean and that the deepest reaches of the latter are warmed by the heat of the earth. "North Pole" became the Arctic's first drifting research station; RUSSIA continues this practice today.

Arctic drift is a relation that is tropological in being climatological. Within the context of debates around the ideological periodization of the ANTHROPOCENE, it enacts a posthumanist relational theory that is at once definitively situated (albeit also mobile) and an attempt to think the "global" in situ (see Tsing 2005 and T. Morton 2013). If the Arctic can be understood as a common energy environment that is the ongoing and open-ended result of atmospheric drift, one that is distributed across a number of actors and scales of responsibility, then Arctic drift is energy's afterlife in the present. Learning to track its movements, to observe its diverse data-based materialities, and to mediate its sociopolitical articulations is a stochastic project that seeks not an act of slowing or reassembly but instead a mode of transduction that can turn open water back into the media of ice without having to claim a new "North Pole" in the process.

See also: ABORIGINAL, ENERGY, MEDIA.

Automobile

Gordon Sayre

The current worldwide dominance of gasoline-powered automobiles with internal combustion engines constitutes a "transportation monoculture" (Sperling and Gordon 2009, 15) with deleterious consequences too familiar and numerous to list. But from 1890 to about 1905, newly-invented automobiles were fueled by steam, electric batteries, alcohol, diesel, and biodiesel fuels, as well as gasoline, and it was not obvious which fuel would come to dominate the market. Rough, unpaved roads, short trips, and low speeds were the norm, which negated the advantages in power and range that hydrocarbon fuels later offered. The 1894 Paris-Rouen horseless carriage competition, often called the first automobile race, was won by a steam-powered vehicle. Electric cars were initially popular with women, as they were quiet, odor-free, and did not require crank starting. Thus today's monopsony of automotive and oil industries was not inevitable, and the barriers to its decline may be as much behavioral as technological.

Horseless Age was the title of an early auto industry periodical published between 1895 and 1918, and it is worth reflecting on the shift from equine- to engine-powered personal transport during this period. The automobile was introduced "at a time which horse populations were reaching their all-time peak: 3.5 million in Britain and 30 million in the U.S." (Pettner and Turner 1984, 36). The spread of railroads had actually made horse-drawn carriages and wagons more essential for carrying people and heavy goods the critical "last mile" from rail stations. The term *station wagon* originally referred to these vehicles. Growing cities faced a huge sanitation problem, as a horse produced about forty-five pounds of manure per day, and therefore "the man in the street one hundred years ago was ready for the car and expecting a great deal from it. He was expecting it to provide a cure for most of the ills for which the car is now blamed: air pollution, congestion, and death on the roads. Cars were judged to be more reliable, safer and cleaner than horse-drawn transport" (ibid., 54). The bucolic image of horses and manure as organic and sustainable is an ideal that did not fit the reality of horse-powered transport in the late nineteenth century.

The first mass-produced cars, notably Peugeot and De Dion-Bouton in France and Oldsmobile in the United States, resembled carriages so closely in design that observers really did perceive them as horseless carriages, albeit lacking the balance and beauty of a horse to pull them. Fake horse heads were sold as accessories to attach to the dash of one's car (Pettner and Turner 1984, 41). The power of motors is still measured in horsepower today, and features and styles of automobiles—such as the dashboard, trunk, brougham, saloon, and landau—are still named for the corresponding parts or types of carriages. A chauffeur was originally a man who fed fuel to a steam engine. Major early manufacturers of cars in the United States included large carriage makers, such as Studebaker, which manufactured electric cars from 1902 to 1911, and bicycle makers such as Albert Augustus Pope's Pope Manufacturing Company, which controlled several key patents for bicycles and expanded into electric cars in 1897. The car, like the bicycle, was at first adopted as a leisure vehicle by affluent urban consumers, and only with Henry Ford's Model T did it become indispensable to the RURAL majority.

Ford is famous (or infamous) for the assembly line and the Taylorization of labor, but the success of his cars also contributed to the hydrocarbon fuel monopoly. Before the Model T (which could run on gasoline, KEROSENE, or ethanol), Ford had worked in the 1890s as an engineer with the Edison Illuminating Company building power plants, and from 1901 to 1903 had built race cars to help promote his first two automobile manufacturing ventures (which both failed). High federal excise taxes on alcohol before Prohibition also helped ensure that gasoline would continue as the dominant fuel source. Concerns about gasoline supplies and prices in the 1910s abated when chemists at Standard Oil and at Columbia University devised new methods to "crack" heavy crude oil molecules and thereby double the efficiency of the refining process for gasoline (T. McCarthy 2008, 17–19, 48–49). By 1920 gasoline could be purchased at an estimated 15,000 stations in the United States, and by 1924 the US census reported 46,904 of them (Melaina 2007, 4922). In 1930, three-quarters of the world's cars were built in the United States and 90 percent of oil was produced there (Sperling and Gordon 2009, 113).

Sociologists of "automobility," including John Urry and Mimi Sheller, have developed the concept of "lock-in" or "path-dependence" through which cars have ensured the dominance of the oil industry and its gas stations. Today, as battery-powered cars again make a bid for the mass market, the problem of range anxiety and the lack of a network of charging stations show how automobility and gas stations are part of our INFRASTRUCTURE and can be changed only with sustained effort and expense. Automobile dependence is highest in the United States, CANADA, and the MIDDLE EAST, is growing fast in India and CHINA (which in 2009 surpassed the United States as the largest market for cars [Russell and

Bradsher 2016]), and threatens to grow also in Africa and the least-affluent nations, where major public investment in mass-transit systems is not feasible.

What, then, are the prospects for breaking out of the path of our dependence on hydrocarbon fuels? In the United States, policy efforts have been ineffective in large part because they have sought to change the automobile rather than the infrastructure and behavior that constitute automobility. It has been politically expedient to impose safety, emissions, and efficiency regulations upon auto manufacturers, even if they at first resist them, and politically difficult to impose high taxes on cars and fuels, even if the taxes might be a more direct and effective method of reducing vehicle miles traveled and car dependence. (In the European Union, taxes on cars and fuels are higher, but the consumer-friendly approach still prevails.) The Corporate Average Fuel Economy or CAFE regulations enacted in the Energy Policy and Conservation Act of 1975 and implemented in 1978 have been frequently delayed and weakened by loopholes, with the result that in 1991 the US fleet was deemed to have met the law's standard for an average of 27.5 mpg even though the true level of consumption was 21.2 mpg.

A second major policy deficiency is that North Americans promise and flatter ourselves with the prospect of transformative technologies for automobile fuels and use this fantasy to ignore or postpone practical, broad-based, incremental changes that would reduce the harm caused by cars. The pastoral dream of the zero-emission automobile, as mandated, for example, by the California Air Resources Board in 1990 and celebrated in the 2006 film *Who Killed the Electric Car?*, promotes electric cars as a conscientious objection against fossil fuels. But the ultimate source of the ELECTRICITY, and the comprehensive network of recharging stations for it, remain invisible and unavailable, rendering the electric car a niche product. The hydrogen-powered car is an even more distant fantasy, promoted by the Bush administration, which ended the 1993 Partnership for a New Generation of Vehicles (an initiative to subsidize the development of prototype vehicles that achieved better than 70 mpg) and replaced it with a vague mandate for hydrogen-powered cars (Sperling and Gordon 2009, 27). Today 95 percent of hydrogen is produced from natural gas or methane, and if it is used to power a hydrogen-fueled car the total carbon emissions exceed that of a conventional gasoline engine. A full 85 percent of the hydrogen produced is used to remove sulfur from gasoline. There are many barriers to using hydrogen in car engines, not least that no safe distribution and storage network exists for the volatile gas, and to build such a network would be more difficult than electricity recharging stations.

John Urry has lamented, "The car . . . is rarely discussed in much contemporary social science, including that of globalization, although its specific character of domination is as powerful as television or the computer normally viewed as constitutive of global culture" (2006, 17). ENERGY studies, as a new approach to social and human sciences, must also address the car, for it is among the most intractable problems facing the global commons in the twenty-first century.

See also: AMERICA, AUTOMOBILITY, EXHAUST, LIMITS, ROADS, RUBBER.

Automobility

Lindsey Green-Simms

In *The Life of the Automobile*, Soviet critic Ilya Ehrenburg writes, "Cars don't have a homeland. Like oil stocks or like classic love, they can easily cross borders . . . . The automobile has come to show even the slowest minds that the earth is truly round" (1929, 167). Ehrenburg's semi-fictional chronicle of the rise of the AUTOMOBILE can help us understand the specific, paradoxical ways that different subjects experience automobility in a world that is increasingly linked through technologies yet profoundly uneven. His interwar tour de force addresses the combined pleasure and violence of the system of automobility as it was experienced not only by modern European and American drivers but also by accident victims, women, immigrant factory laborers, strike leaders, and RUBBER plantation workers across the world. Beginning with the first patent for a gas engine, Ehrenburg juxtaposes discussions of the automobile as a public good, a sign of global prosperity, a worldwide business empire, and an enabler of speed and autonomy, on the one hand, with stories about the new oppression of machines, the destruction of NATURE, gruesome accidents, ecological DISASTER, and noxious gases, on the other. In his dialectical account of the global life of the automobile, he contrasts the speeding cars of New York, Paris, London, and Berlin with the "peripheral" world regions that are mined for the resources that enable modern automobility. Thus, his work implies a link between the traffic created by automobiles and the traffic in consumer goods, labor, and raw material that makes this individual mobility possible. What *The Life of the Automobile* offers is one of the earliest accounts of an ideology and system of automobility in which cars and the pleasures of driving are deeply entangled with forms of oppression and environmental degradation.

Since the beginning of the twentieth century, the term *automobility* has been used colloquially to describe the type of movement associated with the motorcar.[1] Though Ehrenburg had set the stage in the 1920s, it was not until James Flink's *The Automobile Age* (1988) that cultural critics began highlighting the automobile's ideological promise of autonomous mobility. Flink argues that the car engendered an age of "mass personal automobility," stressing the individualized and privatized nature of this collective phenomenon. But as Mike Featherstone (2005) argues, the *auto* in automobile initially referred to the idea of a self-propelled motor vehicle free from dependence on an ANIMAL; the automobile was a self-directed vehicle liberated from the restrictions of a rail track, able to move in any direction (ibid., 1). *Automobility* therefore expresses how the autonomy of the self-moving vehicle enables a freedom of movement for its drivers: the autonomous machine merges with the autonomous human on the open road, producing speed, adventure, and unfettered mobility (Featherstone 2005, 1–2; Urry 2005, 26). John Urry (2005) takes the notion of automobility one step further by demonstrating how not only cars and drivers but also roads, traffic signals, parking, and fuel constitute a self-organizing, *auto*-poetic system. For Urry, *automobility* captures both the humanist notion of autobiography, or self-making, as well as the machine's capacity for automation. The idea of an autonomous human is linked to an automatic machine through a network of ROADS, technologies, and policies.

However, immediately after defining the system of automobility as such, we can see that it involves an auto-contradiction, for it is technically impossible to be both a fully autonomous human subject and one who is dependent on both a machine and the entire sociotechnical institution that supports it. It is for this reason that Steffen Böhm, Campbell Jones, Chris Land, and Matthew Paterson, the editors of the collection *Against Automobility* (2006), have proposed viewing automobility as a disciplinary regime rather than a self-organizing system. They argue that automobility is not a smooth auto-poetic network but rather one characterized by core antagonisms. Pointing to the proliferation of traffic jams, the dependency on non-renewable resources, the ecological devastation entailed in roads and engines, as well as the prevalence of accidents, they assert that within the regime of automobility, "the pursuit of individual mobility becomes collective immobility" (9). For these scholars, automobility is fundamentally an "impossible" system, not simply because it is unsustainable given the finiteness of fossil fuels but also because it is inherently contradictory: it produces as much inertia and destruction as it does mobility. Thus, like the socially determined human subject, automobile and driver exist within a system of dependent relations. These dependencies include, but are by no means limited

1. The *Oxford English Dictionary* defines *automobility* as "the use of automobiles or motor vehicles as a mode of locomotion or travel" or "mobility by means of an automobile or motor vehicle" and dates its usage back as early as 1903. "Automobility, n." OED Online. June 2016. Oxford University Press. http://0-www.oed.com.catalog.multcolib.org/view/Entry/13483.

to, manufacturers, laborers, laws, police officers, roads, signs, advertisements, geographers, oil companies, gas stations, and gas-station clerks.

Automobility can therefore never be a pure expression of autonomy or a manifestation of the free individual on the unconstrained open road, because neither the individual nor the automobile is unencumbered. In this way, the fantasy of unfettered automobility is linked to what Imre Szeman (2011) calls the "fiction of surplus," the belief that there will always be enough fuel and that daily life will never be hindered by a CRISIS of ENERGY, despite the very obvious LIMITS to nonrenewable energy (324). Likewise, automobility persists as an ideal: people cling to the notion of the automobile as a private cocoon that takes its owners wherever they want to go, whenever they want to go there. As Sudhir Chella Rajan argues, the automobile is "the (literally) concrete articulation of liberal society's promise to its citizens" (2006, 112–13). It is what Kristin Ross (1996) calls the "key commodity-vehicles" (38) of a post-Enlightenment order that has not yet given up on the idea of the free and autonomous individual. An ideology of autonomous movement, automobility obscures the fact that mobility is always dialectically related to immobility and that autonomous movement is always dependent on something else, as Ehrenburg intuited as early as 1929. In other words, seeing automobility as a self-making process (per Urry) obscures the fact that automobile ownership is the privilege of the few who depend on both the labor of the many and the energy that has caused social and environmental degradation in places like the Niger Delta, the Persian Gulf, and the Louisiana coast. Moreover, as entire cities and regions are built around the car at the expense of alternative forms of transportation, citizens become locked into a system in which the options for mobility are extremely limited and the dependency on fossil fuels multiplies.

The increased popularity of the Hummer and SUV on American roads during the second US-Iraq War provides an example of the way in which automobility is a system of dependent relations that is paradoxically valued as an ideology of autonomous movement and freedom more broadly construed. Charissa Terranova argues that the Hummer, a brand of SUV based on the US military's High Mobility Multipurpose Wheeled Vehicle (or Humvee), operates within an economy of desire rooted in the false idea that "in owning a Hummer one participates in a vast matrix of martial security" (2014, 189). The Hummer, she writes, aligns its driver with pre-emptive war and participates in a myth "of endless oil reserves . . . American exceptionalism . . . radical atomization . . . national machismo and government-less self-rule" (192). In other words, the prevalence of Hummers in America's suburban parking lots, highways, and popular television shows seemed to propel the fantasy that gas-guzzling vehicles were somehow not the cause of the war but a sign of American freedom, self-determination, and toughness that were only bolstered by the 9/11 attacks. Of course, while Hummers might be icons of self-sufficiency and absolute independence, the car depends on oil reserves abroad, refineries, tankers, the American military that protects oil interests (with their Humvees), workers for General Motors who manufacture the Hummer, and US tax payers who support the military and who, as it turned out, bailed out General Motors. The Hummer, then, like any car but even more so, is in practice far from being a tool for autonomous mobility. It is connected to a global system of people, energy sources, laws, and regulations that are interconnected and entangled and, as Ehrenburg writes, have "come to show even the slowest minds that the earth is truly round."

See also: AFFECT, AMERICA, AUTOMOBILE, ENERGOPOLITICS, EXHAUST, LIMITS, MIDDLE EAST, ROADS, RUBBER.

Boom

Brenda K. Marshall

The first boom was wheat.

In 1873 the Northern Pacific Railroad (NPRR), the recipient of a federal land grant to build a line from Duluth to Puget Sound, ran out of money at Bismarck, Dakota Territory. With newspapers in the east deriding this "wild scheme to build a railroad from Nowhere, through No-Man's-Land to No Place," the NPRR came up with a strategy to advertise the fertility of this No-Man's-Land: "bonanza" farms owned and operated mostly by eastern industrialists who exchanged increasingly worthless NPRR stock for huge parcels of land abutting the railroad.

Stories of the productivity of the bonanza farms in the Red River Valley spread worldwide, drawing immigrants to the new Northwest to homestead 160-acre parcels. The narrative of boom drew the homesteaders in, but it was the pioneer narrative of independence that gave them the strength to remain on the rich, but often brutal, treeless northern plains through seasons of hardship, wind, snow, and isolation. This narrative of independence was never accurate. The NPRR set storage rates for the railroad-owned elevators, as well as transportation rates to ship grain east to Minneapolis millers, who determined the grade, and thus the worth, of the wheat.

The narrative of independence remains as false, powerful, and necessary as ever in the Bakken oil boom. The Bakken Formation spans about 200,000 square miles in western North Dakota, Montana, Saskatchewan, and Manitoba. Although wells have been drilled there since the 1950s, until 2007 the bulk of the oil and gas was locked in a rock formation with low permeability. Horizontal drilling and FRACKING changed that. North Dakota is now the second leading producer of oil in the United States (after TEXAS), with production topping one million barrels per day (US Energy Information Administration 2014). Fracking technology makes the success rate of each well about 99 percent. The formation is predicted to continue to produce for thirty years (Bob Johnson 2014).

The western counties of North Dakota where the Bakken lies were settled during a second wave of immigration in the early 1900s, soon after statehood. The arid climate and topography—short-grass prairies dotted with craggy buttes—are nothing like the Red River Valley along the state's eastern border, the site of the wheat boom. North Dakotans dislike being confused with South Dakotans, but they also insist on a distinction between west and east "of the [Missouri] river," with the east having been the locus of money and power.

But with a new boom—oil—power has shifted.

The oil boom created a surge in jobs and money. At the height of the boom, North Dakota led the nation in population growth. Per capita income in the boom's epicenter was twice the state's average, nearly triple the US average (Bob Johnson 2014). Truck drivers in the oil fields make six figures annually. A local McDonald's franchise offered $15 per hour and a $300 signing bonus (Gruley 2012). One in seven private sector jobs in North Dakota is tied to oil and gas development (not including the demand for food, housing, health care, and government services) (Nowatzki 2014). North Dakota's unemployment rate dipped to 3 percent; in the western counties, less than 1 percent.

A boom is not all sweetness and cash. Housing is scarce and often unaffordable. A two-bedroom apartment that rented for $500 per month before the boom cost $2500 (Gahagan 2014). Thousands of newcomers live in "man camps" that range from relatively cushy facilities with eating, laundry, and recreation on site to clusters of small campers, travel trailers, and skid shacks. Some of the company-owned man camps let rooms for twelve hours. Then the rooms are cleaned and new occupants move in for the next twelve hours. In 2013, North Dakota saw the biggest surge of homelessness in the Union (B. Ellis 2014). An estimated 25 percent of high school students are homeless in Watford City (Bob Johnson 2014), which grew from 1,600 to 12,000 residents between 2010 and 2013 (B. Ellis 2013). Public school teachers (who earn a statewide average annual salary of $47,344) are offered subsidized housing.

Public services are stretched. Nine out of ten vehicles on the highways are trucks. Country roads are crumbling and traffic fatalities soaring. Regulations lag behind the dumping of oil field waste, which includes millions of gallons of salty, chemical-infused wastewater, not all of which makes it into underground disposal wells. There are leaks and spills and illegal dumping.

In Watford City I first heard the rhetoric of boom displaced by more localized talk of "immigration," a seemingly odd term to describe the influx of fellow US citizens. Beyond its suggestion of insularity, the term expresses the hopes (and anxieties) of people who, although deeply unsettled by difference, know that this storm of change is unstoppable.

North Dakotans know something about how to survive a storm. The weak, they tell you, complain; the hardy square their shoulders and get to work while they wait for it to pass. But this waiting is never passive; it is always accompanied by stories. The story they tell now is an old one: there were pioneers before; this was the home of immigrants before. Some will stay and others will move on. Those who stay will be worthy; those who leave will be forgotten.

Recently, in a conversation about someone who had disappointed me, a fellow North Dakotan said, "It doesn't sound like he could have crossed the prairie in a covered wagon." This comment was not a non sequitur—even if few twenty-first-century North Dakotans are up for a transcontinental trip in a prairie schooner. What matters is that the frontier language of competence and rugged individualism still resonates. And so the families, the hordes of young men, the entrepreneurs—the strangers, that is, who pour into the state in response to the oil boom, clog the roads, choke the lines at the grocery stores and doctors' offices—all become "pioneers," following dreams of opportunity. Thus they are known, understood, contained.

In the words of one local resident, "The boom didn't care if we were tired or had other plans . . . . In an instant, I saw people that worked hard their whole lives, work even harder. With their heads down and their shoulders square, in the dead of winter, they met the boom head-on. They had no choice and neither did I" (Ruggles 2014). The newcomers have adopted the language: "Like all the other pioneers making the adventurous trek to the overpopulated small towns of western North Dakota with infrastructures the size of a stalk of wheat, we found ourselves living in a man camp lined with dozens of campers and propane tanks" (M. Smith 2014). This is a communal story of hardship overcome, diligence and patience, working and waiting and setting aside.

The discourse of the brave immigrant and the hopeful pioneer provides a reassuring veneer of control. Another discourse at work, however, is freighted with danger, more likely to tear than mend the social fabric: the discourse of wealth. Talking about money in North Dakota has always been in poor taste. Reticence about personal wealth and income is partly due to a regional suspicion of difference; here, it is better to fit in than stand out. But the newcomers, many of whom will tell you they are in North Dakota to make money fast and get out, arrive without this reticence and with a very different ethos about the display of wealth. Long-time residents who are cashing oil checks, in contrast, tell you that it is really the oil companies that are getting wealthy and that the bulk of the oil money is going back to Wall Street. They are not wrong, although this construction sidesteps the fact that the rising tide of oil production is not raising all boats equally. For example, if you own oil-rich land but not the mineral rights, the income from producing wells flows into other hands. In the words of one county extension agent, "It's the good, the bad and the ugly. The good is the guy with seven wells who's a millionaire in 24 hours. The bad are those who own the land but not the mineral underneath, and whose roads are tore up and they're not getting anything. And the ugly is the disparity, which creates a lot of animosity" (Stone 2008).

Ultimately, the history of this boom will be written on the land. So what do the locals think as they drive, slowly, down crumbling roads lined with bobbing and nodding wellheads, and note yet another hill scraped flat for yet another well? How do they abide the lost vistas, the changing community and culture? Some, I imagine, think about their growing bank accounts. Some mourn the good old days of a few years past when the streets, stores, and bar stools were not occupied by strangers. And some assuage their anxiety by returning to the old story of immigrant pioneers, come, once again, to the land of opportunity.

It is a comforting story that need not be true to be powerful.

See also: CANADA, PETRO-VIOLENCE, RESOURCE CURSE, RUBBER, RUSSIA.

Canada

Kit Dobson

Canada offers a case study in what it means to quite literally fuel culture. Since its colonization by European settlers, Canada has built an economy around the extraction of resources destined to feed colonial capitals. In its early days, Canada simply exported whatever was wanted in Europe (furs, mostly). Economist and historian Harold Innis's now-classic articulation of the staples thesis holds that Canada developed as it did because of its grounding in the export of staple goods like fur, fish, and other raw materials. Innis maintains in *The Fur Trade in Canada* (1930) that "energy in the colony" of early Canada was "drawn into the production of the staple commodity" (1970, 385) and that this production enabled the country to develop rather than simply be exploited. Fast-forward to the present, and we see that Canada has not changed much: it continues to export raw materials, except that the materials have fluctuated along with the markets. Rather than exporting materials to European markets, today Canada exports first and foremost to the United States but also increasingly to developing markets in Asia. The raw materials have shifted too, most markedly away from fish (after the collapse of the Atlantic cod fisheries in the 1990s) and furs (since beaver hats are out of vogue) and toward bitumen from northern Alberta and other resources (lumber, potash, and more). Mel Watkins, whose classic article "A Staple Theory of Economic Growth" provides, in tandem with Innis's thinking, the foundational articulation of the staples thesis, argues that today the "resource-based structure" of the Canadian economy has "actually deepened" (2013).

We can understand the consequences of such an economic model in several ways. One would take us through Marx's writing on colonial spaces that the European metropole required for the expansion of markets (as expounded both in the *Manifesto* and *Capital*). The expansion of Canadian markets, according to a Marxist analysis, relies in turn upon external markets: Canada grows economically by selling off little bits of itself (some RENEWABLE, others less so) to markets that maintain a demand for the resources that it can supply and by exporting manufactured goods as well. In its emphasis on trade through export markets, Canada continually relies upon what geographer David Harvey terms "the spatial fix" in order to resolve the internal contradictions of its own markets (2000, 23): it adopts, in other words, its particular model of resource extraction in order to resolve its economic problems through market impacts elsewhere by selling raw materials—or sometimes completed ones like Blackberry mobile devices—to other markets. Canadian markets, society, and culture, in this view, are developed through extraction and export.

A second way of understanding Canada's situation comes through *Bataille's Peak*, Alan Stoekl's reworking of Bataille's notion of expenditure. Stoekl's invitation to read oil dependence in terms of expenditure might suggest that, rather than merely depleting its own resources in an effort to stave off a faltering economy (which a classical Marxist analysis would suggest all economies do), Canada is perhaps engaged in something quite different. By acknowledging that power is ultimately not infinite (as the sun will one day go supernova, likely long, long after the human species no longer exists), one might conclude that conscious acts of expenditure may more realistically reflect the LIMITS placed on the human and global lifespan. If we take the theoretical Hayflick limit (named for Leonard Hayflick), for instance, to be the upper limit of the possibility of human life, then it becomes clear that the notion of SUSTAINABILITY (criticized by Stoekl) is, over a long enough timespan, insufficient for thinking about our ENERGY use: we will die; we will run out of energy; we cannot sustain our individual lives forever. Human life is finite; the Hayflick limit suggests that human cells can only divide a certain number of times before they become irreparably damaged. This limit would set the maximum human lifespan at approximately 120 years. The idea of sustainability that Stoekl critiques is one that advocates perpetual motion—or, rather, a vision of sustainability that maintains the goal of developing human systems that are indefinitely self-sustaining. In his argument for "postsustainability," Stoekl suggests that consciously choosing particular energies to expend might instead offer a means of making ethical choices about how our energies will be spent rather than focusing on preserving all energies in the spirit of sustainability. And doing so, too, might provide a counter-logic to Canadian resource extraction, as the energies extracted from Athabasca bitumen, for instance, could then be consciously resisted.

A version of this thinking can be traced through Matthew Tierney's *The Hayflick Limit*, a collection of poems on chess, science, personal observation, and the names of various fears that, taken together, convey the fallibility of human action. The economics of perpetual growth appear continually, and derisively, as in the poem "Temperance St.":

> Tomorrow there will be less, the wreckage carted out
> to landfills overflowing with conviction.
> Tomorrow stretches out its arms, embraces the present
> perfect continuous: Have. Have *been*. Emerging.

> Tomorrow is the afterworld, a surface unbroken
> by need: life forms engulfed in never enough, pawing space
> to breathe. (2009, 19)

Here the poet invokes a notion of tomorrow as an imaginary realm, an afterworld overflowing with DETRITUS that fulfills every need yet clogs even the air. Tomorrow is both "less" and more, more stuff that overflows the landfills of Canada and elsewhere. It is a space and time in which, to invoke the title of another of Tierney's poems, the seeming "Perpetual Motion Machine" of the earth and the economy reach their limits.

The poem "The Chess Player" opens with a bleak take on life, using the chessboard as a metaphor:

> Theory:
> the ideal position is the first,
> each move a further weakness,
> a giving way of perfection.
> In this sense,
> the game is meaningless. (2009, 45)

The poem goes on to recount the relationships of both the speaker's brother and chess prodigy Bobby Fischer to the game, yet this "theory" also reads as a gloss on human life and the overriding notion of cellular division with which the book concerns itself. Rather than being founded upon perpetual renewal, perpetual motion, or perpetual growth, here "the ideal position" is that with which we begin, and everything else is a form of degradation. If this "theory" is read as a stand-in for cell division, human life, and what often goes under the rubric of "development," it offers a pessimistic take on all three.

The final irony of *The Hayflick Limit* is that Tierney's book of poetry itself represents a possibility for some sort of development: either the development of a concept or the creation of a book where previously there was none (and hence, possibly, an improvement on the original, "ideal" position). As a book published by Coach House Books in Toronto, it is also a development away from a staples model of economics, as it is printed on paper manufactured in Canada and printed and bound in Toronto (hence the completed product is made domestically). In this sense, the logics of science and economics might nevertheless provide the impetus—fuel—for this particular cultural artifact, demonstrating that the very concepts of the "Perpetual Motion Machine" or the Hayflick Limit can, in turn, create cultural works. Indeed, stereotypical Canadian culture, from Alice Munro and Margaret Atwood to hockey and lumberjacks, is all, with varying degrees of removal, "fueled" by the natural environment. Over a long enough time span, the economic model of perpetual growth under which capital operates today will fail. Canada's economy, still hewing to the staples model of extraction and export, will in turn, reach its limit. What cultural workers decide to do with these limits—how they decide to expend their energies—will demonstrate ways in which an economics of extraction might fuel culture.

See also: EXHAUSTION, FUTURE, LIMITS, RESOURCE CURSE, SOLAR.

Catastrophe

Claudia Aradau

> The electric grid, as government and private experts describe it, is the glass jaw of American industry. If an adversary lands a knockout blow, they fear, it could black out vast areas of the continent for weeks; interrupt supplies of water, gasoline, diesel fuel and fresh food; shut down communications; and create disruptions of a scale that was only hinted at by Hurricane Sandy and the attacks of Sept. 11.
>
> —MATTHEW WALD, "As Worries over the Power Grid Rise, a Drill Will Simulate a Knockout Blow"

Invocations of fear, attacks, and adversaries have long been characterized as security imaginaries. More recently, the prospect of catastrophic disruption has led security professionals across the Western world to draw up new scenarios of the worst still to come and to prepare exercises for inhabiting the catastrophic futures they have imagined. More established threats insidiously morph into unexpected, unknowable, and unpredictable catastrophic events that can erupt anytime, anywhere. Over the past few decades, security has come to be appended to almost everything: human security, food security, water security, energy security, climate security, GENDER security, cyber security, data security, and so on. If security has become the "fetish of our times" (Neocleous 2008, 9), it is in a different guise than the defense of sovereign territory or preventive management of populations.

Security is articulated in a temporal logic in which the threat that must be countered at all costs is that of the unexpected catastrophe to come (Aradau and van Munster 2011). Security experts have relinquished the hubris of taming epistemic uncertainty; disruptive events in the FUTURE cannot be known or neutralized. Instead, we are enjoined to adapt to regimes of unknowability and unpredictability, where "generic events" (B. Anderson 2010) can erupt at any point. In harnessing the imagination toward the future event that punctuates the stability of the present, a future perfect is summoned to contain the event in the indeterminate "as ifs" and "what ifs" of worst case scenarios. Catastrophe after 9/11 entails a double move of bureaucratizing imagination (Bougen and O'Malley 2008) and minimizing discontinuity. The logic of catastrophes to come recasts the circulation of oil, gas, or electricity through the sinews of capitalist markets as continuously prone to interruption and disruption. This generic interruption subsumes risks, crises, hazards, disasters, and catastrophes in a continuum of events whose unexpected, unpredictable, and utterly surprising occurrence fabricates new political coordinates of thinking and acting.

Un/expected Events

In 2012, the Department of Homeland Security declassified a report it had commissioned five years earlier on the vulnerability of US electricity systems to terrorism. Tasked to imagine the effects of a potential terrorist attack, scientists from the US National Research Council unleashed imaginings of the worst to come:

> A terrorist attack on the power system would lack the dramatic impact of the attacks in New York, Madrid, or London . . . . But if it were carried out in a carefully planned way, by people who knew what they were doing, it could deny large regions of the country access to bulk system power for weeks or even months. An event of this magnitude and duration could lead to turmoil, widespread public fear, and an image of helplessness that would play directly into the hands of the terrorists. If such large extended outages were to occur during times of extreme weather, they could also result in hundreds or even thousands of deaths due to heat stress or extended exposure to extreme cold. (National Research Council 2012)

Although not immediately spectacularly catastrophic, the attack gradually becomes so, multiplying dead bodies and intensifying fear. Alongside the politics of resource dependence and neoliberal concerns with privatization and price fluctuations, imaginings of future events with a catastrophic impact on the grid—gas, ENERGY, or transport—activate new logics of security. A potential attack surpasses the geopolitical logic of war, while unexpected disruptions unsettle the logic of efficient, continuous supply. ENERGY SYSTEMS are continually disrupted by events that remain unknowable and untamable. What remains of energy security? A drill, a rehearsal of the un/expected event that disrupts the continuity of supply, the flows of energy, the virtual NETWORKS, and the circulation of money sustaining the grid.

Securing the production and circulation of energy becomes an exercise in insecurity—playing out a future of catastrophic events and worst case scenarios. As security imaginaries insert the unexpected event in the realm of governmentality, experts are asked to imagine that which always eludes them. Understood in its etymological sense of an "overturning" point in a theater play, catastrophe is the moment of radical interruption, the world upside down. It breaks with modern understandings of progress and linear time. It also breaks with the understanding of serial and probabilistic temporality that Foucault assigned to biopolitics. Statistical, probabilistic reasoning discerns the future through the calculability of mass phenomena and the computation of frequencies. As a consequence, the future emerges as "a domain of finite possibilities, arranged according to their greater or lesser possibility" (Koselleck 2004, 18). Disruptive events attend to unlimited possibilities, to a future always open, always emergent, but always already tilted on the brink of catastrophe.

On/off the Grid

From its theatrical origin, *catastrophe* preserved the element of "overturning" while only gradually acquiring negative connotations. As late as the early nineteenth century, uses of *catastrophe* outside the dramatic context indicated the negative sense by adding adjectives, as in "sad catastrophe" ("Coroner's Inquest" 1803), "melancholy catastrophe" ("Melancholy Catastrophe" 1805) or "shocking catastrophe" ("Shocking Catastrophe" 1805). Moreover, *catastrophe* and *revolution* were used interchangeably. *Catastrophe* entered the domain of natural sciences to refer to geological and biological transformations. *Revolution*, which initially referred to astronomic cycles, took the reverse path and became the signifier of "overturn" in human history and political action. The substitutability of *revolution* and *catastrophe* in the eighteenth century derives from the similar temporality of interruption they enact. Catastrophes and revolutions overturn the present and activate security responses of containment and neutralization. Today, catastrophes can be neither prevented nor contained. Yet, the injunction to imagine the worst stumbles at the threshold of the destruction of capitalism (Jameson 1994). One might say that today, embroiled in the security apparatus, catastrophes fall short of revolution. Ultimately, the catastrophic interruption of the power grid does not undermine capitalist processes of circulation and ACCUMULATION.

Yet, it is not in the limitation of destruction that the new logic of security is most insidious. The production of worst case scenarios reproduces a double threshold: that between normality and exceptionality and that between imagination and knowledge. The normality of catastrophe in the processes of capitalist reproduction is effaced by the looming presence of catastrophic disruptions in the future, stripped of conditions of possibility, politics, and history. The injunction to imagine possible catastrophes-to-come is ultimately an injunction to ignore what we know today about already unfolding catastrophes. Or, as Günther Anders warns (2008), it is a sign of a lack of imagination so profound that we are unable to perceive what we see or to acknowledge the catastrophes that take place before our very eyes.

See also: CHANGE, CRISIS, DISASTER, GRIDS, NETWORKS, OFF-GRID, RISK.

Change

Ian Buchanan

> The predominant mood continues to be not indignation, or enthusiasm; it remains a depoliticized quietism.
>
> —PERRY ANDERSON, "Homeland"

It is simple, really. Our dependency on carbon fuels is jeopardizing our only planet home. To avert DISASTER, we must either switch to a more sustainable fuel source or find technological solutions to the environmental problems we have created. But who is this "we" and how can "they" effect the necessary changes? It cannot be done alone; no individual can pull off this miracle herself. It cannot even be done one country at a time. It will require a coordinated global effort, one that changes our very conception of change.

Political science has three main theories of how change of this sort happens: revolution, reform, and revelation. *Revolution* means the overthrowing of a government by violent or peaceable means; *reform* refers to gradual, internal transformation of the way a government rules, not always to the betterment of the people or the nation; and *revelation* is any sudden and widespread political change of heart, possibly but not necessarily religious in character.

In the West, only the last of these three theories retains any credence; its place in contemporary political thinking is nonetheless uncertain because revelation in a secular age is often understood reductively, as a variety of fanaticism. This may explain why it has been so hard to measure the achievements of the Occupy Wall Street (OWS) movement, which epitomizes the revelation model. As important as OWS was in raising awareness about global inequality, it did not force any political changes, which might make it seem a failure. To the disappointment of many, it was reticent about even demanding change; its assertion that the US political system was no longer a legitimate form of government explicitly rejected the reform model's assumption that meaningful change can occur through existing channels of the state.

In this respect, OWS differed significantly from the great social movements of the recent past—the battles for equal rights for women, nonwhites, and gays and lesbians. It explicitly broke faith with reform. It proposed no legislative changes and did not turn to the idea of revolution, which effectively left it in political limbo. It seemingly closed off two of the models for change. Instead of facing a double dead end, however, OWS opened up a new kind of "smooth space" (to use Deleuze and Guattari's useful phrase: the undetermined space between fixed points): revelation. The problem is that, for the time being, revelation can be difficult to distinguish from depoliticized quietism.

I would argue that revelation is the conceptual chasm between reform and revolution where political thinking must operate now. The apparent quietism of the Occupy movement was not a failing, therefore, but rather a symptom of the present impasse: we want change, but we do not know how to achieve it. Those in power appear to be caught in the same paralyzing trap, epitomized by the political and structural constraints the Obama administration has faced. From this situation Perry Anderson (2013) draws three conclusions about the politics of the present: (1) cultural change is tactical rather than strategic; (2) the political party system is paralyzed by its inner contradictions, particularly its dependency on campaign donors, and is therefore unlikely to generate systematic economic change; and (3) because there is no groundswell of opposition—"*the people are missing*," as Deleuze put it (1989, 216)—the government is not compelled to do anything more than offer up cosmetic changes (legalization of undocumented immigrants, gun control, marriage for all, and so on) presented as momentous concessions. In effect, Anderson suggests that reform is no longer viable.

This is the state of politics today: We fight hard to achieve small concessions and we believe they are larger than they are, in a context where it is understood that larger changes, such as a shift to socialist democracy or a post-carbon ENERGY REGIME are out of the question (P. Anderson 1983, 27). Armed revolution is also assumed to be out of the question, a conclusion that most Western political theorists have come to accept in the post-Vietnam era.

Even someone as consciously inflammatory as Slavoj Žižek is cautious about advocating armed revolution. In an essay about mass protests in Turkey sparked by the government's decision in 2013 to allow a portion of Istanbul's Gezi Park to be turned into a shopping mall, he writes: "Just because the underlying cause of the protests is global capitalism, that doesn't mean the only solution is directly to overthrow it. Nor is it viable to pursue the pragmatic alternative, which is to deal with individual problems and wait for a radical transformation" (Žižek 2013, 11). What option is left, then? Žižek's answer is that because global capitalism is inconsistent—it gives with one hand and takes away with the other—the thing to do is "demand consistency at strategically selected points where the system cannot afford to be consistent . . . . Such demands, while feasible and legitimate, are de facto impossible" (ibid.). This reworking of the old utopian "demand the impossible" slogan is fine neither in theory nor in practice because demanding the impossible is not the same thing as obtaining the impossible.

This "demand" strategy is doubly flawed: not only does asking not amount to getting (we might call this the *Oliver Twist* fallacy), but sometimes getting what one asks for creates its own problems (the *Great Expectations* fallacy). Consider Fredric Jameson's argument about the utopian demand for universal employment. On the one hand, it would require major changes in economy and society, not least the belief in the structural necessity of some amount of unemployment. On the other hand, even to consider the idea begins to generate the opposite demand, namely the end of work itself and a so-called right to laziness (Jameson 2005, 147–50). Some petro-states achieve a unity of these opposites; Kuwait guarantees jobs for all its citizens even though there is not enough nonmenial and nonstrenuous work to go around. The net effect is a fully employed, wealthy citizenry that nonetheless lacks purpose, caught between a postmodern fascination with commodities and a reactionary (but no less postmodern) fascination with religious orthodoxy.

Alain Badiou unequivocally rejects this "demand" option: "Just as our states and those who vaunt them (parties, trade unions and servile intellectuals) prefer governance to politics, so they prefer demands to revolt and 'orderly transition' to any rupture" (Badiou 2012, 107). His alternative is mass uprising to overthrow the state, which is equally problematic, not least because in advanced capitalist states like the United States it is inconceivable. In part this is a failure of the imagination, along the lines of Jameson's famous statement that "it seems to be easier for us today to imagine the thoroughgoing deterioration of the earth and of nature than the breakdown of late capitalism" (F. Jameson 1994, xii). Perhaps we can be forgiven this collective failure of imagination, just a little. Hollywood is probably correct to assume that only a cataclysmic event—say, an alien invasion or global natural disaster—could galvanize the people to effect regime change in the world's most powerful state.

Even that might not be enough, considering the weak response to the slow-motion global CATASTROPHE that is climate change. So what is left? We need to create new forms of "smooth space" where revelation can take place; we need to cultivate a climate of thought in which the truths of our situation compel us to act rather than leave us paralyzed. To bring about change, we need to change how we think about change: even if revolution and reform are off the agenda, we are not necessarily condemned to the status quo. But if we do not exercise our imaginations, we will be spinning our wheels until we run out of gas.

See also: CRISIS, FUTURE, SUSTAINABILITY, TEXTILES, UNOBTAINIUM.

Charcoal

Caren Irr

While grilling outdoors is a nostalgic leisure activity pursued by many Westerners, WOOD is the primary fuel of the poor throughout the developing world—especially in sub-Saharan Africa where it is mainly used for cooking. The World Future Council estimates that 80 percent of Africans rely on biomass (wood and charcoal) for their energy needs. The bulk of biomass energy involves combustion of unprocessed fuelwood, but a significant and growing percentage results from charcoal burning in urban settings. Producing charcoal requires burning several times as much per unit of energy as one uses when burning fuelwood directly; charcoal is inefficient and expensive to produce, in addition to being unsustainable. Regions where the lucrative industry of producing charcoal for cookstoves has taken hold see rates of deforestation that vastly exceed rates of tree planting; Africa is losing its forests twice as quickly as other continents. Meanwhile, even though burning charcoal releases less harmful smoke than burning fuelwood, it is also a high polluter that creates serious health effects for producers and consumers alike and releases methane and carbon dioxide in amounts that demonstrably contribute to climate change.

When Western environmentalists talk about charcoal, they usually try to imagine alternatives: cow dung, biogas, oil, reforestation, new kilns, new cookstoves. They celebrate the efforts of activists such as Nobel Laureate Wangari Maathai, because they think about African forests as a "global carbon sink" necessary for cleansing the atmosphere of the earth as a whole, and they hold up politically unlikely efforts to plant a Great Green Wall of trees across central Africa as a talisman against the prospect of desertification on a massive scale. However, for skeptics like Binyavanga Wainaina, these idealistic initiatives join the $100 laptop, windup radios, and other upstanding, well-intentioned but shortsighted development projects (Wainaina 2007). Wainaina consigns these efforts to a dustheap of eccentric experiments, and in so doing he speaks for any number of East Africans who see charcoal burning as a development opportunity that is here to stay.

For many African business and government leaders, charcoal is an important growth industry and one that has generated its own structures of regulation, management, and taxation, as well as new smuggling and black-market routes. Charcoal is prized because it substitutes for expensive imported energy sources—especially oil—and provides needed economic opportunities for vulnerable populations with few options. In this context, the problems raised by charcoal circulate mainly around transport issues and future prospects for the industry when massive swaths of land are "grabbed" by foreign investors who limit foraging and forest access. In the African press, global environmental worries connected to the ongoing use of charcoal usually take a back seat to pressing social needs in the present.

Western environmentalists need to recognize the real social concerns of human populations that rely on charcoal burning and biomass fuels more generally to survive. Neither a Druidic romanticism about trees nor a hypocritical demand that Africans immediately change their energy consumption practices at a pace that Europeans and Americans have not been able to manage should guide action on this issue. Those working to mitigate the impact of charcoal burning on the climate need to consider political and economic as well as environmental priorities. After all, cooking with charcoal is just as aesthetically enjoyable in Kenya as it is in Kansas, and if charcoal could be produced in a sustainable manner (e.g., by making briquettes from fast-growing bamboo rather than slower-growing rain forest species) it might well prove a more enduring fuel than petroleum products. Charcoal is an ancient form of fuel, and if stewarded well its reputation as an environmentally unfriendly resource could and should improve. But with or without changes to the way charcoal is produced and consumed, it needs to be taken seriously if we are to foster an environmentalism of and for the poor.

See also: ACCUMULATION, ANTHROPOCENE, LIMITS, SUSTAINABILITY.

China 1

Arif Dirlik

"China Is Choking on Its Success" (Pesek 2013)
"Forget the New Air Pollution Plan, GDP Growth Is Still King in China" (Nan 2013)
"How to Fix China's Pollution Problem? It May Not Be Able to Afford It" (Vanderklippe 2013)
"China's Response to Air Pollution Plan Poses Risk to Water Supply" (Huawen et al. 2013)
"Intense Smog Is Making Beijing's Massive Surveillance Network Practically Useless" (Riggs 2013)
"Beijing Slashes Car Sales Quota in Anti-pollution Drive" (2013)
"China Faces New Car Explosion" (Holloway 2013)
"China's Soviet-Style Suburbia Heralds Environmental Pain" (2013)
"The Coming Age of Coal" (Magstadt 2013)
"Newest Pollution Concern: 'Ugly' Sperm" (Deng 2013)
"China's Smog Threatens Health of Global Coal Projects" (Wong 2013).

These headlines date from a short period of no particular eventfulness in late 2013. They are typical of reports that are increasingly streaming out of the People's Republic of China (PRC). Attention to the social and environmental toll of development in China now claims nearly equal time to celebrations of development and its yet-to-be fulfilled promises. Such gloomy prospects are of deep concern among the leadership and the population at large, especially the latter, who have to live with the negative consequences of development even as they benefit from it.

Headlines are intended to lure the reader. Even so, these headlines and their reports offer cause to wonder whether energy consumption and policy in the PRC signal the end of the world as we have known it. Not just the end of the world constructed under US domination over the last half century, a way of life that the PRC ardently desires. Nor "the demise of the capitalist world economy" of which the PRC is theoretically a nemesis but in practice an integral part (Minqi 2008). I mean the world as earth, a livable habitat for its denizens. So rapid has been the PRC's rise to a world economy, so voracious its consumption of resources, and so far-reaching the environmental destruction wrought by development, that it evokes metaphors of apocalypse—as with the chronic "airpocalypses" in North China (see, for instance, Duggan 2013). We have become inured to photographs of coal waste billowing out of towering chimneys, automobiles barely creeping along highways jammed as far as the eye can see, ghostly shadows shrouded in gray smog in a seemingly futile search for their destinations, and pig carcasses by the thousands floating past glittering skyscrapers. The pollution China generates is not confined to the domestic space of the PRC. Meteorologists have detected mercury from Chinese coal plant emissions on Mount Bachelor in Oregon (which had purportedly been the cleanest place in the United States) (Kirby 2011). The "rise" of the PRC seems to confirm environmentalist fears about the threat that unbridled development poses to the FUTURE of the earth as a habitable environment.

The PRC is presently the largest energy consumer in the world and since 2007—when it surpassed the United States—the number one polluter. If it continues along its trajectory of the last two decades, consumption of COAL, oil, natural gas, and other resources will approach current levels of production worldwide—levels that are not likely to increase in proportion, if at all. Already, it imports resources that can no longer be provided by domestic production. The PRC is also a world leader in the production of RENEWABLE—including GREEN—energy, but these still make up only a small portion of its needs.[1] One solution to China's growing energy DEMAND might be the expansion of NUCLEAR energy; given the dangers, as witnessed in the recent meltdown of the Fukushima plant in neighboring Japan, how far the PRC is willing to expand in this direction remains to be seen.

The search for energy is a major determinant of foreign, trade, and military policy for the PRC in relation to its immediate region and across Central Asia and also extending to Western Asia, Africa, and the Americas. Aggressive claims on the Southeast Asian Sea and the Senkaku/Diaoyutai Islands are motivated at least in part by securing oil and natural gas deposits. While the possibility of military conflict is greatest in these areas, concern over protecting sources of energy is a major motivation behind the PRC's naval militarization. The inevitability of resource wars is already part of the chatter among strategic planners in the PRC and the United States.

Given this situation, it is remarkable that the PRC's "development" continues to be the subject of hope and celebration. Improvement of popular welfare is good cause for celebration everywhere. The impressive rise in the standard of living over the last two decades has lifted hundreds of millions out of poverty and enabled unprecedented freedom of self-expression and mobility. Hopes for continued economic progress and upward mobility also continue to nourish support for Communist Party leadership and its policies, despite widespread dissatisfaction with sharpening class divisions, pollution, abusive officials, arbitrary relocation of populations, and repression of dissident intellectuals and minority populations. Even though the Communist Party has turned its back on the social promises of the revolution that brought it to power, and the PRC shows little hesitation to engage in imperialist and colonial practices "with Chinese characteristics," the memories and language of the revolution persist in official as well as popular assertions of national power. Memories of the revolution also surface in popular grievances against the state, but

1. See US Energy Information Administration (2013). The projections come from the Earth Policy Institute, cited in Richard Smith (2013).

to a much lesser extent, and are less likely to question development itself than unfairness in its execution.

Despite worries about markets, democracy, Tibet, pollution, repression, class inequality, militarization, and so on, many outside observers also celebrate the PRC's development. For Euro/Americans in particular, it would be heartless and arrogant, if not outright racist, to deny the Chinese people—or anyone else, for that matter—the right and privilege of living the life of consumption they themselves enjoy. Anyone who is critical of the colonial past and in sympathy or solidarity with the people of the Global South has reason to celebrate the PRC's development for the benefits it has brought to the people. For many, this development also brings some ideological relief that the people of China are on their way to becoming "like us." One analyst of China's flourishing AUTOMOBILE culture has suggested that "it fosters a postsocialist subjectivity anchored in individual autonomy, agency, and freedom" (Seiler 2012, 358).

Yet it is estimated that if car ownership reaches the same per capita level as in the United States, in a matter of decades there will be more than one billion cars on China's ROADS. Admiration for China's development ranges far beyond concern for the welfare of the people, the majority of whom still live poverty; the kind of poverty they experience, however, will likely become increasingly similar to that in so-called developed societies. Their example should give pause to expectations that poverty, inequality, and pollution will cease if and when China reaches a level of development comparable to the United States. Nonetheless, for all the dangers posed by poverty and pollution, the PRC remains still far behind the United States in the consumption of energy resources and generates only slightly more pollution, despite having a population four times the size.

Unlike during the Cold War, when "communist" societies sought to establish alternative spaces of their own, the PRC is an integral part of the global capitalist economy, and its problems are those of capitalist societies in general. The environmental damage the PRC inflicts at home and abroad is, in large part, due to its function as a global factory. Its complicity is not limited to tacit consumption of its commodities. Euro/American companies, including producers of energy and energy-consuming contraptions, have invested heavily in the Chinese market and are among the most conspicuous cheerleaders of Chinese development, consequences be damned. Questions of energy and environment in the PRC bring us back to capitalism as a global problem. In the midst of a broad (if still politically ineffectual) awakening to the ravages of the globalizing capitalist economy, the developmentalism that has taken hold in the PRC has opened up new frontiers for capital, significantly exacerbating the environmental damage already inflicted by so-called developed societies. The damage is most severe on those—the vast majority of the world's population—who are on the wrong side of an increasingly inescapable class division of global scope. The more socially conscious among critical environmentalists are already well aware of the interrelation of environment, social (in)equality, and democratic politics. What is equally essential is overcoming divides of developed and developing, or north and south. There are no doubt significant differences in need and capability, but the problems of one are the problems of the other if we are to take globality seriously. It is necessary to question modes of development that—against all accumulating evidence—falsely promise sustainability, plunder human and natural resources, and offer the allure of limitless consumption to disguise oppressive inequality, precarious existence, and fatal marginalization. There is nothing north or south or east or west about the fact that human welfare and survival require an equitable and just development that recognizes LIMITS to human desire and nature's bounty. Whether humanity is capable of achieving it is another matter.

See also: ACCUMULATION, COAL, FRACKING, MEDIA, NECESSITY, SUSTAINABILITY.

China 2

Amy Zhang

Demand for ENERGY requires a constant search for new territories of extraction. China's ever-expanding economy has turned to refuse and debris as energy sources. Waste-to-energy (WTE) incineration, which burns garbage to generate energy, carries the promise of transforming trash, previously a constant reminder of the crisis of consumption, into a new power source. What WTE has produced, however, is less an alternative ENERGY REGIME than a new rationale for current consumptive practices.

In spite of a long history of burning waste, modern incinerators now function as an emblem of progress through technological engineering as they transform and reorient matter in service of human needs. In addition to generating ELECTRICITY, modern incinerators often house entertainment facilities—such as swimming pools and revolving restaurants—to showcase how technology overcomes the problem of waste management and can improve lifestyles. The Amagerforbraending WTE incinerator in Copenhagen, which takes heat generated from boilers to make snow for a ski slope, exemplifies what the designers refer to as "hedonistic sustainability," which promises to reward increased consumption, material wealth, and pleasure with more and cheaper energy (Eriksen 2011). Once a lingering reminder of our disposable culture, garbage is now sublimated into fuel for affluence. This irony is lost on planners as they enthusiastically highlight the contradictions: use heat from burning your garbage to produce snow, come for a swim or to enjoy a nice meal with your family atop the trash your wasteful consumption has produced. These projects not only absolve people of guilt; they celebrate the act of consumption and consequent waste as part of a utopian, optimistic modernism where all social challenges can be defeated by clever engineering.

China, meanwhile, is bringing incinerators online at an astonishing rate, thereby provoking widespread contention. Official plans call for 35 percent of municipal waste to be treated by incineration by 2015, up from 1 percent in 2005.[1] Environmentalists have dubbed this decade China's "Great Leap Forward" in incineration, a skeptical allusion to the unrealistic scale and pace of earlier state-led attempts at transformative CHANGE. Ironically, to justify building incinerators so quickly, the state has mobilized the language of CRISIS and frequently refers to a coming "waste siege." In this analogy, waste is decontextualized and the new garbage-producing culture of consumption and disposal of the past thirty years is obscured. The piles of garbage that encircle the city are imagined as autonomous agents; their "siege" is tantamount to attack by an external force. Deploying technological weaponry is the only defense. Behind every high-rise apartment building, luxury shopping mall, and expansive banquet restaurant sits an unruly pile of garbage tended to by countless migrant workers. As the city awakes to clean sidewalks and emptied bins, waste has been expelled to villages, neglected fields, and gutters now filled with colored glass, paper cups, and crushed containers of half-eaten dumplings. In socialist China, the dominant political ideology demanded the demystification of capitalist symbolisms, while in the postreform era of "capitalism with Chinese characteristics" a profound remystification obscures the social and material consequences of daily practices of disposal in the name of consumption-driven growth.

The conversion of waste to energy is not, at the end of the day, a novel innovation of technological modernity. For centuries, people in China have transformed waste into all sorts of fuel and other products. Agricultural manuals from the twelfth century recorded at least four different processes for turning human feces into fertilizer and fuel, and by the seventeenth century there were about eighty conventional waste products (Needham 1984, 295). During the Maoist era, state-run recycling depots were filled with citizens lined up to sell everything from old blankets to empty toothpaste containers. Itinerant migrants roamed back alleys singing and calling out for chicken and duck feathers. In an era of material scarcity, the objects of daily life were repurposed for immediate reuse. In each of these instances, however, reprocessing began with fine-grained sorting focused on separating and disaggregating objects. Recycling, refurbishment, and the transformation of matter focused on the integrity of materials—steel was melted to forge more steel, paper pulp hardened to produce more paper.

Contemporary WTE incineration, however, burns heterogeneous waste—vegetable peels, plastic wrappers—to generate homogenous and interchangeable units of energy. Waste is reduced to caloric value, how much heat an object can generate as it burns. All matter can now be mobilized toward production; the effect is an illusion of consumption

1. By 2014 the state had planned or begun construction of incinerators in many of China's largest cities. In many cases, protests have forced municipal governments to stall or to consider resiting their projects.

without consequence where nothing ends up wasted. This shift necessitates an ethical re-orientation. David Graeber argues that using the cyclical or equilibrium model for understanding recycling leaves out the property regimes that make these circulations possible as objects pass from production to consumption (2012, 287). The idea of an energy cycle, where nothing is lost, prevents us from questioning the ethics of our relationship with the material world and provides a justification to perpetuate existing regimes of spending and disposal. The ability to generate "green energy" by incinerating waste has the potential to displace previous environmentalisms that called for a re-evaluation of consumption.

While the state and industry portray WTE incineration as a new clean energy source, the material consequences of burning waste for fuel present significant challenges in China. Jane Bennett asks whether "patterns of consumption [would] change if we faced not litter, rubbish or trash or 'the recycling,' but an accumulating pile of lively and potentially dangerous matter?" (2010, viii). In China today, matter is increasingly viewed as unreliable, deceptive, and deadly. There is growing doubt about the ability of technology to contain toxicity, and WTE incinerator has come to be viewed as a noxious machine likely to emit poisons including mercury, lead, and dioxin. Surrounded by a carcinogenic landscape, local communities perceive incinerators as hazardous sites. The vital uncertainty of material transformations animates questions around the logic of burning for energy.

In China, WTE incineration is also fueling an emerging urban politics. Left out of the decision-making process regarding large INFRASTRUCTURE projects, citizens, villagers, and environmental activists searching for alternatives to WTE incineration organize forums, lectures, and education initiatives and participate in protests and public dissent. In the process, they raise pressing questions: Who should be included in the decision-making process on the direction of development in Chinese cities? What institutions of oversight and regulation exist to monitor the technologies deployed by the state? Finally, how can we trust the science behind technologies when political influence limits what data count as facts? For many, the very machines intended to symbolize urban modernity in China have come to highlight the limits of the state's vision of progress. Villagers living in close proximity to incinerators perform their own risk assessments with homemade maps. Meanwhile, urban citizens cite the specific regulations and enforcement measures of pollution-monitoring standards in Japan and Germany as they argue that China lacks its own suitable standards of transparency and oversight. In response to a lack of accountable and responsible government in the face of an impending urban environmental crisis, WTE incineration is also actively producing new forms of engagement, cooperation, and knowledge-making practices that challenge the seemingly totalizing role of the state.

See also: ACCUMULATION, COAL ASH, DETRITUS, RENEWABLE, RISK, SUSTAINABILITY, URBAN ECOLOGY.

Coal

Ashley Dawson

Coal is the big dirty secret of our time. Although coal-fired power plants generate more than 50 percent of ELECTRICITY in the United States, few Americans think about coal when they stop to reflect on where their power comes from (Bob Johnson 2010). The tense geopolitics of oil attracts many more headlines than coal, yet 35 percent of the world's electricity is currently generated by coal power, and developing nations such as CHINA and India bring hundreds of pollution-belching, coal-fired power plants online each year. When we turn on our sleek iPads and MacBooks, we seldom consider that the ENERGY used to power these totems of the global economy is derived from fiery combustion of the fossilized remains of massive Paleozoic plants. Nor do we often think about the human labor or environmental toll associated with the consumption of coal today. Our failure to consider the origins of the coal-based power we consume is a particularly extreme version of the broader ignorance about energy that characterizes the present.

To bring coal's pivotal role in contemporary bioenergetics back to light, we need to overcome its association with the grimy industrial past. The film *Brassed Off* (1996) links coal miners in Thatcher-era Britain to extinct creatures like dinosaurs and the dodo. The film narrates the struggle of a north English village brass band to play on in the face of the impending closure of the coal mine where all of the band members work. The Grimley Colliery Band battles feelings of helplessness and futility in the face of the coal board's harsh redundancy policy. Why play on when the occupation that defines their band is no longer viable? *Brassed Off* captures the miner-musicians' double bind as they struggle to maintain their masculine pride as workers in a job that consumes their bodies yet is no longer recognized by the public as the foundation of Britain's prosperity.

The fate of the Grimley Colliery Band is particularly bitter since coal did indeed power Britain's rise to global hegemony. Although the country had been exploiting its abundant reserves of coal since the late medieval period in order to offset a scarcity of timber supplies, it was the early-eighteenth-century invention of the steam engine and the later development of viable methods for producing iron using coke that established the great triumvirate of coal, steam, and iron.[1] This combination catalyzed the industrial revolution, thereby unleashing a chain of epochal changes, including the divorce of labor from human and ANIMAL body power, the transformation of perceptual experiences of time and space, a massive demographic boom and migration of Britain's populace to cities, a hitherto unimaginable rise in the nation's productivity and military power, and a concomitant growth in imperial control of the colonies, which provided raw materials for British factories. As Timothy Mitchell suggests, coal was vital to the rise of modern democracy since the logistics of coal extraction and transport allowed workers to control chokepoints in the national economy and thereby to force political concessions from governing elites (2011, 19). Perhaps most significant, coal also sparked a reorientation of consciousness around a culture of ceaseless, exponential growth freed from ecological LIMITS, a mindset whose perilous folly we are only now beginning to confront in spiraling concentrations of atmospheric carbon.

As the power of coal miners in the modern economy grew, middle class representations of miners' culture and corporeality became increasingly anxious. From early on, miners, many of whom ended up in the mines after being displaced from their land by great waves of enclosure, were seen as dangerously different. Along with factory workers, they were initially perceived, E. P. Thompson argues, as a "fresh race of beings" (quoted in Freese 2003, 78). By the late nineteenth century, this attribution of cultural difference had transformed into a eugenically tinged moral panic over the degeneration of the working classes in general and miners in particular. In H. G. Wells's *The Time Machine* (1895), the protagonist travels to a distant future in which humanity has evolved into two distinct races, the effete and childlike Eloi and the twisted and brutal Morlocks, who dwell in subterranean caverns beneath the bright sunlit world inhabited by the Eloi. Wells's cannibalistic Morlocks represent, in particularly grisly form, elite fears of corporeal and social degeneration imputed to coal miners.

Similar fears were articulated in the United States, which had made the shift from WOOD to coal power by the 1880s. Writing at the acme of coal's influence in 1922, when the nation was almost entirely dependent on coal for everything from domestic heat to industrial power, novelist Sherwood Anderson lamented the impact of coal in his article "My Fire Burns." Coal, Anderson opined, had catapulted Americans from an age of organic lyricism where autonomous rural producers ensured the republic's Jeffersonian democracy to an age of mass production in which people, subsumed by machines, had become slaves

1. For a history of coal's role in Britain's rise to global dominance, see Freese (2003).

to the corporate oligopoly of King Coal (cited in Bob Johnson 2010, 268). Fears of racial degeneration akin to those articulated by Wells were replayed in the United States as coal miners gained more political power. Eugene O'Neill's play *The Hairy Ape* (1921), for instance, features the Neanderthal-like protagonist Yank, who works as a stoker in the hold of a transatlantic steamer. When Yank meets the play's other protagonist, the young steel heiress Mildred Douglass, whom O'Neill characterizes as "sapped of energy . . . a waste product of the Bessemer process," tragedy ensues (quoted in Bob Johnson 2010, 277). The figure of the coal worker in O'Neill's work is synonymous with a deathly dialectic that augurs the destruction of modern civilization rather than its transformation into the kind of workers' paradise imagined by Marx and Engels.

With the rise of oil, the haunting figure of the energy worker was effectively banished from the world stage. The Oil Encounter was, as Amitav Ghosh observes, remarkably mute, with the exception of a few texts such as Upton Sinclair's *Oil!* and Abdelrahman Munif's *Cities of Salt* (Ghosh 1992). This silence was, in significant part, a product of the altered material INFRASTRUCTURE of oil, a change that made oil far more slippery than coal in terms of both aesthetic and political representation. As Timothy Mitchell argues, the shift from coal to oil after 1945 was part of a conscious strategy to smash the power of miners and other key sectors of the working class mobilized around the coal economy's chokepoints. With its shift from nationally rooted coal production to the oil economy's transnational flows, the energy sector might be seen as a prolegomenon to the new international division of labor that characterized the neoliberal era from the mid-1970s, a site where the new transnational material infrastructures and flows that helped disaggregate existing working class organizations and strategies were experimented with and perfected.

Today we reap these bitter seeds, in the weakness of the global working class and the increasingly palpable reality of catastrophic climate change. Our situation is clear: we must cease consuming fossil fuels if we are to avert more disastrous forms of climate chaos. Of the fossil fuels currently in use, coal is by far the dirtiest and most dangerous. Campaigners for climate justice should therefore make the elimination of coal consumption one of their primary targets. There is some evidence of success in this regard: in recent years, the anti-coal movement has stopped the construction of one hundred new coal-fired power plants in the United States (Nace 2009). But coal continues its disastrous forward march internationally, with support not only from governments in developing nations such as China and India but also organizations such as the World Bank.

To stop coal, the climate justice movement must make coal's environmental and political toxicity visible once more. The campaign against coal can draw strength from historical memory, not simply of the importance of miners in the struggle to deepen democracy in industrialized nations but also the specific weaknesses of the coal industry's dendritic NETWORKS. Climate and environmental justice movements need to join with workers in the energy sector to choke the infrastructure of coal. However, as *Brassed Off* reminds us, workers will fight to retain their jobs in a dirty industry that kills them, unless they are offered a clear alternative. The campaign against coal must therefore demand a just transition to a RENEWABLE, decentralized energy infrastructure (Abramsky 2010). We have powerful imaginative resources to mobilize in this regard. After all, the figure of King Coal reminds us that the energy infrastructure created by what Lewis Mumford calls "carboniferous capitalism" has bred rampantly undemocratic forms of corporate oligopoly (2010, 156). Taking power thus implies a radical democratic transformation of both global ENERGY SYSTEMS and governance. Let us dethrone Old King Coal.

See also: CHARCOAL, COAL ASH, ENERGY REGIMES, FICTION, PETROREALISM.

Coal Ash

Susie Hatmaker

To live in a coal-fueled culture is to live in the time of ash: a time of irrationality, unpredictability, and unanticipated events that reveal not the work of an angry god, but the limits to human progress and scientific planning. Power generation and electrification emerged in the twentieth century as core elements of modernization and development on a global scale. The failure to account for the corresponding production of waste is neither mistake nor oversight, but inherent to this logic.

To fuel culture, COAL must burn. We convert prehistoric organic sediment into an invisible, world-shaping force of ELECTRICITY. Coal powers modern life, from the quotidian task of plugging in and charging up, to the future-oriented task of technological innovation. Industrial machines process and burn coal in landscapes where the terrain and the lives of people constitute a material basis for power production. From the power plant, tentacles of connection extend. Smoke rises into the air. Railroad tracks reach across the land and connect power plants to coal mines. Payments flow. Coal moves. Multiple landscapes intertwine the interests of their human populations. Beyond tracks that contour the landscape, power's reach extends down rivers and across seas. In our era, we know coal is "bad" or, at least, not so good. Yet we continue to seek new, more complex ways of turning coal into power, despite widespread acknowledgement that the sky is changing and the water is rising and it is getting hotter. Were there always this many extreme natural disasters?

Beyond these new anxieties lies a little-acknowledged effect of coal consumption. As coal burns, it accumulates coal ash, a physical byproduct. As electricity flows constantly and invisibly, coal ash piles up across acres of land with every kilowatt-hour. This waste changes the look and feel, and the ecological interconnections, of the earth on which it sits. Coal ash landfills exist everywhere coal is burned, yet they attract relatively little attention. These landfills hold the millions of tons of fly ash, bottom ash, and boiler slag produced in coal burning. Early power plants released noticeable ash into the air. As the United States began to legislate environmental protections in the 1970s, atmospheric concerns led to the addition of electrostatic precipitators and, later, scrubbers, which capture most of the particulate matter in the smoke and redirect it into the growing solid waste ponds.

Coal ash is currently unregulated in the United States; environmentalists want the Environmental Protection Agency to label it toxic or hazardous, while industry lobbies against such designation and seeks to minimize disposal costs and maximize profits by selling the waste for reuse in concrete, wallboard, bowling balls, home appliances, etc. Beyond this political opposition, I suggest that coal ash is itself a material force, antithetical to the logic of progress that creates it. Born of modernity's world-shaping epistemology, coal ash also evidences its LIMITS. Its destructive potential invites a rethinking of how we understand time, knowledge, power, and progress.

How might this waste come to matter in a culture that mostly ignores it? To better understand this material component of the energy life cycle, I trace the life story of one body of coal ash.

On December 22, 2008, the largest SPILL of coal ash in US history occurred in Kingston, Tennessee, flooding the adjoining river in thick gray sludge and destroying a small lakefront community (Dewan 2008). Held by a dike of earth and ash, an enormous "pond" had contained the waste of over half a century's power production. On that night, the ash suddenly burst forth and excretions of mercury, arsenic, and selenium gushed into a river already polluted with nuclear waste from Manhattan Project facilities upstream.

We can specify some of the events that this power plant fueled; we can trace the policies that influenced its generation. But the waste knows no time. Ash deposits mix together and blur distinctions between political eras. This murky gray body constitutes a physical plane of existence outside of history, politics, and modern epistemology. Both before and after the flood, dry ash drifts into air, liquid waste seeps into river and earth, and this giant mass shifts form and composition in ways that exceed human abilities to know, control, and contain.

Numbers and measurements remain the default method of understanding the event: over 1.1 billion gallons; 100 times larger in volume than the Exxon-Valdez oil spill; ash piled up over 55 feet above the water level; 5.4 million cubic yards rushed into the adjoining river; this plant burns an average of 14,000 tons of coal per day (Matheny 2013; Dewan 2008). Other notable information: the holding cells have always been (and remain) unlined. Engineers designed the holding pond in the 1950s with a specified maximum capacity, which was reached and exceeded a few times over. Each time, the walls were raised with more ash and dirt. Reasons for the spill according to one investigation: a culture of mismanagement, worker isolation, tendencies to not report problems, unauthorized fixes, lack of funding for waste management, lack of oversight, confused bureaucratic channels (Ide and Blanco 2009). In short, numerous "lacks." What becomes clear: the waste itself is the lack. It is the accumulation of lacks, absences, oversights and un-thoughts, of our electrified culture; the flood, an opening and a chance to know differently.

I grew up in the shadow of this coal-burning power plant. Twin pale gray smoke stacks rise into view like skyscrapers over rolling hills. The tallest buildings in this rural landscape, over time they became a natural part of the vista—the landmark for a town dominated by power production. Beneath the tall stacks, a row of much shorter, brick red stacks from the plant's earliest days. When in use, they dusted nearby residences in soot that burned trees and killed gardens. The tall stacks were built as a corrective, but they increased exponentially the area of ash dispersal. Most recently, a rotund white scrubber stack emerged. Half the height of the tallest twin stacks, it constantly exhales a thick man-made cloud. Each stack stands as a monument to the prevailing energy science of its time.

The generating capacity of the Kingston Steam Plant exceeds DEMAND from local household use. It powers the NUCLEAR weapons facility in neighboring Oak Ridge, as it did throughout the Cold War. It sells excess power for profit. It is one of several power plants in the South operated by the Tennessee Valley Authority (TVA)—one of the largest national development projects in US history, which has been imitated around the world.[1] This project, begun under the New Deal, seized land from thousands of subsistence farming families and erased their communities. TVA built over twenty DAMS in twenty years, creating a vast system of lakes. Then came the shift to power generation (hydroelectric, coal-fired, nuclear), today the primary function of TVA. While TVA originated in utopian dreams of agricultural and social development, these ideals faded in value once ENERGY became the most profitable operation. Residents receive ample cheap power and managed recreation in lakes and parks; waterfront property is today considered a luxury.[2] In Kingston, a town economically and physically centered on the steam plant, everyday life unfolds within a landscape given over to power generation; coal ash builds up as an afterthought.

In the 2008 flood, the ash displayed its force and reminded witnesses that things given life through rational, scientific planning can far exceed human control. A TVA/EPA report on the cancer risk of coal ash in the wake of the Kingston spill contains an unintentionally absurd section on "uncertainties." It notes three "key areas of uncertainty": "data uncertainties," "exposure scenario uncertainties," and "toxicity value uncertainties" (Tennessee Valley Authority/US Environmental Protection Agency 2012). These unknowns

1. TVA collections at the National Archives in Atlanta, Georgia, provide numerous documents of visits from abroad. Leaders, scientists, engineering students, and planners visited from places including CHINA, India, Burma, Jordan, Egypt, Brazil, and South Africa. TVA tracked dam and power projects based on its model that were built in other nations: a link between the methods and epistemology of development in the American South to global developmentalism in the mid-twentieth century.

2. A "government-owned corporation," TVA maintains its own police force that patrols its properties, including the rivers, lakes, and parks.

mark knowledge's absence where processes (always) already underway push back against desires for infinite power and vitality. In this infrastructure-landscape arrangement, we produce power and are in turn produced by its generative force. So we must acknowledge the existence of coal ash, a physically enormous cultural inheritance forged in the desire for a particular style of modern life. As long as coal is burned, coal ash will pile up by the ton, contouring the landscape of the culture it fuels. Perhaps industrial "recycling" will increase, if the coal industry continues to prevent the EPA from labeling it hazardous. But for the most part, we continue to hoard coal ash in gray landscapes of waste, hoping that the walls will hold as we build them higher.

See also: CHINA, COAL, DETRITUS, NETWORKS, STATISTICS.

Community

Sara Dorow

Oil's binding of production–social reproduction–consumption binds us, in turn, to it. One shape that such a collective truth takes is "community." As it has underwritten new arrangements of global capitalist ACCUMULATION over the last century, oil has offered up community by reconstituting the possible forms and imaginaries of collective social life. Petroleum products have promised public and private spaces of leisure, freedom, and urbanized modernity; oil value chains have produced transnational and cosmopolitan NETWORKS of IDENTITY; flows of oil have invited us into affective spectacles of consumption; and oil has extended and/or compressed the time-space experiences of humans. Oil and community have been naturalized together, the life-giving properties of each conjured via the magic of the other.

At the same time, the co-naturalization of oil and community has served to distract from the very forms of inequality and violence it entails (Huber 2011). This sleight of hand is enabled by a range of sliding geographical and temporal scales of community. For example, oil-based national and global accumulation is accomplished via the sponging up of social and environmental "externalities" by out-of-sight, out-of-mind local communities. For another, *our* energy is secured over against *their* backward unfreedom (see Levant 2010 on "ethical oil"; Bauman 2001; Maass 2009). And not least of all, oil's extractive logic redraws lines of familial, ethnic, and national identification (Watts 2004) and differentially integrates populations into global divisions of labor and nature (Coronil 1997; Shever 2008; Joseph 2002; Bougrine 2006).

Naming these spatiotemporal transformations and disjunctures is part of the task of understanding the assemblages and reassemblages of community that are created and perhaps necessitated by oil. These are sites where social reproduction intersects with oil production and consumption in context-specific ways (Joseph 2002; Schofield 2002), cathecting subjects to nation, region, city, or village and calling them into shared modernity, tradition, development, kinship, or history in ways both delimiting and productive. Put another way, if community is a powerful coinage in the mediation of inclusion and exclusion (Creed 2006), then we might say oil is one of its dominant global currencies. This is more than a matter of oil's importance to capital accumulation along the production-distribution-consumption value chain. Oil affectively produces socio-spatial and political assemblages of community through its own distinctive material properties—as an energy-intensive, easily transported, and seemingly endlessly commodifiable version of "fossilized sunshine" (Black 2012; Hornborg 2013; T. Mitchell 2009)—and through intimate forms of everyday use and barter (Rogers 2012). In other words, the geographical, social, and political relationships between oil and community constitute a rich arena of mediated exchange.

What are the various cultural technologies that do the mediating WORK of codifying, assembling, and imagining "community" (Tsing 2005) in the late oil age? Jody Berland (2009) uses the term "cultural technology" to designate the intersections of spatial materiality, governmentality, subjectivity, and symbolic labor (12) that constitute culture. Such technologies situate "us" in relation to oil (or other forms of ENERGY) via the two-sided coin of community. This shift in perspective turns inside out Leslie White's (1943) energetics formula of sixty years ago, in which energy, harnessed by technology and yoked to human-serving social systems, advances culture. Rather, we might pose a different energetics, where the cultural technologies of oil effect social arrangements of humans into "communities." Apprehending these arrangements is increasingly important in an era when most humans lack visceral knowledge of increasingly specialized and technologized energy production (a form of alienation that Laurie Shannon contrasts with the embodied intimacy between personhood and grease in the age of TALLOW [2011; see also Black 2012]), and when the temporal and spatial arrangements of that production seem to circumscribe political imaginaries and actions (T. Mitchell 2009).

For these reasons, oil production zones, as intense and intimate sites of the cultural technological work that articulates oil and community, are particularly "good to think with" (Amit 2010; Dorow and O'Shaughnessy 2013). While scholars have more commonly presumed existing forms of community and then studied how the extraction of fossil fuels affects them, we can and should ask how community is affectively and culturally produced in sites of oil production. Doing so might also foreground the social reproductive activities of extractive capitalist accumulation (see Mitchell, Marson, and Katz 2004).

Cultural technologies in oil production zones are perhaps most obviously at work in some of the spectacular displays of oil wealth found in oil cities (Dubai, Dallas, Abu Dhabi) and oil nations (VENEZUELA, NIGERIA). Of course, one spectacle is not the same as another; how these articulate oil production to social reproduction and consumption differs by context. Some are simulated arenas of community and national participation that package cultural pride, tradition, or the people themselves (Apter 2005) into a display of progress, modernity, globality, or even populism (Coronil 1997; Elsheshtawy 2012). Here cultural technologies convert oil to community via the latter's high exchange value and the "magical state" made possible by it (Coronil 1997). There are also somewhat less spectacular but no less magical versions of such cultural technological work. In Fort McMurray, the city at the heart of bitumen production in northeast Alberta, CANADA (and the largest industrial mega-program in the history of earthlings), oil wealth is being transmuted into the promise of a GREEN global city of the near future via bold urban development plans and participatory campaigns that effectively displace a cold, hard economic calculus with one based on multicultural wealth, state-of-the-art leisure, and feminized familial bonds (Major and Winters 2013; Dorow 2015). Bitumen production, and the consumptive excesses required and produced by it, get rewritten as community life.

Cultural technologies in oil production zones thus also build affective bonds of loyalty to the singular economy of oil production through direct appeals to kinship and community. But perhaps "singular" is a misleading term, since cultural imaginaries of belonging are enabled, constricted, and refigured along with the changing productive and political infrastructures of oil. With the privatization of oil in Argentina, collective claims to oil as a family heirloom have had to adjust to, and have struggled under, a new system of local and kinship-based microenterprises (Shever 2008). And in the Perm area of RUSSIA, forms of petro-barter have been enabled by a distinctive regional cultural identity carried over from the previous era of centrally planned oil production; this social intimacy with the exchange of oil has contributed to the "*unimagining* of the post-Soviet federal state" (Rogers 2012, 14, emphasis in original).

If the cultural technologies of "oil community" can work both top-down and sideways, they also have unintended consequences, including refigured claims on "community oil" and the emergence of nascent forms of civil society. Hugo Chavez's version of the magical state, which "channeled oil revenues into the sponsorship of culture" as a tool for Venezuelan national cohesion, spawned both oppositional alliances among class and cultural elites as well as counter-logics of SPIRITUAL and cultural community identity in the barrios (Fernandes 2011, 106). In northern Alberta, ABORIGINAL bands find themselves exchanging cultural and historical rights to land for monetary and market rights to the business of oil extraction; the latter, in turn, are to be converted into the social and economic rejuvenation of their communities. These are complex processes of both cooperation and conflict (Schiller 2011) where the use of oil wealth to "activate community claims" amplifies subnational political entities and generates new crosscurrents of inclusion and exclusion (Watts 2004, 210).

Cultural technologies reproduce *Homo energeticus* (see Kowalsky and Haluza-Delay 2013)—embodied subjects for whom the given alignments of oil and community make sense—but they also disrupt or erupt. The transmutation of resource wealth into community—chained to the chaos of commodity and speculative markets and bound up in irreversible transformations of NATURE—creates lasting violence and exacerbates new and old inequalities. Sometimes these unbearable costs mean imagining an altogether new kind of community, a full turning away from the oil complex. The subaltern "phantom citizens" of Ecuador who leapfrogged the impossible chain of broken responsibilities for their health and environment by suing a subsidiary of Texaco in the United States rather than in Ecuador created "an alternative anatomy" of transnational subjectivity (Sawyer 2001). In Nigeria, a women-led "social anatomy of coordinated global actions by producers and consumers" deployed nakedness and direct action to insist that oil multinationals leave and never return (Turner and Brownhill 2004, 63). In these anatomical shifts, cultural technologies work both within and beyond oil production zones to create an alternative politics of community and do so by reimagining the material and symbolic body of community. They enact a fundamental shift in the local and trans-local relations of social reproduction, production, and consumption through which the cultural technologies of oil constitute community.

See also: AFFECT, EMBODIMENT, GENDER, IDENTITY.

Corporation

Andrew Pendakis

An ambiguity has clouded our understanding of the corporation ever since Stewart Kyd's 1793 treatise distinguished between the corporation as a stable juridico-material entity and the process (of incorporation) by which the corporation gains its unique legal prerogative. Incorporation, however, is never mechanically anterior to the corporate form, per se, but instead its very essence: a dynamic, precarious activity characterized by a desire for what Kyd called "perpetual succession," an existence projected aggressively onto a temporal horizon limited only by the criterion of profit (Kyd 1793, 6). The corporation is a "living being" that combines Spinoza's *conatus*—the impulse to self-preservation that he sees in all entities—with an interminable, indeed legally binding, will-to-profit (2009 [1677]). Its core fantasy derives from Hobbes's mechanist materialism, a system constructed around the continuous, incorrigible circulation of emplaced and desiring monads (2009 [1651]). Survival—but also a deeply un-Spinozan erasure of rational limit, a total openness to the abstract infinity of money—characterizes the corporate formula. The corporation is a being without precedent, which smuggles into its DNA a measureless dimension at odds with the very concept of appetite so crucial to the Western philosophical tradition. Appetite, though restless and perennial, is never insatiable; it always reaches choke points, gray zones, places where pleasure flips into its opposite, a bad fullness, nausea, or pain. The danger posed to life by the corporate form occurs precisely at the juncture between this "indefinite duration"—its dream of immortality—and its evolution of a material body shorn of the capacity for satiation: the corporation's juridical claim on time is matched only by the full tenacity of a spatial prerogative that wants nothing more or less than expansion forever (Kyd 1793, 6).

What happens when the substance most indispensable to the character of the existing world system is genealogized through the historico-ontological nexus of the modern corporation? It is no small surprise that the era of the full-blown modern corporation—the cruel efficiencies of Rockefeller's monopolistic Standard Oil—is coeval with the birth and hegemony of Darwinian naturalism. The unveiling of the economic value of oil takes place only within the instrumentation and activity of the historical corporation, while the latter only really takes flight in the aftermath of the technical and infrastructural conversion of oil into historically actionable fuel. The modern corporate form—its scale, integrated geographical reach, and logistical intricacy—only becomes thinkable in the context of the ENERGY unleashed by petroculture, just as oil itself could only become the fuel of our age on the condition of economies of scale generated by the corporation's vertical and horizontal evangelism. Oil before Rockefeller was a chaotic weave of small-scale operations, which combined on site the processes of extraction and refinement. Rockefeller stabilized and perfected the concept of the modern high-flow refinery, situated within reach of railroad NETWORKS then ramifying across AMERICA. However, it is not just a question of connection but of exemplarity. The paradigm of the modern corporation, its very template, originates in Standard Oil's success, as if the germ of the corporate form were linked to the substance itself. The modern corporation assembles itself specifically on the back of the substance that lays down the rules for its subsequent generalization, its universality. Twentieth century immensity was born and reproduced via the nexus of the oiled corporate form; the proclivity for mass, from the unprecedented monumentality of the modern highway (without question the biggest object in history) to the semi-planned gigantism of World War, requires both the bureaucratic infrastructure of the managerial corporation and the cheap combustibility of oil as much for fuel as telos. One should remember the discontinuous phenomenologies of this process, not a technical auto-poesis of forms assured in advance but instead a conjunction of embodied strategies, ranging from Rockefeller's infamous underselling of competitors to the violent repression of oil workers and unions.

The corporatization of oil is one of the great destinal events in the history of life on our planet. Out of it swirl the suburb, the fibrillating culture of consumerist AUTOMOBILITY, the lineaments of our PLASTIC ecosystem, but also, and perhaps most critically, the structural de-linking of oil's destiny from every collective, recognizably human decision. Henceforth, the paradigmatic dream of republicanism shared from Rousseau to Jefferson to Mao—that of a socius in the grip of its own general intelligence, a life-world vetted dually by reason and civic AFFECT—will be forgone for the decentralized and amoral sovereignty of DEMAND. The latter is the belief that anything that *can* exist *should*: its axiology grafts precisely onto the body of what is, sanctioning only that which can be engendered within the coordinates of the profitable. Oil directly fueled the concatenation of privacies that now systematically forecloses the kind of politics necessary to theoretically avert the fatal environmental cycles today engendered by corporatized oil itself. Markets grew, technologies were innovated to better and more efficiently extract the oil from the earth, and institutional (governmental) conditions were (often violently) emplaced geopolitically to ensure the continuous flow of oil to the West from its volatile, postcolonial margins. Out of the corporatization of oil emerged an imperial privacy never before encountered in history, titanic entities so deeply inserted into the productive machinery of nations, so strategically indispensable to their status quos, that they, along with the substance they peddle, stand to the ground of what exists like an *arche* or first principle, a fully operative social metaphysics.

Political and scientific transparency were always cultural fantasies, but we will never know to what extent the erosion of our capacity to believe in either was abetted by the colonization of both by the financial, electoral, and legal prerogatives of unaccountable corporate privacy. At issue in this predicament is not merely the ostensible exception of monopoly, its excessive deviation from the otherwise good rule of proportionate corporate order. Although the monetary and political resources of the mega-monopoly often outweigh those of nation-states, explicitly belying the consistency of liberal equality and rule of law, the corporate form itself, its very modality, has deposited an injunction to secrecy, obfuscation, and the effective manipulation of appearances into the very fabric of postmodern experience. In the domain of advertising, the lie is a truth we openly concede as harmless second nature; far more troubling is the corporation's discursive heart—its scientific and legal activity, its internal memos, its public relations output—in which dissimulation is a logistical imperative coextensive with survival itself. Whether it is the situated suffering produced in Ecuador or NIGERIA or the risks posed to planetary life by climate change, the Darwinian discursive production of (oil) corporations is continuously fissuring issues into the false open-mindedness of two sides. "Spin" is the euphemism by which our culture transforms lying into a state of nature, a slight bending of the truth, a gentle distortion, one proper to an order the basic social ontology of which was already fully adumbrated by Hobbes.

The results of statist attempts to interrupt the nexus linking oil and the corporate form have been mixed. Where states like RUSSIA or Mexico have used oil to reinforce hierarchy, reproducing inequality through layers of aristocratic oil bureaucracy, only socialist VENEZUELA has been able to channel oil revenues into functional social programs and the reduction of poverty. This effort notwithstanding, state oil monopolies, insofar as they remain within the liberal productivist paradigm, have not been able to deter the ecological effects of burning fossil fuels or to de-link oil's virtual social value from the consumerist imperative of eternally increasing levels of disposable income (what the twentieth century called "development").

The octopus: this is the animal called upon by critics to concretize the scenario of corporate oil at the turn of the twentieth century. Again and again the motif returns, the corporation's essence pictorialized as somnambulant body, pure reflex, an encroachment without limit or measure. This IMAGE has been replaced today by corporate oil commercials featuring flows of scientists or sprawling wind farms. Yesterday's oil leviathan is today's astonished artist/scientist, a nimble, responsible, socially oriented "energy provider." The postmodern oil corporation is pure restless will to knowledge, its telos nothing more than the vexed solving of sociotechnical problems of interest to us all. Jargons of COMMUNITY, innovation, and discovery replace an older establishment rhetoric built around the concepts of herculean efficiency and cheap energy: if Standard Oil could not disguise its troubling vastness, it could at least recalibrate it as consumer value and modernity. Sometimes, comically, Rockefeller's ascetic head would be placed on the body of the octopus, one tentacle drawn building a church while another strangled Congress. Dressed up as curiosity, creative expertise, today's oil scenario cuts a thinner but no less exigent and dangerous profile. Those of us who continue to discern beneath this new face that old acquisitive body, the insatiability and violence of the corporate form, are inevitably framed by the technocratic center as partisans of an imaginative conspiracy, animists, premoderns, speakers of a language still filled with animals and metaphor. Circumventing this representation of critique as antiquated exaggeration is one of the key political imperatives of our moment.

See also: ADDICTION, CRISIS, EMBODIMENT, ENERGY, EVOLUTION, FUTURE, PETROVIOLENCE.

Crisis

Jason W. Moore

A funny thing happened on the road to a theory of environmental crisis and environmental degradation in capitalist development. It never happened.

I do not mean to suggest that questions about the role of ecology in crises of ACCUMULATION have not been posed. But strikingly little movement has occurred in socio-ecological thinking about capital accumulation and its crises, a quarter-century after James O'Connor's groundbreaking theory of the second contradiction (1998), which finds in the expanded accumulation of capital an exhaustion of the relations and conditions of (re)production. Radical thought today has settled on a language of crisis that is arithmetic rather than dialectical: today's "epochal crisis" represents "the *convergence* of economic and ecological contradictions" (Foster 2013). Nature + Capitalism = Crisis. The two are seen as related yet distinct systems, each prior to the other: earth-system and capitalist system.

The irony of the two-systems theory of crisis is that nearly everyone agrees that humans are a part of NATURE. Marx repeatedly makes the point: humans are a "natural force," experience "natural limits," and are linked to nature as if to themselves (K. Marx 1973, 612; 1977, ch. 10; 2007, 74). This philosophy of humanity-in-nature has largely won the day in twenty-first-century GREEN thought. Even mainstream advocates of the ANTHROPOCENE perspective pay lip service to it (e.g., Steffen, et al. 2011). The relational ontology of humans as natural force has informed regional political ecology and environmental history (e.g., Kosek 2006; R. White 1995). Elsewhere in the historical social sciences, however, this coproduction of nature perspective has gained little traction. In global studies, nature remains what it always was: a resource zone, a rubbish bin. In short, an object (J. Moore 2013).

What happens to our thinking about crisis if we shift our premise from the separation of humanity and nature toward the dialectical unity of humanity-in-nature? From environment as object to environment-making? These questions underwrite a unified theory of capitalism encompassing the accumulation of capital, the pursuit of power, and the co-production of nature. This unified theory is the point of departure for the world-ecology perspective, where the modern world-system is a capitalist world-ecology through which emerge successive configurations of capital and power in the web of life (J. Moore 2011b, 2013; Niblett 2012; Deckard 2012; Weis 2013). World-ecological thinking shares common ground with post-Cartesian social theory, but its insights extend into new narrative strategies, methodological premises, and theoretical frames in the history of capitalism. If nature matters, it matters not least for how we study and narrate world-historical CHANGE.

The what and why of systemic crisis in the twenty-first century have been relatively easy to identify: climate change, financial volatility, biodiversity loss, food insecurity, rampant labor-market informalization, the decline of American hegemony, etc. But how do these processes work through manifold and multilayered bundles of human and extra-human relations? This question illuminates the limits of a Cartesian frame premised on a mechanical, rather than dialectical, relation of humanity-in-nature. The conventional approach is to think ecological crisis in terms of diminishing flows of substances: not enough food, not enough oil. It may be more productive, however, to think crisis as a process through which new ways of ordering the relations between humans and the rest of nature take shape.

The Green critique is perhaps best known for sounding the alarm about impending "limits to growth," a phrase instilled in popular consciousness by Donella Meadows and her colleagues in the early 1970s (1972). Their thesis is simple enough: modern economic growth creates relative scarcities in nature. LIMITS to growth are found in external nature first and social contradictions second. The problem with this way of thinking is that there is no such thing as natural limits—not because there are not limits but because the natural cannot be divided from the social in this way.

Limits and crises are fundamentally about histories of relation involving both humans and extra-human natures. Substances do matter. But their historical meaning forms within specific relations—human and extra-human, organic and inorganic. To treat them as self-evident facts in isolation from the relations that transform geological substances into resources is to enact a series of violent abstractions and to retreat into a new form of environmental determinism. In the "peak everything" thesis (see Heinberg 2003), for example, the messiness of human enmeshment with the rest of nature is abstracted in the interest of narrative coherence. But history matters to how we think and narrate crisis. COAL in sixteenth-century Europe was entirely different from coal in the nineteenth century. To paraphrase Marx, coal is coal; it becomes a fossil fuel only within certain relations.

This relational approach to historical limits orients us toward moments when the strategic relations governing a civilization reach a qualitative and sometimes fundamental impasse. (The present impasse of neoliberalism's financialized accumulation regime is one example.) These moments are turning points in the life of capitalism. They are world-ecological crises—not crises in an exogenous nature that constrains social development but instead crises in civilizational ways of organizing nature, humans included. It is not merely soils and species, forests and fuels, that make world-ecological crises but the relations of power, markets, labor mobilization, and reproduction. There is no distinct ecological crisis operating alongside other crises, since the mosaics of constitutive relations (power, capital, science, etc.) are themselves messy bundles of human and extra-human natures.

It's no accident that conventional theories of crisis in classical political economy were formulated in the late eighteenth century, a period of generalized agricultural deceleration marked by stagnant labor productivity, rising cereal prices, and sharply rising inequality. This agrarian depression reached from the Valley of Mexico to Scandinavia, but its significance for theorizing crisis lay in the threat that sharply rising food prices (relative to industrial prices) in Britain and across Europe after 1750 posed to the emergence of industrial capitalism. The contradiction posed by this signal crisis in the ecological regime was that land productivity could only be increased through labor intensification, a solution that would have contracted the reserve army of labor at precisely the moment it was most needed for industry and empire. (A similar contradiction exists today.) A solution was ultimately found in two great commodity frontiers that yielded windfall profits. The first frontier was vertical, moving into the earth to extract coal. The second was horizontal, moving across the earth to produce wheat, particularly in North America. When another great depression arrived in the 1870s, together the coal and wheat frontiers produced cheap food that enabled rapid industrialization—notwithstanding mass starvation elsewhere in South Asia and CHINA and genocide in North America (J. Moore 2010).

Underlying the emergence of these commodity frontiers was an earlier shift in the valuation of nature that came to characterize modernity: "cheap nature" did not just appear, it had to be created (J. Moore 2014a, 2014c). In the transition from land productivity to labor productivity as the civilizational metric of wealth, European states and capitals in the long sixteenth century came to see time as linear, space as flat and homogenous, and "nature" as external to human relations. This broader epistemological shift—known as the Great Frontier—amounted to the invention of Nature itself and the subjection of biospheric reality to the rationalizing and disciplining logic of successive scientific, botanical, cartographic, metrical, and other revolutions. These knowledge revolutions identified and facilitated the appropriation of those cheap natures necessary to sustain rising labor productivity and rising volumes of material production (J. Moore 2014b). Without a widening sphere of appropriation—including the appropriation of cheap human nature and its transformation into labor-power—production costs would have risen and profitability faltered, as Marx warns in his general law of underproduction (1967, 119–21) and his account of the working day (1977, ch. 10).

The various crises of the present—financialization, climate change, the erosion of the capital-labor relation—represent the closing of the Great Frontier (Webb 1964; Balakrishnan 2009). No longer can obstacles to accumulation be overcome by appropriating cheap natures. Yet the end of the Great Frontier and the obsolescence of cheap nature strategies do not represent the exhaustion of an abstract and external nature, since frontiers are not simply "out there": they continue to be actively produced by bourgeois knowledge, from Linnaean taxonomy in the eighteenth century to genomic mapping today. Instead, it is the frontier strategy that bears ever less fruit—low-hanging or not—in the twenty-first century. The salient crisis is within the relational matrix that we might call *historical nature* (J. Moore 2014b), whose specific form is the capitalist world-ecology. The so-called global ecological crisis is indeed real enough, but the way that our civilization knows this crisis is through value (abstract social labor), the historical vitality of which lies in the appropriation of massive volumes of unpaid work, performed by "women, nature, and colonies" (Mies 1986). As Marx understood, to say that the ecological crisis is a crisis of value relations is to say that we live at a moment in human history when the absurdity of value relations as a way of regulating the web of life becomes apparent. At such moments, the possibility of a civilizational world-ecology premised on different, more participatory, holistic, and egalitarian values also comes into view.

See also: ACCUMULATION, ANTHROPOCENE, CORPORATION, LIMITS, NATURE, RISK.

Dams

Peter Hitchcock

To understand the dialectical relationship between human species being and NATURE, consider water. Water is so implicated in our substance and our needs as to be axiomatic both in our forms of socialization and their contestation and in how we articulate material being. Yet its apparent ubiquity masks contradictions involved in its measure. Hegel, for example, invokes water in his discussion of measure to explain the leaps between quantity and quality (2015, 288–93). Marx subjects this Hegelian antinomy to a materialist injunction: in the *German Ideology*, he writes that "the 'essence' of the fish is its 'being,' water" (Marx and Engels 1988, 66). He immediately qualifies essence, however, by specifying its material conditions. Thus, the essence of a freshwater fish is the water of a river, but this river ceases to be the essence of the fish when it is made to serve industry, polluted, navigated, diverted, etc.

But here Marx faces his own dialectical challenge. What would constitute real liberation beyond idealist conceptions of consciousness? Real liberation in the real world can only be achieved by real means: "the *development* of industry, commerce, agriculture, the conditions of intercourse" (Marx 1988, 61, emphasis added). The development of productive forces must be sufficient for qualitative change. For Hegel, water is a determinate content, therefore always already an objective principle for philosophy, irrespective of productive force. The real measure of the qualitative in water is independent of development. Marx's question for Hegel is: What happens to qualitative quantity if water is not just an impetus to thought but remains material, matter? Hegel's question to Marx might be: What happens to quality in the meantime? What about the water? What about the fish?

Humans have worked out how to store or redirect water so that the ravages of natural cycles and deep, otherwise inaccessible water tables or aquifers could be cheated or circumscribed in all but the most inhospitable places. The history of dams is consanguine with such ingenuity, but dams do not just store water; they provide irrigation, mitigate seasonal flooding, and, most important, convert the force of flowing water into distributable power and usable ENERGY. The earliest dams were constructed over five thousand years ago in West Asia. Damming in India does not date back as far, but its systems were among the most extensive in antiquity. Ancient dams were relatively small, save for those of the Roman Empire, which benefited immensely from the development of hydraulic mortar. If droughts were a bane of human settlement, the dam's role in water management might be seen as developmental succor, the advance over nature that permits humans to settle in places otherwise largely uninhabitable, like Las Vegas or Phoenix.

These days, dialectics is probably in as much trouble as the environment, and for the same reasons—the more humans labor in the negative, the more negative labor subtracts from the regulative vicissitudes of planetary homeostasis. This is hyperbole that nonetheless usefully accentuates a specific mode of dialectical thinking, Hegelian Marxism, which is intimately tied to the transformations of modernity, including its crises and catastrophes: laboring in the negative is overdetermined by the productive logic of modernity that it both challenges and confirms. Figuratively, this is modernity's dam, for what preserves and expands production through irrigation and power is out of joint with water's flows, which are neither exactly rational nor absolutely chaotic. We might say that water is modernity's purloined letter: by addressing it, we also ensure that it does not reach its intended addressee. Modernity sets this dialectical contradiction in motion, to which dialectics responds by motioning its exception. Yet the problem of dams takes the form of its own solution: dams engineer progress in such a way that they can never slake our thirst. Indeed, they produce the conditions they are meant to remedy: they can demonstrably generate droughts and floods that yet preserve thirst. Modernity, not nature, fashions this negativity, which is not to say that we can return to nature but that we articulate alternative ACCUMULATION and distribution strategies in our socialization for which dams can only ever be an arena of contestation.

Dams are now a primary scene of economic engineering and geopolitical strategy. A 2011 majority staff report for the US Senate Committee on Foreign Relations articulates the stakes: "The Obama administration has recognized the critical role water plays in achieving our foreign policy goals and in protecting our national security interests" (2011, 1). Indeed, water is a national security issue precisely because it does not comport with normative claims that secure the borders of the nation state. If the United States wants, say, to shape the geopolitics of Afghanistan and Pakistan (the subject of this senate report), then its funding of water projects must be integrated with the demands of neighboring countries, particularly those in Central Asia, Kyrgyzstan, Tajikistan, Turkmenistan, and Uzbekistan. In the Soviet era, regional water and energy needs were centrally controlled and distributed, so at least in theory those downstream countries that benefited from increased water volumes in the summer for irrigation and hydropower would then send COAL and gas upstream to help countries with power needs in the winter. To tamp down instability in Afghanistan and Pakistan, the United States now realizes that it must, without replicating Soviet policy, encourage management of the entire regional water basin. Dams, planned or under construction, like the Kambaratinsk, Rogun, Sangtuda I, and Sangtuda II will raise hydroelectric production to 30 percent of total energy output in the region.

The Pacific Institute's "Water Conflict Chronology Timeline" has, since 1994, shown increasing rates of belligerence over water resources, symptomatic of the disjunction between national developmental desires and regional realities. Such conflicts over water and its unboundedness by political boundaries will significantly inform the shape of water ontology to come. Together, privatization and geopolitical desire constitute the *arche* against which basic socialization breaks. Both elements inform the prospects of environmental DISASTER and are quite willing to manage them to their own advantage.

As the global environment is degraded by our priorities, so the social effects of dams in modernity are concomitantly intensified. The writer-activist Arundhati Roy is a dialectician of this aspect of modernity, not just because she measures the "small things" against the "big things" that are paradoxically diminutive by contrast, but because for her the modernization of India is a litany of contradictory symptoms in which determined national prestige is factored by equally determined national privation. "The Cost of Living" is the last chapter of Roy's novel, *The God of Small Things*, where Ammu and Velutha's lovemaking by the river refuses to yield to any cost that would deny it. *The Cost of Living* is also the title of Roy's subsequent nonfiction book that featured essays on the emergence of India as a NUCLEAR power and its related desire for mega-dams. It would be unfair to say that Roy discovered her social activism in these two developments, but she nevertheless finds her responsibility predicated on the immediate and pressing concerns of India's contradictory self-image a half century after independence.

Roy asserts that like India's nuclear deterrent, India's dams are weapons of mass destruction. Refusing the false dichotomy between Nehruvian developmentalism ("dams as the temples of modern India") and Gandhi's "romanticized village republics," Roy asks how one can measure progress without knowing what it costs and who has paid for it (2007, 16). Measuring progress must be qualified by the question of qualitative quantity, for measuring by quantity will miss the qualitative leaps that are water's essence for human ontology (Hegel continually thinks of this dynamic in terms of water's changing states, like melting glaciers or evaporating lakes). Since independence, India has embraced big dam projects for their prestige as much as their practicality: the argument goes that a large and growing population needs massive infrastructural water management programs to ensure food and energy production. Yet after over three thousand large dam projects alone, India is now more prone to droughts and floods than before independence, and cheap power seems to be the monopoly of business interests. The Narmada Valley projects are symptomatic of such a miscalculation or missed calculation, involving huge population displacement and polluted silt water. Roy does not object to the principle of water management but rather to the costs that too often go unmeasured: the larger the dam, the greater the propensity for social and environmental harm, especially over the long term.

Even if we say that a dam can provide clear benefits for agricultural irrigation systems, power, and water storage, increases in a dam's scale do not necessarily magnify the benefits sought. True, powering larger turbines tends to favor larger electrical output, but this relation is dependent on maintaining sufficient water behind the dam wall (and also, to some extent, creating adequate efficiencies in electrical transmission). The modern history of dams suggests that rational tendencies cannot guarantee effects in a predictive way; indeed, the variables are such that even a single dam involves unpredictable changes in circumstance, like the effects of global warming, which might have seemed negligible at the moment of proposal. Beyond cyclical variations, small changes can be magnified by the size of a dam and its relationship to other dams and instances of management in the water basin. Small changes can also be exacerbated by the form of development: compressed modernization is more likely to produce sudden critical changes than more modest and deliberate expansions still consistent with social need. Yet, in the Global South what might seem like a regulative discrepancy is understandably risked in the face of the Global North's obvious historical abstention from the same question about the real costs of modernization. This is now the globe's most pressing water test, one that forces a distinction between quantity and quality in water's capacity for CHANGE.

See also: CHINA, COAL ASH, ELECTRICITY, INFRASTRUCTURE, RISK.

Demand

Elizabeth Shove and Gordon Walker

For physicists, ENERGY is the ability to do WORK. The demand for energy depends on the type of work to be done and how that work is organized. This simple observation opens the way for some less than obvious lines of thought.

In borrowing the physicists' definition in order to understand the historic and ongoing dynamics of energy demand, we need a broad interpretation of *work*—one that includes practices of sociability, having fun, growing up and growing old, and eating and sleeping, along with work in the more conventional sense of production, employment, and labor. Most social practices generate some form of energy demand or, in more technical terms, most call for a conversion of resources (fuels) to provide services (such as heat, light, movement, power) integral to the ongoing reproduction of those practices.

Within the established literature on energy-society relations, the central goal has been to understand and analyze the energy intensity of different societies by comparing the quantities of energy (measured in horsepower or some analogous unit) required to achieve a set of standardized outcomes, for example, to provide heating or lighting (L. White 1943; Fouquet and Pearson 1998; Sørensen 2012). Such studies compare systems that depend on human and ANIMAL labor with those "powered" by COAL and steam, or by other types of fuel, and seek to explain historical transitions from one system to another. Not surprisingly, interpretations of the social and political significance of such transformations differ widely.

For example, Leslie White (1943) argues that societies progress by harnessing more and more energy, which allows them to move from "barbarism" to "civilization." By contrast, Ivan Illich (1974) contends that social divisions and inequalities increase within societies that depend on increasingly intensive uses of energy.

Either way, debates of this sort miss the recursive relation between technologies (including those involved in harnessing and converting energy) and the constant transformation of the sum total of practices that constitute the work—or culture—of which societies are made. In other words, energy demand is not simply a consequence of the efficiency with which the same service is delivered. Taking a step back, energy demand depends on the entire range of practices that are, or are not, enacted in any one society (Shove, Pantzar, and Watson 2012). This range is, in part, an outcome of previous conjunctions of energy + work, conjunctions that are sedimented in knowledge and technology and that make some forms of social action possible while precluding others. In this way, and in contrast to exercises in comparing demand in standard units across societies, one generation or era of energy + work sets the scene for the next. Put simply, demand has a material and a cultural history.

Accordingly, no simple metric can be used to describe the transformations involved in heating a home with coal rather than WOOD or coming to rely on gas-fired central heating. In one move, the home is detached from one system of provision and labor (wood/coal supply and distribution, storage, stoves, skills in lighting and maintaining a fire) and repositioned within another (gas, pipes, remote storage, new systems of billing, different skills, time, and attention). The forms of work involved in delivering and using energy, as well as the forms of work that energy makes possible, are simultaneously reconfigured.

In short, it is impossible to talk about energy demand as if it could be separated from the life of the home, the rhythms of the day, and the patterns of practice that go on within. In the example of home heating, one key move is from point sources of heat (the fire) to entire space heating: rooms change character and function; new activities—like watching TV in the bedroom—become pleasurable; new styles and distributions of work are enabled. New routines of dusting and hoovering (less) and servicing (chimney sweep, boiler maintenance) are established, and time previously spent fetching and carrying fuel becomes time devoted to other practices.

As this example demonstrates, energy systems do not simply provide the means to do work; they are also implicated in constituting the work itself and vice versa. Crucially, there is no inherent direction to this relationship. For instance, portable power (in batteries) enables new distributions of practices that were previously locked into a specific location precisely because of the type of energy they required. Equally, as practices like mobile phoning and laptop computing become established, they generate different patterns, and especially timings, of demand: power taken from the office is used at home and the other way round. In this instance, the link between energy and doing is wirelessly separated in space and time. This is one of many examples in which spatial rearrangements in the consumption and provision of power map onto—and perhaps enable—the reconfiguration of place and the changing meaning of *home*, *work*, and *leisure*.

Some of these dynamic relationships are outcomes of deliberate intervention. ELECTRICITY is difficult to store; from its earliest days, providers have sought to build loads and orchestrate demand to suit the needs of the supply side, including the need for constant operation (C. Harrison 2013). As Adrian Forty (1986) explains, demand for nonlighting uses (such as trams) was actively constructed by the electricity industry to fill the daylight hours. More of this management of demand occurs than one might suspect. Building intermittent renewables into energy supply calls for intermittent practices, hence the concerted attempts to shift time-negotiable "needs" or forms of work to off-peak hours through peak-time pricing or via devices such as washing machines that turn themselves off and on to suit the grid.

These few observations on the constitution of demand underline the importance of what we might think of not as embodied but rather as entwined energy. Embodied energy is that used in making a specific item or delivering a particular service. A building consequently embodies the energy used in producing the concrete, glass, and plastic of which it is made (Cole and Kernan 1996). Entwined energy, by contrast, involves the items and services that structure the potential for future instances of energy demand. For example, railway systems embody energy, but they also facilitate a more complex web of arrangements encompassing the means to move, the construction of destinations, and the potential for objects and people to circulate and travel—all of which generate other FUTURE possibilities. The concept of infrastructural entwining is useful in understanding how it is that past and present patterns and systems of provision (partly) generate future demand.

To summarize, energy demand is best understood as an outcome of how people spend their time and what work is done. Systems of energy provision are implicated in shaping the types of work and the range of practices enacted in society today. On both counts, energy demand and the dynamics of social practice are inextricably interwoven.

See also: ABORIGINAL, ENERGY SYSTEMS, GRIDS, INFRASTRUCTURE, TALLOW.

Detritus

Sharad Chari

Oil is literally rot. As biomass decomposing over millions of years, oil is the rot of ages, and yet it has become the fuel we cannot yet do without. This dialectic of protracted ruination and fatal promise crystallizes the ethos of our time. If the late-nineteenth and early-twentieth-century moment that inaugurated the age of oil was also, at least in some places, a time of modernist hope for human liberation against the specter of annihilation, the present appears more obviously marked by proliferating decay, desperate walling-in from inequality, political discourse utterly disengaged from arts of survival, and painful archiving of creative effort mired in the muck of the murderous twentieth century.

We might turn, as many have, to Walter Benjamin's Angel of History, not just in relation to objects clearly marked as waste—the great rubbish dumps of our ignoble globalism—but also in relation to monuments of productivity (Benjamin 1977, 259). Still counted in the metrics of growth, praised in the bibles of "emerging markets," and foisted with niche precision through media new and old, the truisms of capitalist "development" ring more hollow than ever and just as loud. In a time of what Michael Denning (2010) calls widespread "wageless life," hopes of social reproduction—for instance, the mirage of global Fordism—have become works of futures past. Karl Polanyi famously called the idea of "the self-regulating market" utopian, which seems more true than ever (2001, 141, 150). And yet, new frontiers of capital accumulation are lauded for producing piles of plastic, despite ruining environments and degrading working bodies. These forms of destructive creation are also carbon-fed, in form and fuel, as are the flickering hopes for a world of plenty they are presumed to keep burning.

If the commodity was the collective hope the nineteenth century bequeathed to the twentieth, ever-piling detritus is the reality of our necropolitical present. The challenge remains how to think with detritus other than through the language of lack and CATASTROPHE, or through longing for a set of twentieth-century verities borne of rose-tinted views of the "Golden Age" of North Atlantic capitalisms (pity about the racism and environmental suffering). The monumental PHOTOGRAPHY of Sebastião Salgado and Edward Burtynsky marks the presence of wasted objects, but it is the poetry of Derek Walcott and the prose of Patrick Chamoiseau that marks their enduring weight in wasted places and ruined lives (see Stoler 2013).

Two lines of thought converge in a productive theory of this strange category, *detritus*. The first emerges from the materiality of oil and its political effects. Timothy Mitchell proposes that democracy's basic presumptions were engineered alongside the COAL commodity chain, from the militancy of semiautonomous miners, to the invention of sabotage, the threat of general strike, and the spread of democratic rights and biopolitical government across North Atlantic societies. The material dynamics of oil, by contrast, circumvented and undermined popular political power. Reading the development of oil in the MIDDLE EAST against Whig histories, Mitchell posits Western states and corporations as concerned with limiting not only oil exploration but also the spread of democratic claims into the colonial world; the idea of "self-determination" became the instrument through which democratic impulses were forestalled (2011, 68.)

Mitchell's narrative pivots on the minerals boom and the South African War of 1899–1902 it provoked and the subsequent rapprochement between Afrikaner and British settlers in the remaking of a "self-governing" (he does not say white-supremacist) Union of South Africa in 1910 (2011, 70–72). What happened in South Africa would prove to be of global significance, particularly through the work of imperial statesman Jan Smuts, architect of segregation and proponent of "self-government" and "self-determination," concepts that fundamentally hollowed out the "self" that was to be empowered (ibid., 81–82). Mitchell takes a concern of late-twentieth-century postcolonial studies, the failure and afterlives of anticolonial nationalism, and places this problematic squarely at the dawn of the century. This move potentially disrupts our understandings of the great anticolonial struggles for self-determination, including the antiapartheid movement led by the African National Congress. Mitchell's argument is suggestive for a theory of detritus premised on the material and infrastructural damaging of politics as we know it but also as condensing, in fossilized form, the detritus of decolonization.

As I write from Johannesburg, a city built on the minerals revolution central to Mitchell's book, I am forced to consider the decaying dream of decolonization in active terms. After all, *detritus* names a process of wasting that is not willing. This deindustrializing City of Gold (*eGoli* in isiZulu) is fundamentally uneven, but it has also become the command and control center for capital in extractive industry across the continent (Chari 2015).

However, we cannot simply recycle concepts like "sub-imperialism" to account for the reality that Johannesburg is also a vibrant Afropolis, whose residents each day rewrite their place in the world in myriad ways (Mbembe and Nuttall 2008). Moreover, Johannesburg is a densely forested city with the most securitized urban forest in the universe: nobody is going to steal our trees. Not least, Johannesburg is a city of ongoing political struggles; the transformation of the liberation movement into a highly securitized machine for ACCUMULATION has not meant the end of politics.

A political theory of detritus would posit the ruins of racial capitalism and decolonization as protracted, if frustrated, sites of struggle. A second line of thought for a theory of detritus is a genealogy that emerges through a different geohistorical lens, focused on the interconnected critical traditions of the Black Atlantic, the Caribbean diaspora, and the Americas. Of numerous figures and texts, I signal three. First, C. L. R. James's dramatic rendition of the Haitian Revolution (1989) emphasizes the damaging of the foundations of James's own political hopes for Black Liberation, and also, in this Hegelian spirit, the importance of holding onto the idea, even the damaged idea, of liberation. Second, Michel-Rolph Trouillot's (1995) meditation on the ruins of the Haitian Revolution turns forgotten sites and silenced narratives into a politics of the ruin in which damaged sites might be read as crucibles of political possibility. Fernando Coronil's postcolonial Marxism can be read as taking these questions of ruined politics and the politics of the ruin further still, by asking how they have been shaped by a colonial ontology that continues to partition the world into bounded regions with separable, indeed hierarchical, histories.

In *The Magical State* (1997), Coronil argues for a global understanding of capitalism in terms of the various triadic configurations of land, labor, and capital that have shaped multiple, interconnected formations. Crucial to his understanding of postcolonial VENEZUELA is the fueling of national capitalism through oil rents that enable a fetishism of politics and the state. Written while his body was beginning to be colonized by cancer, Coronil's late work aimed against the inevitability of this fetishized politics of "carbon democracy." Drawing from Walter Benjamin, Coronil (2012) points to popular traditions of conserving the capacity for critique from within the detritus of capital and decolonization.

I have followed this route in my research on neighborhoods surrounded by oil refineries in the Indian Ocean city of Durban, where people are forced to live with various kinds of detritus: the stench of pollution, its protracted effects in uncounted rates of leukemia and respiratory illness, the stigma of postapartheid racialization, and the poverty of underemployment (Chari 2013). While documenting the remains of segregation and struggle that people are left with, I have found that residents refuse to *become* detritus. However, while these areas have been crucibles of political creativity before, during, and after apartheid, what distinguishes the present is the deep abstraction of Left political discourse from lived suffering in places such as these. When political speech fails, I have found the most powerful critics of ruination to be documentary photographers Peter McKenzie and Cedric Nunn. I am constantly led back to the power of Nunn's photograph of football players in the shadow of the oil refinery, a work of great insight that, in the powerful asymmetry

FIGURE 3. Playing Soccer. Highbury sports ground, Wentworth. 1995. (Cedric Nunn)

of its iconic contrasts, frames in black and white what I take to be the cultural politics of detritus (see Figure 3).

Coronil ends his last piece with damaged forms of critique as the fragmentary poetry of our possible FUTURE. What else could a cultural politics of detritus be, if not bittersweet?

See also: CHANGE, CHINA, EXHAUST, RESILIENCE.

Disaster

Claire Colebrook

Disaster films frame life and death situations in such a way as to sharpen the meaning of life. In small-scale films such as *The Poseidon Adventure* (1972), *The Towering Inferno* (1974), or even *Jaws* (1975), a local threat drives characters into familial or tribal collectives, all focused on an exit strategy, with key characters playing out both romance and villain plots. In more recent end-of-the-world films like the viral pandemics *28 Days Later* (2002), *28 Weeks Later* (2007), or *Contagion* (2011), the very survival of humanity and civilization is at stake. Today, in a world of widespread CATASTROPHE, disaster provides narrative fodder for shoring up a sense of who we are and what our world means to us.

But disaster has also been understood as evacuating all meaning from everything. Ordinarily, we live as if in a world of meaning—of projects, purposes, narratives, and community, as beings-in-the-world. Martin Heidegger (1996), the great philosopher of being-in-the-world, also opened new ways for thinking the end of the world. Once my world of meaning and projects breaks down, I may be left without world, without a horizon of sense and intentionality—without a FUTURE toward which I am thrown. Maurice Blanchot's *The Writing of the Disaster* (1986) captures disaster's evacuation of meaning: something about disaster precludes writing. Yet one can only begin to write after losing all that one has accepted as one's world.

In this way of thinking, disaster is not a mere mishap or disturbance within my world. With a disturbance, meaning remains and perhaps even deepens. Maybe the power goes out; this disturbance may prompt reflection, RESILIENCE, and a stronger sense of self. I become aware of how writing and researching are constitutive of my being (and how these activities, in turn, depend on technology and external ENERGY sources), but I still have my world and a sense of who I am.

Disaster, by contrast, is not even just death or my loss of world; it is widespread world-less-ness. Maybe the power goes out, the water supply is interrupted, and flooding renders the water in the streets and the ground level of my home toxic and dangerously close to fallen power lines. Food deliveries stop and my town becomes isolated. People begin to panic and lose sympathy; panic is exacerbated by fear that reinforces itself without the ordinary communication and transport technologies that might ameliorate this disaster. My world is not simply disrupted, for my projects of writing and researching are not only halted, they are no longer my projects. My horizon of meanings, purposes, and intentions has contracted; I need to stay alive. "Life" is no longer the rich *Lebenswelt* of phenomenology: I do not move, breathe, speak, touch, and think with the thick past of humanity coursing through my veins. "Life" is not the living present with a retained past and anticipated future. I must stay alive, and in that surviving all that is left is what remains. Like Heidegger's stone, rather than Heidegger's animal, I am world-less and not simply "poor in world" (Heidegger 1996, 185).

This version of disaster has three stages: first, there is world into which we are thrown, with meanings, projects, and others. Second, that world of meaning and "being-with" is destroyed. Suddenly, the things and relations that gave my life meaning, connectedness, IDENTITY, and forward momentum are gone. There is only one way of being: to live on, to survive. For Blanchot, it becomes necessary to write (the third stage). Writing would not follow from the meaning of the world, or from the expression of a self, but from the absence of either of these terms (Blanchot 1986, 52).

Blanchot pleads: "May what is written resound in the stillness, making silence resound at length, before returning to the motionless peace where the enigma still wakes" (ibid., 53). Perhaps Blanchot's sense of disaster is more authentic than twenty-first-century disaster films: rather than emphasizing what is self-evidently worthwhile and meaningful—friends, love, family, heroism—disaster in his sense exposes the utter fragility or impotence of these human motifs. There was a time when disaster destroyed the *Lebenswelt* and threw us back upon a life without pregiven sense; now disaster is the very stuff of marketing, fantasy, and intimacy (for disaster now melds just as happily with romance plots as it does with pornography).

So we might conclude with a glum diagnosis of the present: we have lost a noble sense that disaster might awaken us from our cozy human slumbers. Far from destroying the homeliness of world, self, and meaning, disaster seems to heighten the demand for spectacle and consumption. We have disaster capitalism, where opportunists cash in on grand catastrophes with no fear of choice or competition (N. Klein 2007), and we have disaster porn and disaster tourism too. We have become happy with disaster, as yet one more spectacle to be purveyed, commodified, and rendered affectively homely.

Perhaps, though, we should not be too quick to see the present as fallen into simple ideology and spectacle, and we should question what Heidegger and Blanchot take to be the authentic condition of disaster as world-destructive and therefore world-disclosing. Heidegger refers to the breaking of a single object, whereas today's planetary crises indicate the disruption of worlds at a far more systemic level. Rather than an "end of the world" that occurs for an individual and her projects, we might think today of how the end of a certain material world (by way of climate change, resource depletion, and concomitant disasters) results in an end of the liberal polity. Rather than seeing the present everydayness of disaster as a loss of authentic intensity, we might look back upon the brief, localized, cozy, industrialized period of comforting worldliness as the exception to a general rule of disastrous human existence.

With the assistance of fossil fuels, NATURE came to appear as stable and nondisastrous for a privileged portion of humanity. Drawing on nonrenewable and toxic resources made it possible for a manufactured "humanity" to protect itself from the volatility of life. In other words, in the two centuries of intensive fossil fuel consumption, the otherwise chronic disaster of vulnerability to the elements and the environment seemed to be mitigated. For some humans this was a moment of ease, when the disaster enterprises of warmongering, fear-campaigning, and sovereign-terror were put on hold and experienced as exceptional. This brief period of industrial capitalism, now understood to have precipitated the disaster of all disasters that is the ANTHROPOCENE, is the same brief period in which we could imagine (with Heidegger and Blanchot) that disaster was exceptional enough to be existential. Now it is disaster itself—the spectacle of something that might come in and destroy "us"—that, far from "exposing a hurt which can no longer be endured, or even remembered" (Blanchot 1986, 52) prompts "us" to huddle together and regroup in fantasies of a "we" and of "our future" and in a language of survival, adaptation, and mitigation. If disaster has become a way of life (see Buell 2003), then we also need to forge ways of life that confront the myth, spectacle, magic, and religion of disaster that reason cannot vanquish.

See also: ANTHROPOCENE, CATASTROPHE, CRISIS, INFRASTRUCTURE.

Ecology

Timothy Morton

When we divide the world into the categories *nature* and *culture*, we perform the quintessential gesture of modernity. But modernity is predicated on the ecological emergency that has given rise to a new geological epoch: the ANTHROPOCENE. "Modernity" is how the Anthropocene has appeared to us historically thus far. Dividing the world into NATURE and culture is precisely anti-ecological insofar as it participates in the logistics that enabled humans to act as a geophysical force on a planetary scale. The Anthropocene is the moment when Western philosophy restricted itself to the (human) subject-world correlate (Meillassoux 2008, 5). This self-imposed ignorance of the real seems to go hand-in-hand with direct intervention in the geological real.

Ecological thinking must, like premodern cultures (particularly indigenous ones), lack the concept *nature*; therefore, it must also dispense with the concept *culture*. However, the gap between phenomenon (subject) and thing (world) cannot be breached without metaphysical (and physical) violence. Instead it must be universalized beyond the gap between the (human) subject and everything else. A truly postmodern—that is to say, ecological—thinking is alive to the uncanny difference between what things are and how they appear.

We should simply acknowledge that social space is already teeming with nonhumans; it was never fully human from the start. This means that we must drop the metaphysical boundary between humans and nonhumans—or as Marx puts it, the difference between the worst of architects and the best of bees (1990, 284).

The concept of Nature (capitalized to stress its artificiality) is an ideological and physical function of the particular logistics of agriculture that propagated throughout the Fertile

Crescent about ten thousand years ago.[1] Since then, Earth's climate cycles have been smoothed out by human action on the planet. This smooth periodicity is called Nature: "Season cycle moving round and round, / Pushing life up from a cold dead ground" (XTC 1986). Modernity's concept of the "medieval worldview," cyclic and seemingly ageless, is a product of human activity as a physical, planetary force.

In modernity, the Nature concept begins to assert itself in contradistinction to the accelerating logistics of agriculture that gave rise to the Agricultural Revolution and subsequently to the Industrial Revolution. There is an often-noted cultural irony here. Using Nature to combat industry is like using a fake medieval weapon against a modern machine gun. This flimsiness should not lead us to see the Nature concept as ineffective. The idea of a "return to Nature" has underwritten fascism and other forms of political violence. Nazism infamously deployed concepts of health, pathology, and the German *Volk* in order to purge society of its putatively decadent pathogens. Nazis were keen on animal rights and reforestation, which makes it difficult to talk about ecology in some left circles (see Ferry 1995). Yet the Nature concept is by definition atavistic, evoking a nonexistent, more fully present, "organic" past. Thinking oneself or acting oneself into that illusory past can only involve violence. Why? Because it is impossible to return to something that does not exist. Attempting to do so pushes against the grain of how things are.

Yet an even more profound irony lies within the originary irony of Nature itself, a concept born of human activity on a planetary scale, the unconscious of human activity as such. The Nature concept is not separate from humans, but until we could think concepts such as *species* in a successful way, we remained ignorant of it. This thinking emerged in modernity's first century—the nineteenth—a moment when the acceleration of the logistics that resulted in the Industrial Revolution began to generate what is now called the Anthropocene by emitting carbon compounds and radioactive materials (and so on). It started in the later eighteenth century, when, for instance, the patenting of James Watt's steam engine as a general-purpose machine in 1784 catalyzed both massive fuel use and massive carbon emissions. In 1945 there arose the Great Acceleration, in which carbon emissions and other human geophysical activities (such as the production of NUCLEAR materials) began to spike. People love dramatic data spikes, and the graphs certainly look as if we have gone from the natural to the disturbingly artificial in record time, as if Nature really did exist before 1945. But 1945 was predicated on 1784, and 1784 was predicated on the no-weeds, no-pests logistics of agriculture from the Fertile Crescent, which for convenience I call *agrilogistics*. The deep, physical irony, inscribed within the very crust of Earth, is that the smooth periodicity before 1784 was inherently unstable: it was already a kind of violence, whose expansion via Jethro Tull and Enclosure, steam engines and factories, nuclear bombs and computers (with their PLASTICS and rare earths, fiber optic cables and lithium batteries) was only a logarithmic increase within agrilogistics.

Nature is a modern term for a less complicated, less obvious, smoothed out violence, its inner logic muted so as to appear uncomplicatedly nonhuman, policing a thin and rigid

1. I am grateful to the geologist Jan Zalasiewicz for discussing this with me.

boundary between the human and nonhuman worlds. The very concept that establishes this boundary, then, is predicated on a slow, (until very recently) invisible war of humans against the nonhuman. This war takes place on a planetary scale in which there is no "away," resulting in the immiseration of entities that have the not-so-good fortune to share Earth with us. Moreover, this war is waged with the best of intentions (civilization, fighting disease, predictability). It is just that the emergent effect of these intentions is precisely that war. Humans have been creating a sort of geophysical pointillist painting whose dots are perfectly reasonable, often benevolent decisions. Yet from a distance—at a planetary scale where one part of the painting cannot be separated from another (there is no "away")—the whole thing is war. Nature is war.

The immiseration of the nonhuman is also the immiseration of humans. Jared Diamond has described agrilogistics as the worst mistake ever made by humans (1987). Very shortly after the introduction of agrilogistics, humans in the Fertile Crescent underwent a series of changes from which they have not yet recovered: patriarchy, rigid class division, luxury for a very few and poverty for almost everyone else, environmental disasters such as epidemics that are based (another irony) on keeping one's agrilogistical space free of disease. And agrilogistics has spread, like a successful virus, throughout Earth, from the hyperbolic sheep fields of New Zealand to the Dust Bowl of the 1930s American plains.

Thus a fuller recognition of the nonhuman, and a more comprehensive inclusion of the nonhuman in social and philosophical space, must be achieved by dropping the Nature concept. So too with the agrilogistical project that gave rise to the Nature concept in the first place. A truly post-modern (rather than postmodern) age will be an age of ecology without Nature but also without agrilogistics.

Modernity has given us a strange noir awareness of our deep implication in geophysical reality. It is now impossible to unthink the irony by which our very attempt to achieve escape velocity from the nonhuman and the physical has resulted in our quite literally drilling down ever deeper into them. The way to think ourselves out of this bind is to allow for beings that can be self-contradictory. Agrilogistics is a social, physical, and philosophical attempt to smooth out contradiction, to create a predictable, consistent plane of thought and action. We must throw some effort into allowing things like meadows and whales and clouds to be contradictory, lest we eliminate them, in both thought and deed. Ecological awareness is necessarily uncanny. The life sciences have worn away at the human-nonhuman boundary and the life-nonlife boundary. A life-form is made of other life-forms (EVOLUTION): it contains parts that are strictly not members of it, in glorious violation of Russell's set paradox. Ecological entities are fuzzy, inconsistent, and contradictory. They cannot be smoothed out without violence. Since at least Plato, such smoothing out has resulted in a philosophical war to eliminate the gap between being and appearing.

Philosophically, politically, and physically, if Nature is war, then ecology is nonviolent coexistence.

See also: ANTHROPOCENE, NATURE.

Electricity

Alan Ackerman

Electrical charges are positive and negative. That polarity is crucial to the physical and figurative power that electricity generates. "It is impossible to read the compositions of the most celebrated writers of the present day," writes Percy Bysshe Shelley in "A Defence of Poetry," "without being startled with the electric life which burns within their words" (1973 [1821], 762). Poetry is "a sword of lightning," a revolutionary force "which consumes the scabbard that would contain it," while a fascination with "electricity and galvanism" leads Mary Shelley's Frankenstein to his "utter and terrible destruction" (2002 [1818], 43). Is our ambivalence about electrical ENERGY any different than that of people two centuries ago, when electricity first fueled a culture of individual freedom and consumerism? As the basis of modern culture, electrical generation is inherently unstable and contradictory, both empowering and dangerous.

A vital metaphor of Romanticism, electricity charged works of individual genius and bound (or incinerated) the social body. The idea of harnessing it for human use caused a paradigm shift in early-nineteenth-century cultural production. In *Leaves of Grass*, Walt Whitman would "sing the body electric," imagining himself in a circuit of electrostatic attraction with readers that would "discorrupt them, and charge them full with the charge of the soul" (1982, 250). As physical force and SPIRITUAL idea, electricity generated modernity, like the lightning bolt Benjamin Franklin channeled through a key on a kite, thereby illuminating new ideas, transmitting voices, and cooking turkeys, while also driving and epitomizing the creative-destructive dynamic of capitalism. "Scientific and technological investigations into electricity gave rise to its metaphoric and symbolic use to represent the human potential to harness the natural world and to free humanity from the chains of the past," writes literary historian Paul Gilmore (2009, 6). From the telegraph to Twitter, electricity has been linked to language, transforming human communication with radical socioeconomic consequences. A medium for goods and ideas, electricity has generated the materialist/idealist dialectic that defines the relation between modern culture and commerce.

Since Whitman's death in 1892 (the year General Electric was incorporated), the generation and use of electrical power has caused and undergone enormous CHANGE. In *The Ecological Thought*, Timothy Morton argues that, ideologically speaking, "nothing much has changed" since the Romantic period; the intensification of science and technology, economic growth, and the expansion of both democracy and social alienation "are quantitative differences, not qualitative ones" (2010, 11). However, the "mass" aspect of contemporary society, with a global population seven times that of Shelley's day (and growing), generates questions about electricity as source and limit of culture. Quantitative change becomes qualitative. Looking back to Romanticism in a 1954 essay, "The Question Concerning Technology," Martin Heidegger makes this point in contrasting the windmill to a hydroelectric plant in the Rhine that changes the essence of the Rhine itself, from river to water-power supplier or standing reserve. Though not opposed to technology as such, Heidegger critiques the "monstrousness" of modernity's divorce of techne from *poiēsis* and asks us to "ponder for a moment the contrast that is spoken between the two titles: 'The Rhine' as dammed up into the power works, and 'The Rhine,' as uttered by the artwork, in Hölderlin's hymn by that name" (Heidegger 1977, 321). Heidegger's ecological conservatism anticipates the leftist politics of environmentalists today, who object that our instrumentalist approach to the planet's resources has altered its very essence. Thus, suggesting a new name for our dwelling-place in *Eaarth: Making a Life on a Tough New Planet*, Bill McKibben predicts, "The next decade will see huge increases in renewable power; we'll adapt electric cars far faster than most analysts imagine. . . . It will be exciting. But it's not going to happen fast enough to ward off enormous [climate-induced] change" (2011, 52). Electricity is at the nexus of our well-being and our undoing, both culturally and materially.

The crux of electricity's power is its unique relation to the materiality of cultural production and distribution, even to the very matter of culture itself. Found in NATURE—a key point for the Romantics—electrical potential energy is unlike solid or liquid forms of fuel, which can be stored and released. It is both invisible and ubiquitous. Electricity is not a thing but a set of phenomena produced by the flow of an electric charge; unlike a lump of coal (the epitome of inertia) electricity is never at rest. Though it may be "static" rather than "current," the it-ness or material basis of electricity is difficult for the average person to comprehend: electricity is the balance or imbalance of charges on the surface of something else that is material. Franklin and Frankenstein looked to the skies; we depend on massive power plants. COAL, natural gas, water, wind, and the sun generate the electrical power we need. It is not simply that our electrified culture is substantively different than that of the candlelit Romantics but also that generating electricity has transformed the socioeconomic order and the physical planet. The emergence of forms of culture that are transmitted and stored electronically has changed the nature of both techno-utopian dreams and dystopian nightmares, if not the very nature of being human.

Generating the electricity to power this culture has produced new catastrophic scenarios. On the one hand, we are connected by massive electrical GRIDS. On the other, breakdowns can plunge fifty million people into darkness. The consequences of such crashes fill the screen in apocalyptic movies like *American Blackout* (2006). But when the power is on, electromagnetic fields surround us. What is the effect of "electromagnetic smog" on our health? Cell phone companies suppress research into "dirty" electricity, though studies of its carcinogenic impact and of electric hypersensitivity have prompted France, Germany, and England to remove wireless networks from schools. Coal-fueled power plants alter the climate. In 2012, Hurricane Sandy, which caused electrical failures from Jamaica to New York, epitomized the vicious circle: power production contributes to weather events that result in power failure. Meanwhile, global DEMAND for electricity grows exponentially, but how to generate it without damaging the world and ourselves? "Now, if we only had a superhero who could stand here and turn the generator real fast," a *Schoolhouse Rock!* video comments, "then we wouldn't need to burn so much fuel to make . . . electricity" (*Schoolhouse Rock!* 1979). That is the problem.

To address it, we might start with another imperative drawn from the mechanics of electricity: the need to ground ourselves, just as measuring the voltage of an electrical circuit requires a reference point in direct contact with the earth. Gilmore argues that the revolutionary impact of Samuel Morse's invention of the telegraph was to separate communication from bodies, discourse from the material world. Telegraphic discourse, writes James Carey, "located vital energy in the realm of the mind, in the nonmaterial world" (1983, 307). Yet the anti-empirical bias of postmodernism has contributed to the current environmental CRISIS. Lawrence Buell writes of the need to make discourse "accountable to the object-world" (1995, 91). Perhaps electrical engineering suggests a lesson: grounding protects us from shocks and prevents hazardous voltages from overloading our circuits.

The second electrical metaphor that may prove instructive is the circuit; electricity works by connecting components, flowing between. The language of electrical connection suggests experience of near-unbearable intensity. Love may be "electrifying," but electricity also powers a culture of distraction. Lovers gaze momentarily into each other's eyes before turning back to iPhones. Facebook expands and flattens friendship. For Whitman, electricity charged body and soul, linking the one and the many. Yet, sounding like Henry David Thoreau, Jaron Lanier, the father of virtual-reality technology, argues that electronic communication has produced a "postpersonal world" in which "life is turned into a database" (Lanier 2010, 69). He critiques techno-utopianism that has turned culture into advertising. For the Romantics, electricity bridged nature and technology, but our twenty-first-century digital reality, in Lanier's terms, is alienated and flat.

Nonetheless, as Benjamin Franklin found, electricity can energize creativity. This was the vision of my favorite television show in childhood, *The Electric Company*, which opened with the song, "We're gonna turn it on/We're gonna bring you the power." In 1973, they added a lyric that equated the power "coming down the lines" with language: "We're gonna tell you the truest words that you heard anybody say" ("*The Electric Company* Theme Lyrics"). The variety form of the show eschewed the Romantic yearning for wholeness. Yet, evoking Friedrich Schiller's *On the Aesthetic Education of Man* (1794), electricity powered the show's *Spieltrieb* or "play drive." Language/electricity is our place of being, always in between, mediating form and content. The medium is much of the message, a circuit that overrides the polar opposition of art and nature, self and world.

See also: DAMS, ENERGY, GRIDS, UTOPIA.

Embodiment

Bob Johnson

Breathe in the warm swell of COAL.

Empty your mind of chatter. Recline into child's pose. And let the quiet textures of petroleum caress your body. Congratulate yourself for being here. This is your hour. This is your truth.

Hot yoga emblematizes who we are as a people—it is both metonym and exaggeration of the modern condition. The wild excess of steam heat, the span of vinyl-wood flooring, the plate glass walls and mirrors, and the textures of spandex, polyester, and PVC mats immerse the modern body in a festival of tactile and visual sensations that trace back to the pleasures of combusted carbon. Here, as in our other rituals, modernity's carbon culture is materialized—the body acclimated to the surfeit of coal's heat, conditioned to the touch of congealed oil and natural gas, and joined to a subterranean INFRASTRUCTURE of carbon that delivers these sensations to the skin, ear, and eye.

We inhabit a peculiar moment when even the search for the self delivers us back to the coal mine, the oil well, and the boiler room.

The Eroticism of Fossil Fuels

Our erotic attachments to fossil fuels are "ultradeep," to repurpose a phrase used by critics of deep-water drilling (LeMenager 2014). They are rooted in the intimate quotidian rituals of such unlikely spaces as the boudoir, the lavatory, and the yoga studio. This embodiment of fossil fuels neither begins nor ends with the pleasures of the gas-fueled AUTOMOBILE and the transoceanic flight; nor is it limited to the sexualized imagery of vehicular mobility and the luminance of the modern city. Such things shape our erotics, but they are only the transparent manifestations of a much deeper phenomenology of carbon.

Put briefly, the embodiment of fossil fuels concerns how the modern soul is disciplined by and made present through the combustion of prehistoric carbons—how it is fleshed out in everyday carbon rites that imbue its sinews with muscle memory, provide material for its imagination and senses, and shape its expectations about being fully human in the twenty-first century.

This structure of embodiment occurs through various modalities.

First, it derives from the body's saturation in a surfeit of mineral heat that carries us beyond the organic forest.

This surplus heat, a fossil heat rather than a traditional organic heat, is the precondition of modern life. It arises from the large, controlled burn that has been going on in boiler rooms throughout the West, and elsewhere, for two hundred years. Today we combust in fossil fuels the equivalent of 21 billion acres of sustainably harvested forest, or more than double our global reserves.[1] Whereas heat in the premodern world had purchase—it was often scarce, limited, hard-earned, and frequently visible to the eye—heat today is simply ubiquitous, assumed, and invisible. According to Dennis Skocz, we are daily conditioned to this surfeit heat that our bodies absorb unconsciously and that appears to us without origin: modernity's "warmth and comfort," he writes, "are not perceived as the effect of anything. They simply are experienced phenomena without a history or an anchor in anything outside themselves" (2010, 18). That heat calms the body when we step into a hot shower, generates the savoriness of soup on a kitchen stove, and provides the security of entering a warm home. It teaches us, Skocz says, "that climate is not an issue and is controllable" (ibid., 20).

The hot yoga studio makes this heat visible while erasing its source. Steam rises from ventilators to provide a sensory contrast to the waterless air of the desert, and hot blasts, exceeding 104°F, invite the body into a ritualistic sweating that is grotesque, therapeutic, and perfectly modern. In this space, modernity's heat is institutionalized and elevated to totem.

Second, fossil fuels are embodied through what Victor Hugo called the "'consubstantial flexibility of a man and an edifice'" (quoted in Bachelard 1994, 91).

The body, phenomenologists tell us, is co-penetrated by its material environment; for us, an ARCHITECTURE of congealed carbons—ENERGY transmuted into glass, cement, and

1. The US Energy Information Administration estimates 2010 global energy consumption at 524 quadrillion BTU, of which approximately 84 percent was from fossil fuels, the equivalent of 16 billion tons of coal. In the preindustrial world, an acre of forest could sustainably produce about .73 TCE, so our fossil fuel consumption is the equivalent of 21.7 billion acres of sustainably harvested forests (US Energy Information Administration 2013). For the energy yield of a premodern forest, see Vaclav Smil (quoted in Pomeranz 2000, 308–9).

steel (Bachelard 1994, 3–37). Today, our eyes, ears, and fingertips anticipate a material environment that differs from the world of WOOD and stone that the body once navigated, grated against, and enjoyed. We carry inside us the rough memory of knees hitting cement, eyes lingering on the curve of steel arches, and a sensibility of light refracted through plate glass. Floor-length mirrors and a terra firma of concrete are energy incarnate—and they signal a materiality beyond the forest (Bob Johnson 2014, 21–26).

In the yoga studio, mirrored walls reflect the body back to itself while open steel rafters, floor-length sheet glass, and recessed lighting provide invitations to the eye. This house for the soul might exude the aura of naturalness, but almost nothing in this space does not depend on mineral heat for its fabrication. In the yoga studio, we encounter a harmony of congealed carbons: coal and enlightenment go hand in hand.

Third, we embody carbon through fabrics we wear on our bodies as accoutrements of the self.

Through polymerization—wherein monomers coaxed from hydrocarbons are strung out into chains of polymers and heated into resins—we acquire the elasticity of modern life in the form of spandex, polyester, and various other synthetics (Huber 2014, 302–7). This carbon that makes contact with the flesh, that drapes down our walls, and that is assembled like idols around our bodies, finds its way consciously and subconsciously into the poetics of daily life as memories, nerves, and muscles become "encumbered" with these little quotidian facts (Bachelard 1994, 57).

This other embodiment of fossil fuels is physical and intimate in the yoga studio. It embraces the lissome stretch of spandex across the thigh, the cushion of high-density foam on knees, and the wet touch of polyester wicking sweat away from an overheated body. The tree of life might stand boldly at the entrance to the yoga studio, giving to this space the feel of the unprocessed, but we too are polymerized here, bound in body and mind to these manufactured chains of carbon.

Finally, our embodiment of fossil fuels derives from the unspoken leisure that carbon accords us as it performs disembodied labor behind our walls and beneath our streets.

Today, mineral heat signifies more than simply fuel and materiality; it also implies energy turned into motion, power, or, in short, WORK, with each American burning sufficient carbon throughout the day to generate the power of nine horses or eighty-nine human bodies (Nikiforuk 2012, 71). This vast kinetic work subsidy—an infrastructure of empowerment—is not always visible to the eye, but it leaves its bold signature on our leisure, productivity, and psyche (Bob Johnson 2014, 14–21).

At the close of yoga practice, I anticipate a shower that was not drawn by my hand—nor any other—and I am entitled to a heat not borne of wood nor delivered by foot. Electric pumps move rivers of water (8.3 pounds a gallon and 25 gallons a shower) across deserts to deliver streams to my feet; natural gas lines imbue me with heat and kinetic energy at the flick of a switch; and oil PIPELINES condition me to a universe of horsepower that emancipates me whenever I prefer not to walk, lift, or hew without help. Although the body takes center stage in the hot yoga studio, that body rarely arrives there on its own and never leaves without accruing to itself this heavy attachment to kinetic carbons that generate and circulate the staggering quantity of water, people, resources, and heat that defines our modernity. Even here in the yoga studio the body is moved by carbon.

But for now—sweet *savasana*.

I take a final drink from a polyurethane water bottle, strip off a pair of nylon-blend prAna pants, and turn the plastic handle of this plastic basin to release the steam of a hot shower. Copper pipelines hiss with gas behind a wall, electric water pumps churn silently, and a burst of water rains down in the desert. As a warm cataract gushes over me, there is only the faintest trace of the FRACKING in California, the drilling in North Dakota, and the strip-mining of Wyoming that gives me pleasure.

I needed this. Soaked in coal and oil, I am now fully and finally in my body.

See also: ADDICTION, ARCHITECTURE, COAL, GRIDS, INFRASTRUCTURE, SPIRITUAL, TEXTILES.

Energopolitics

Dominic Boyer

In Iceland, in the dead of winter, you can hike across barren fractal plains of volcanic rock, covered with crusts of ice and lashed with snow. But go with a guide. Because beneath these plains churns water-wishing-to-become-steam, containing ENERGY enough to power an entire country, with plenty heat left over to scald an errant traveler. The whole landscape simmers and plumes. A geothermal spring in the wilderness is one of life's miracles. Lush moss thrives improbably in the space where bubbling water ends and ice begins.

That barrens can be more than they seem is the premise of this essay. I offer a brief discussion of contemporary theories of power, especially those that are thriving in the human sciences and clinging like verdant Icelandic moss to rock between two hard places. Through the wars and cosmic-ideological struggles of the twentieth century against its sibling titans, fascism and socialism, liberalism managed to discipline thinking about power in its own image, excoriating any alternative to its political ontology of "autological" individuality (Povinelli 2006) either as impossible fantasy and a violation of human nature or as possible and within the bounds of human nature but nevertheless a thousand times worse than liberalism.

The topology of mainstream political thinking today thus sometimes looks scarcely different than in the time of Hobbes and Locke. Power- and rights-bearing individuals and even more powerful states and sovereigns are sometimes in outright conflict, sometimes uneasily bound to one another by contracts. In our time, we hear of the mediating power of "the market" and "market forces" as well. But, upon closer examination, "the market" more often appears as a vast constellation of autological self-interested individuals than as an autonomous force to which individuals are fundamentally beholden.

The monopolization of mainstream politics has not entirely evacuated political thinking. Legacies live on, and new ideas and practices germinate in out of the way places. In the human sciences, the two principal counter-analytics of power in the twentieth century have revealed dimensions and operations of power that run counter to the common wisdom of liberalism. First were the elaborations of an older Marxian critical field (e.g., concepts like capital, hegemony, ideology); second, the emergence of a new conceptual field that alloyed epistemics drawn from semiology, cybernetics, and systems theory. Michel Foucault's "biopower" and Brian Massumi's "ontopower" are exemplary concepts, both indebted to Deleuze's weaving together of the earlier immanentism of figures like Spinoza and Bergson with more contemporary threads of psychoanalysis and information theory. These critical fields intersect in the revelation of the historicity and multiplicity of power. Both speak of forces uncontained by states and subjects yet highly consequential for the constitution of social formation and subjectivities through time.

Capital, for example, invokes a dialectical, historical re-envisioning of power. Capital was human labor objectified and transmogrified into a power-unto-itself in a divided and specialized labor ecology incorporating a monetary system. As capital, the objective form of labor power progressively alienated humanity from its epistemic and material creations and from itself. But the power of capital too was transient. Marx predicted that capital would possess a logic and power of its own only until the social apparatus of production attained such a degree of sophistication, efficacy, and universality that it could satisfy human needs without the need for a divided, specialized labor ecology. In the *Grundrisse* (1973), Marx schematized that horizon as a world of technological advancement in which a society could generate useful things almost effortlessly. Once the universal working class seized this production apparatus away from the owners of labor, the empire of capital would be broken and labor could be freed to explore its true capacities, not just for some but for all, as a species being. Marx regarded the individual and the state, at least in the bourgeois-liberal understanding of these terms, as transitional alienated political forms that would be negated as humanity exceeded them.

The analytics of biopower and ontopower reframe power as *pouvoir*, modal "enablement." As Foucault noted in a 1972 conversation with Deleuze, "everywhere that power exists, it is being exercised" (1977, 213). The analytics of biopower examine two force clusters, the anatomo-politics of the human body and regulatory controls on the species body (Rabinow and Rose 2006, 196). But Foucault's more foundational insight was in the domain of political mechanics. He called attention to the complex organization of relays within power, a capillary network of transversal forces with a capacity for resistance and creation. Massumi's ontopower elaborates this latter quality and describes a priming force, something that shocks us into action, helping us acquire "new propensities," bringing a "moreness to life" (McKim 2009, 10). Ontopower's force can emancipate existing lines of power or reinforce them; it can be exercised by the military as well as the artist. Concepts like biopower and ontopower challenge the idea of power as a function of human will. The liberal individual and state must be viewed as particular assemblages of enablement, not things-unto-themselves but instead continuously undulating things-becoming and things-unbecoming.

These counter-analytics have been extraordinarily generative and will remain so. But I believe that the circumstances of critical thought in the early twenty-first century—when concern about the ANTHROPOCENE is continuously mounting—demand further thinking about power. Humanity has proven capable of reshaping planetary ecology through its population and habits of energy use. This is a new magnitude of power, comparable only to human mastery of NUCLEAR energy and weaponry. It prompts us to think about power beyond the realm of the human, not in a posthuman register, because our presence and agency cannot be denied, but rather in a transhuman register adapted to questioning the political interrelationship of humanity, energy, ecology, and INFRASTRUCTURE.

I have proposed energopower as a conceptual rubric around which to organize these analytics (Boyer 2011, 2014). Energopolitical analysis re-examines our understanding of political power through the twin analytics of fuel and ELECTRICITY, the sources enabling nearly every index and practice of modernity from artificial illumination to transportation to electronic media and computation.

These analytics have a deeper history in critical thinking: the interrelationship of energy and power has haunted us since before the nuclear age. Freud used electricity as his model for intra-psychic energy flow, the medium of the operation of desire—a line of thinking retrieved by Deleuze and Guattari (Boyer 2013, 152–56). But the first truly electric philosopher was Marshall McLuhan, who saw electric instantaneity revolutionizing human thought and culture, relocating the "servomechanisms" of mind increasingly outside our skulls.

The analytics of energopower remain cryptic in such formulations, just as they do in the political insights of poststructuralism. Luciana Parisi and Tiziana Terranova brilliantly counterpose a thermodynamic disciplinary society with the entropic energy and turbulence of "post-disciplinary power" (2000). More recent work by Hermann Scheer (2004) and Timothy Mitchell (2009, 2011) charts even clearer pathways for understanding energopower. A German politician and RENEWABLE energy visionary, Scheer sees the long, inefficient supply chains intrinsic to carbon and nuclear energy systems as the backbone of a centralized energic infrastructure through which political authorities can oppress or ignore communities and individuals. Instead, Scheer imagines a "solar economy" based on a decentralized renewable energy supply, which enables a rescaling of energy infrastructure and unprecedented opportunities for political autonomy.

Mitchell analyzes the inseparability of modern democratic power and political expertise from carbon energy systems; the materialities and infrastructures of COAL allowed miners to choke political power's energic basis until greater rights and protections were offered to labor (2009, 407). He also explains how the biopolitical norms of twentieth-century Keynesian welfarism were authorized by a regime of expertise concerning oil as an inexhaustible resource capable of fueling endless economic growth. This understanding of oil was enabled by a geopolitics of imperial control over the MIDDLE EAST and its subsoil resources. Once that control was ruptured with the formation of OPEC and the oil shocks of the 1970s, the magic of Keynesian biopolitical thinking was disrupted. "Growth" declined radically across the Global North and different powers of life exploited this CRISIS to rise to dominance: this politics we normally gloss as "neoliberal."

Energopolitical analysis is the effort to discover what role fuel and electricity have played in the constitution of political power and expert authority. It is also the effort to imagine new configurations of energy and power/knowledge. Energopower offers no handbook, no definitive statement; we welcome all imaginative explorers and designers. To return to my point of departure, the landscape of political imagination only looks barren from a distance. Close up, it is complex, fissured, steaming. Its hot springs remind us of the abundance of alternative energy sources and the potentiality of alternative political forms. Carbon democracy is not only not inevitable. It is also increasingly vulnerable. Was it not precisely geothermal Iceland that fought most decisively against (neo)liberal truth and power after the 2008 collapse? A self-identified "anarcho-surrealist" was even elected mayor of Iceland's capital. The optics of energopower take Scheer's lesson that the project of regenerating political alternatives is wholly combined with new projects of energic sovereignty, which may relieve us of the infrastructure and exercise of centralized authority.

Today we await the geysers.

See also: CHANGE, GREEN, INNERVATION, PETRO-VIOLENCE.

Energy

Vivasvan Soni

In Aristotle's unusual and tragic sense of motion, as the "actuality of a potential, as such" (Kosman 1969, 57; see also Sachs n.d.), we are on the way to our own destruction, even if in slow motion beyond a human scale of time. The hope that animates Patricia Yaeger's urgent injunction to examine our "energy unconscious" (2011, 306) is that, by bringing energy into cultural awareness, we might be able to fashion forms of collective life and patterns of energy consumption that avert this calamity.

Still, there is a limit to how thoroughly we can probe our energy unconscious, because this concept assumes that *energy* itself has a stable meaning determined by modern physics. *Energy* in this sense refers to the material source of power needed to perform any activity: "a force for making things happen, for driving every kind of process" (Coopersmith 2010, 5). This energy takes many forms (kinetic, potential, electrical, mechanical, heat, chemical, NUCLEAR), each of which is convertible into all the others, according to one of the most fundamental principles of modern physics, the conservation of energy. But the concept of energy, if it can be called that, is of very recent origin. Historians have traced modern use of the word to Thomas Young in the early nineteenth century (P. M. Harman 1982, 51; C. Smith 1998, 8), although Leibniz already had an understanding of kinetic energy, which he called *vis viva* (Harman 1982, 36, 60; Coopersmith 2010, 31–32, 42, 103). Not until the work of William Thomson, Peter Tait, and William Rankine in the 1850s did energy become the central concept of physics (Harman 1982, 5–6, 58). The manipulation of energy and energy equations has since become a staple of modern physics, and it seems that there is little we understand so well or so precisely (Coopersmith 2010, 355). And yet, the concept remains as opaque and mysterious as all matter, eluding our efforts to say what it is: "It is important to realize that in physics today, we have no knowledge of what energy *is* . . . . It is an abstract thing in that it does not tell us the mechanism or the *reasons* for the various formulas" (Feynman 1995, 71–72; see also, Coopersmith 2010, 1, 355).

Beyond these two forms of "energy unconscious"—culture's silence about the energy that powers it, and the unsignifying opacity of the modern concept—I want to call attention to the unconscious of the word *energy* itself, the untapped resources or latent potentials that need to be reactivated. Concealed in its history are clues that could help change our unreflective and indiscriminate habits of energy use. *Energy* derives from the Greek *energeia*, yet a nearly unbridgeable gulf separates these words (Heidegger 1977b, 160). Though not opaque like *energy*, *energeia*'s meaning is difficult to grasp because its conceptuality is lost to us.[1] It is translated variously as activity, operation, actuality, or a "complete" action, to which we might add "the effective," activation, and even, following Heidegger, the real (*das Wirkliche*) (1977b, 160). Tellingly, *energeia* is never translated as energy.

Looking past the confusing list of partial synonyms, what does *energeia* mean? Most distinctively, in Aristotle's *Metaphysics* it refers to a kind of praxis in which the end is implicit at every moment (1984, 1048b 22–28). *Energeia* describes that which is effective or at work in the world (Sachs n.d., sect. 2) not because of the magnitude of its power but rather because its end is attained in the very doing. By contrast, modern energy physics describes the motion or transformation of a system without reference to ends or purposes (Coopersmith 2010, 146). The absence of teleology in physics is not a problem but an important aspect of its explanatory power; thus, attention to ends must come from outside physics. Similarly, the principle of the conservation of energy is indifferent to the form that energy takes: clean, GREEN, dirty, explosive, inert, kinetic, potential, chemical, thermal, etc. Again, the problem is not the principle, which has extraordinary descriptive power. The problem would be to think that the principle is sufficient to discriminate between preferable and undesirable, better and worse forms of energy. These judgments can only come from outside the system, with reference to our own ends and purposes. Aristotle's energeia compels us to remember that, though there may be motions, happenings, events of all kinds, *action* has no meaning without reference to the ends that animate it. *Energy* and *energeia* are key words or articulation points for entire conceptual clusters in two radically different worldviews, one the modern, physical worldview that is still largely our own and the other the Aristotelian worldview in which human action was meaningful, relevant, and effective.[2]

A second difference between *energeia* and *energy*: *energeia* refers to what is actual, whereas the dominant sense of *energy* is potentiality. When we speak of fossil fuels, for example, we mean stores of energy; more informally, when we say that a child has a lot of energy,

1. On the different senses of "conceptual loss," see C. Diamond (1988).
2. On the modern obsolescence of the category of action, see Arendt (1958).

we mean that she has pent-up reserves with no outlet. To be sure, physics has a concept of kinetic energy, but potential energy dominates our attention. When Heidegger characterizes the essence of modern technology as "standing-reserve" (1977b, 14–18), he points to this primacy of potentiality. But the problem is not simply standing-reserve or instrumentalization, which allows us to treat NATURE as something to be exploited; exploitation is unavoidable if we act thinkingly in the world. The problem is rather that "potential" energy is not a potential for anything at all. In its strict definition, potential energy is conceived in relation to position, not actualization (Feynman 1995, 76). This understanding of potential does consider what the potential is good for, what ends or purposes it promises to actualize, or how it might be actualized in better or worse ways. Modern freedom becomes associated with this infinite potentiality rather than with the capacity to actualize itself in particular ways.

Whether we compare energy and energeia in terms of ends-orientation or actuality/potentiality, the Aristotelian conception of energeia demands that we discriminate among ends and consider what a particular potential is good for. Such judgment is at best optional and at worst avoided altogether within a modern understanding of energy, leaving us in a nearly Parmenidean stasis as energy changes from one form to another while remaining always exactly the same. In our endless hunger for energy, what vanishes is that specifically human mode of acting with a view to ends, energeia. Around the time when the modern notion first emerges in the late seventeenth century, the third Earl of Shaftesbury is among the last to preserve the urgency of the older meaning of the word, what Saree Makdisi and Michael Ziser call "human energy" (Yaeger et al. 2011, 318–20, 321). For Shaftesbury, any "philosopher" who attempts to explain a person's motion through purely physical means, without considering their internal ends, makes a profound category mistake, "since . . . he consider'd not the real Operation or *Energy* [emphasis added] of his Subject, nor contemplated the *Man* . . . as a human Agent; but as a *Watch* or common *Machine*" (Shaftesbury 2001, 1.181). If we are to become discriminating in our patterns of energy use, if we want to act energetically together to avert calamity rather than simply converting energy thoughtlessly from one form to another, then we need to remember that it is energeia and not energy that gives meaning to our actions. Energeia turns our movements and muddlings into actions in the first place.

A final meaning of *energeia* is relevant here. In Aristotle's *Rhetoric*, *energeia* is also a rhetorical figure, referring not merely to an especially vivid or forceful expression but to a species of metaphor that suggests "a notion of activity" (1984, 1411b 29). Let us deepen Aristotle's insight into the literary dimension of *energeia*. Ends or purposes, we have seen, are not implicit in physical descriptions of phenomena; they come from beyond. But from where? The WORK of fashioning, fabricating, forging such ends is always a work of fictioning; it is impossible without the literary capacity to imagine and make what was not there before. We undertake this making daily, in the simplest ends-oriented actions. But one genre of writing explicitly takes as its task the work of crafting ends to energize collective action: utopia. In utopias, ends are always implicit, animating and active, making them the exemplary instantiations of the literary figure I would call *energeia*. Orienting our actions toward ends we constitute for ourselves, utopian imagining, rather than the ACCUMULATION of energy reserves, is our energy future. In it, "we [will] find our happiness, or not at all!" (Wordsworth 1979, 399).[3]

See also: DAMS, DISASTER, ELECTRICITY, FICTION, OFF-GRID, UTOPIA.

3. Aristotle describes the highest end for human beings, happiness, as the "activity (*energeia*) of the soul in accordance with virtue" (see Soni 2010, 126–27, 136–41).

Energy Regimes

Michael Niblett

Energy regimes are difficult to define but easy to recognize in a CRISIS. The crucial role of dominant ENERGY sources in structuring everyday life tends to impress itself most strongly on popular consciousness at moments of potential disruption: think of the "shrill cries" of pamphleteers warning of a timber shortage in seventeenth-century Europe (Williams 2006, 166), the "coal panics" that swept through Britain in the late nineteenth century as commentators prophesized the EXHAUSTION of the resource that powered the country's ascent to global hegemony (Podobnik 2006, 1), or the social experiments in sustainable living sparked by oil shocks in the 1970s.

For J. R. McNeill, energy regimes are "the collection of arrangements whereby energy is harvested from the sun (or uranium atoms), directed, stored, bought, sold, used for work or wasted, and ultimately dissipated" (2000, 297). Like many environmental historians, McNeill locates in the Industrial Revolution a fundamental shift in patterns of energy use. He posits a transition from a "somatic energy regime" in which human and ANIMAL muscles are the most important prime mover, directly converting the SOLAR energy absorbed in plants, to an "exosomatic energy regime" based upon fossil fuels (ibid., 10–13). Echoing McNeill, Edmund Burke posits "only two major energy regimes in human history: the age of solar energy (a renewable resource) from 10,000 B.C.E. to 1800 C.E., and the age of fossil fuel (a nonrenewable resource) from 1800 C.E. to the present. . . . This unprecedented transformation lies at the heart of any history of humans and their relationship to the environment" (2009, 35).

The problem with viewing the Industrial Revolution as a historical rupture is that it tends to abstract industrialization from the systemic logic of capital ACCUMULATION and the dynamics of class, state, and power in which it was embedded. Indeed, this perspective risks reifying energy, thereby obscuring the definite social relations that determine how energy is internalized and experienced within civilizational histories. The use of WOOD as fuel in Roman copper smelters was very different from the use of wood as fuel in the copper smelters at, say, the Great Copper Mountain of Falun, Sweden, in the seventeenth century. In the context of an expanding capitalist world-economy and the deeper penetration of commodity relations, the logic dictating the consumption of energy resources was radically transformed from the logic in the precapitalist era.

Bruce Podobnik's linkage between energy regimes and the long wave dynamics of capitalist accumulation offers a framework within which to grasp energy sources not as inputs but as simultaneously structured and structuring relations of the patterned totalities of historical systems. For Podobnik, an energy regime "refers to the network of industrial sectors that evolves around a particular energy resource, as well as the political, commercial, and social interactions that foster [its] expanded production and consumption" (1999, 155). In this view, the transition from the coal-dominated energy regime of the long nineteenth century to the oil-dominated regime of the twentieth century (or from wood to COAL in the late eighteenth century) was not simply the substitution of one fuel for another. Rather, it transformed the world-economy and reorganized "the system of industrial production and consumption" (Podobnik 2006, 49). Indeed, the transition to oil was inseparable from the development of new forms of (Fordist) mass production and consumption. It also entailed the production of new kinds of space (suburbs, urban sprawls, expanded road networks), the reorganization of global natures, a new, oil-soaked cultural apparatus, and new bodily investments and modes of AFFECT "materialized in particular types of vehicles, homes, neighbourhoods, and cities" (Sheller 2004, 229). This is not to say that with the rise of oil, coal disappeared as an energy source. But along with residual or subordinate energy sources such as CHARCOAL or hydroelectricity, coal was rearticulated within an energy regime oriented around the gravitational center of petroleum.

Energy regimes cannot be studied in isolation or in the abstract; as a concept, they have analytical purchase only insofar as they are understood within particular historical systems or regimes of accumulation. Discussing the "great energy transitions of the modern world . . . from peat and charcoal (1450s–1830s), to coal (1750s–1950s), to oil and natural gas (1870s–present)," environmental historian Jason W. Moore cautions against "the conventional view that sees a 'structurally invariant' capitalism (or industrial society) incorporating new external resources." These energy sources "did not make capitalism so much as capitalism remade itself through their incorporation" (J. Moore 2011a, 22). Moore's line of argument is inseparable from his theory of capitalism as "world-ecology," a perspective that joins the accumulation of capital and the production of NATURE in dialectical unity and links historical developments ostensibly located in Europe (like the Industrial Revolution) to developments elsewhere. For Moore, capitalism does not act upon global nature so much as it emerges through the production of nature-society relations; it is a "world-historical matrix" that "knits together humans with the rest of nature . . . within a gravitational field of endless accumulation" (J. Moore 2012, 3). In this understanding, energy regimes represent the cluster of vectors through which capitalism remakes the web of life in accordance with its self-expansionary logic. As the provisionally stabilized configurations of nature-society relations that once enabled accumulation become exhausted, capital seeks to recombine human and extra-human natures in order to revive profit rates. This process necessarily entails restructuring the energy regime through which those natures had been interwoven. Moore emphasizes how the periodic reorganization of energy regimes helps to generate cheap energy windfalls capable of facilitating increased labor productivity without a corresponding increase in capital intensity, thus creating the conditions for new long waves of accumulation (J. Moore 2011b, 130).

The degree to which energy regimes pattern social experience underscores why their presence registers in popular consciousness during periods of regime change. We might turn to cultural production and ask whether there are any trends to be observed in the registration of such transformative moments. Recent work on world literature by critics like Franco Moretti and Stephen Shapiro maps the flow of long waves of capitalist accumulation in relation to cultural manifestations. What happens if we map the flow of energy regime transitions in relation to cultural manifestations?

In an essay on Gothic periodicity and the world-system, Shapiro argues that "generic inscriptions of Gothic narratives and devices tend to re-emerge in swarms at certain discrete periods" (2008, 30). If capitalist commodification produces an intrinsically Gothic experience, he writes, then the intermittent clustering of Gothic tales—at moments like the 1780s/90s, 1880s/90s, and 1950s—indicates that "the oscillating pump of Gothic emissions reveals a more specific, albeit recurring representational purpose beyond its application as a general thematic for describing capitalist-induced phenomenology" (30). Specifically, Gothic's periodicity suggests that "these narrative devices seem particularly to sediment during the passage between two phases of long-wave capitalist accumulation" (31).

Shapiro's argument can be extended to transitions between energy regimes insofar as these are often enfolded in the passage between long waves of accumulation. Indeed, if we map the spikes and swarms of Gothic narratives on to the key moments of CHANGE in energy regimes, an interesting correlation emerges. Take the transition from wood to coal in the late eighteenth century, an era of crisis in the world-economy during which "the progressive exhaustion of the English agricultural revolution" was conjoined with "an agrarian depression that reached from the Valley of Mexico to Scandinavia" (J. Moore 2011b, 125). Crucial in resolving this crisis was the transition to coal and the windfall profits it generated. At the same time that these changes were destabilizing the world-system, a wave of Gothic narratives appeared, starting with Horace Walpole's pioneering *The Castle of Otranto* in 1764 and peaking in the period from 1790 to 1815 (Moretti 2007, 14). Moving forward one hundred years or so, something similar occurs during the transition from coal to oil: a swarm of Gothic texts (e.g., Bram Stoker's *Dracula* [1897]) clustering in the era that "coal panics" swept England and "the expansionary cycle of the global coal system" came to an end (Podobnik 2006, 39). Then again, in the 1970s, during the shocks to the oil regime, there appeared a further swarm of Gothic narratives, with Gothic romances flourishing alongside a horror film BOOM (for which *The Texas Chain Saw Massacre* [1974], a film soaked not just in blood but in oil anxieties too, might be paradigmatic) (Radway 1981; Jackson 2008).

What could account for such correspondences? One response might be that Gothic devices and motifs are well-suited to express feelings of dislocation, anxiety, and strangeness engendered by the social transformations attendant on the reorganization of energy regimes. The existence of such correspondences requires further investigation, as does the possibility that other irrealist literary modes might cluster around energy transitions, functioning as an equally apt register for the anxieties they foster. This preliminary speculation suggests that the rubric of energy regimes has much to offer not only environmental history and political economy but also literary and cultural studies.

See also: ANTHROPOCENE, CRISIS, ENERGY SYSTEMS, UNOBTAINIUM.

Energy Systems

Frederick Buell

The history of oil and its cultures cannot be separated from the more sweeping history of energy systems dating back to the earliest human societies. More proximately, the history of oil cultures cannot be separated from the earlier phase of fossil fuel history that began in the mid-eighteenth century, when the energies unlocked from COAL reshaped and replaced the medieval energy system of wind, water, animal, and people power. Over the next century, a qualitatively new energy system materialized, dubbed coal-capitalism by Jean-Claude Debeir, Jean-Paul Deléage, and Daniel Hémery (1991). Like all previous and later ones, this energy system came together through the dynamic and complex interactions of myriad factors: namely, energy sources, the technological arrangements necessary for their use, and the social arrangements that surround these material components (Debeir et. al. 1991).

Energy systems are interactive—or *intractive* (Barad 2007). Their relationships materialize as technological and social "converter chains" take shape around and "switch on" a particular energy source, so that its material properties come to matter and, in turn, constrain the converter chains proliferating around it (Debeir et al. 1991). These include the technological, social, economic, political, and cultural arrangements that make energy usable and embed it throughout society. Culture, in other words, is an essential (although until now almost completely unnoticed) component of energy systems.

The coal-capitalist energy system was profoundly disrupted in the later nineteenth century with the appearance of oil. A new fossil fuel energy system, which I call the modern oil-electric-coal capitalist system, soon assembled itself (Buell 2012). That system was dominant until World War II, when another burst of dynamic growth, coupled with a new, large-scale conversion of natural gas and oil into petrochemicals, again changed everything. In all these changes, culture was both agent and effect, alternating between exuberance and catastrophism (Buell 2012).

The modern oil-electric energy system appeared culturally as a motive force augmenting peoples' powers: fueling their activities, propelling them forward, accelerating their bodies, energizing them, awakening their faculties, sanitizing their environments, and enhancing their lives. In the postwar period, however, new cultural understandings appeared. The new system replaced its modern predecessor's promises of augmentation with the possibility of systemic metamorphosis. Rather than representing an improvement or acceleration of life, oil-energies became life's dwelling place, even the foundation in which society was embedded. And these new foundations in oil and gas quickly became what their material forms implied: not the architectural fixity of a "foundation" but the fluidity of oil and the volatility of natural gas. Petro-saturated developed-world societies assimilated future shock into a permanent expectation of rapid, exuberant CHANGE. They were also deeply unsettled by a new sense of undelimitable, possibly civilization-ending, social and environmental RISK.

That there was a fundamental shift in the energy system after World War II is clear from several indices. Oil became a key factor in the miraculous acceleration of growth in the postwar period. Graphs of oil use show a sudden spike in the 1950s that matches graphs of world GDP, world population, chemical production, pollution, and other environmental problems—all of which were enabled by oil (Speth 2008). Culture was deeply entangled in these transformations, with three main aspects.

First, the modern energy system fetishized speed. The character of this speed changed in the postwar period, moving far beyond the AUTOMOBILE and the acceleration of bodies. The sites of speed today include consciousness (thanks to information acceleration and pharmaceuticals), computer processing, MEDIA (narrative structures, jump cut edits, television's continuous program streams), technological innovation, social change (and attendant changes in identities, values, and ideas), and even evolution (via genetic engineering) and geological change (Chakrabarty 2009; Gleick 1999; Kelly 1994; Toffler 1970, 1980). The old bodily acceleration fetishized by the Futurists became as retro and downscale as NASCAR, while speeds of integration and invention became cutting-edge, promising not mere augmentation but systemic change of everything.

This accelerated pace of change was particularly evident in social geographies. Rooted in colonialism, this acceleration dates back to coal, when coal-fired ships and rail helped link colonies to coal-fueled industrialism at home and made Britain the workhouse of the world. Coal-capitalism helped construct wider social, economic, and cultural gulfs between colonial peoples and metropolitan ones even as it brought them closer together. Colonial mercantile relationships were transformed—with the help of speedier steam transport—into tighter imperial ones. This change is legible in literary history, in the shift from *Robinson Crusoe*'s redeemable "savage" (1719) to Joseph Conrad's gothically enhanced brute at the *Heart of Darkness* (1899), even as a correlative "heart of darkness" lay beneath the veneer of civilization in the late-Victorian invention of the unconscious. Cultural distancing was accompanied by its opposite, the erasure of geographic distance. With the emergence of oil-electric capitalism, modernization added a new level of difference and distance between core and periphery. Cultural modernism both erased and maintained distance by opening up new avenues of curiosity about (rather than revulsion at) "primitive" art (e.g., Picasso and the Harlem Renaissance) and the "primitive" as a positive element within the civilized unconscious (e.g, T. S. Eliot's "mythical method" [1975, 178]).

After World War II, global spatial reorganization again accelerated from the three worlds theory, where even the undeveloped Third World was poised to modernize, as decolonization gave way to multiculturalism and postcoloniality and then to contemporary globalization. Distances vanished; the whole world began to seem interwoven at every point, in real time, thanks to the globalization of the economy, travel, and media. Motion even became a dwelling place rather than the exceptional acceleration of a body between fixed places; as roots gave way to routes and multiple diasporas, motion supplanted fixture as a lifeworld norm (Hall 1996).

On the other hand, techniques of distancing became more sophisticated (as with what Rob Nixon calls "superpower parochialism" [2011, 33]). While global capital enjoys unprecedented flows across borders, new barriers constrain flows of people. Oil companies today engage in multinational collaborations that effect industrial-cosmopolitan erasures of distance, and even former enemies like OPEC and the International Energy Agency cooperate (Yergin 2011). But this oil integration also amplifies distancing as systemic inequity, from the increasing export of toxic waste or oil-related environmental injustice to the spread in oil-producing nations of what has been called "the Dutch disease" or "the RESOURCE CURSE" (French 2000; Karl 1997; Yergin 2011).

A second aspect of accelerated postwar cultural change involves dwelling rather than speed. In *The Technological Society* (1964), Jacques Ellul remarks that, in the wake of the first Industrial Revolution, society dwelt alongside its "machinery." In the modern (oil-electric-coal) world, by contrast, people dwelt increasingly *in* their technology (in networks of railways, PIPELINES, gas and electric conduits, plumbing). This shift fostered notions of modern identity in which autonomous individuals (and nations) dwelt within infrastructural NETWORKS that augmented their lives.

In the postwar period, the relationship between INFRASTRUCTURE and IDENTITY changed again: from merely augmenting social existence, technology has begun to constitute it. The prior distinction between human and machine gave way to notions of organic life thoroughly fused with technology (Haraway 1991; Kelly 1994). People have become elements within multiple (hard) technological and (soft) social infrastructures. In the United States, the postwar period saw the closing of the infrastructural frontier with the expansion of modern electric, telephone, and sanitary systems throughout the country. Globally, it saw the creation of global networks of oil pipelines and supertankers; always-in-the-air fleets of cargo planes; and money flowing around the globe 24/7 in real time. It also saw the dissemination of screens, from television to the Internet and cellphones. Countless cultural texts imagine peoples' cyborg and chimeric fusion with these technological environments. Daily life now depends on this fusion; bodies and minds are wired into infrastructure.

Today's infrastructure is and must be always on. There is no gas pedal to let up, no rheostat to turn down. Identities, memories, and consciousness now reside, in multiple ways, in human-machine networks. In this sense we can say that many humans dwell within the electric grid. What happens when what is always on is suddenly switched off, when electronic life goes dark? This anxiety about the precarious foundation of hyper-technologized existence predates the Internet age; both Daniel Galouye's 1964 novel *Simulacron-3* and its 1999 film adaptation, *The Thirteenth Floor* (Dir. Josef Rusnak) depict people who do not realize that they are actually computer simulations. Their discovery of their unreality is upsetting, but the ultimate terror is the possibility of being turned off by someone in a world above them. This predicament of being at the mercy of the all-encompassing grid is the inverse of being without reliable access to it, a predicament endured by millions around the globe.

A third aspect of accelerated change after World War II is oil's signature new postwar application: the creation and widespread dissemination of petrochemicals. Oil has not merely provoked transformations in economy, society, and culture; its material substance has metamorphosed and metastasized throughout consumers' lifeworlds and bodies. Cultural transformations attendant upon these changes are visible in differences from their modern precursors, particularly around notions of health. The petrochemicals that once promised sanitary protection from the environment have become a source of toxic threat. This reversal occurred subtly, as the modern sanitary culture that followed Victorian public health reform drew upon the chemical surpluses and homeland-defending anxieties of the postwar period in an attempt to make home and homeland as antiseptic as possible.

First came PLASTICS. These metastasized quickly into an array of consumer goods, personal care products, synthetic TEXTILES, waxes, dyes, automobile bodies, camera film, garbage bags, and artificial hearts. Science and medicine began building themselves on plastics, which line our hospitals to cradle us at birth and swaddle us again at death (Smil 2008). When plastics first appeared on the scene, they were experienced, albeit uneasily, as separating people from the infectious biotic world. Other petrochemicals seemed to save mankind from famine in the Green Revolution, even as pharmaceuticals rescued people from disease.

Soon, however, plastics and petrochemicals were revealed to be not just sanitary protections from or improvements upon biotic NATURE but a royal road to ecosystem and human poisoning, disease, and death. Rachel Carson (1962) exposed the dangers of pesticides, insecticides, and fungicides; Theo Colborn, Dianne Dumanoski, and John Peterson Myers (1996) exposed petrochemicals as endocrine disrupters that led to deformity, disease, and developmental disorders. Some now wonder if the Green Revolution has led us to disaster by spreading a petrochemically dependent factory farming system around the world. As critics like Michael Pollan (2006) and Oliver Morton (2007) demonstrate, the food stream is deeply compromised. We now literally eat oil.

In these thumbnail sketches of today's carbon-saturated world, I have not even mentioned combustion-related CO_2 emissions and the climate change that has resulted from them—the most sweeping systemic metamorphosis of all.

One final observation. The contemporary ENERGY era of metamorphic system change poses a dilemma for cultural analysis and activism. Even though an oil-saturated society may be completely unsustainable, it also seems impossible to walk back from. Is it even possible, theoretically or practically, to walk back from a metamorphosis? Chaos theory offers no way of turning around and walking things back. Entropy allows no return; evolution opens no pathways back; disequilibrial change forbids reversals. One cultural response to this condition has been decisive: the great majority of speculative fictions today do not even raise the question of walk-backs but instead mine the bizarre, postapocalyptic possibilities of dwelling within surpassed limits and uncontainable risks (F. Buell 2013).

See also: ANTHROPOCENE, CATASTROPHE, CRISIS, ENERGY REGIMES, FICTION, METABOLISM.

Ethics

Joanna Zylinska

Life comes to the fore most explicitly when it is under threat. We think about life (instead of just living it) when faced with the prospect of death, whether of individuals; of ethnic or national groups in wars and other forms of conflict; or of entire species, human or nonhuman. My concern here is with the prospect of the impending death of the human species. In the mainstream MEDIA, human extinction is usually presented as something inevitable. In *Ten Billion*, British computational scientist Stephen Emmott remarks, "we are fucked":

> Earth is home to millions of species. Just one dominates it. Us. . . . Our cleverness, our inventiveness and our activities are now the drivers of every global problem we face. And every one of these problems is accelerating as we continue to grow towards a global population of 10 billion. In fact, I believe we can rightly call the situation we're in right now an emergency—an unprecedented planetary emergency. (2013)

My aim is to outline a different response to the discourse of planetary emergency while taking seriously what science has to tell us about the FUTURE of energy, life, and everything else. Philosopher John Gray is no doubt correct when he writes, "The planet does not care about the stories that humans tell themselves; it responds to what humans do, and is changing irreversibly as a result" (2013, 6). Yet we humans do care about such stories. More important, stories have a performative aspect: they have the capacity to make things

Adapted and excerpted from Joanna Zylinska, *Minimal Ethics for the Anthropocene* (2014).

happen rather than merely describe them. Even if we agree with the postanthropocentric mantra that *it is not all about us*, we must also acknowledge the singular human responsibility in responding to both the current threats to life's continued existence and the narratives about life's impending demise. My response takes the form of a proposal for a minimal ethics for a time of a supposed planetary emergency, shaped into twenty-one philosophical propositions for nonphilosophers.

The injunction to outline a "teaching of the good life" when life itself is under threat comes from Theodor Adorno's *Minima Moralia*, a slim volume subtitled *Reflections on a Damaged Life* (2005, 15). Adorno's diagnosis may seem similar to Emmott's, yet the context of his reflections, presented in a series of shards of thought, is distinct: amid the catastrophic violence of the Holocaust, Adorno considers the possibility of philosophizing, against all odds, about what might constitute a "good life." My proposal for minimal ethics draws inspiration from Adorno's persistent search for signs of life amid apocalypse, even if the existential threats we face differ from his. In turning to ethics, I also aim to contest ontology as a dominant mode of knowledge production, whereby a theorist conjures a certain image of the world as truth. By calling this ethics "minimal," I want to avoid the moralism of many environmentalist positions that can deter the emergence of effective and responsible ethics and politics. I also hope to open up critical reflection on how *not* to think about "the environmental/energy/fuel/population crisis" and to contest some of the vocabularies presently at work.

The underlying philosophical framework for this proposal can be described as "critical vitalism," premised on rethinking and remaking "life" and what we can do with it. Taking life as a nonvalorized minimal condition, critical vitalism remains attuned to stoppages in life, seen as both a becoming and a fracturing process. Attuned to "the difference of difference," critical vitalism considers how differences endure and matter, whom or what they matter to, and how. This minimal ethics is not just an updated environmental ethics: it does not pivot around any coherent notion of an environment but instead concerns itself with dynamic relations across various scales among entities such as stem cells, flowers, dogs, humans, rivers, ELECTRICITY pylons, computer NETWORKS, and planets. It therefore becomes "an ethics of life" (Zylinska 2009), with life understood both philosophically and biologically. Its starting premise is that we humans are making a difference to the arrangements of what we are calling "the world." We are not the only or even the most important actors making such a difference—it would be naive and short-sighted to assume this—but we are perhaps uniquely placed to turn the making of such difference into an ethical task. Thanks to our human ability to tell stories and to philosophize, we can not only grasp the deep historical stratification of values through an involvement in what Gilles Deleuze and Felix Guattari call "a geology of morals"; we can also work out possibilities for making better differences across various scales. While our participation in the differentiation of matter is ongoing, frequently collective or distributed, and often unconscious, ethics names a situation when those processes of differentiation are accounted for: when they occur as a cognitive-affective effort to rearrange the solidified moral strata, with a view to producing a better geo-moral landscape.

The ethics proposed here is nonsystemic (i.e., not rooted in any large conceptual system) and nonnormative (i.e., not resting on any prior values or postulating any firm values in the process). One key assumption underlies this project: that we have a responsibility to engage with life—materially and conceptually—because, as we know from Socrates, "the unexamined life is not worth living." What counts as the examination of life goes beyond the Socratic method of inquiry conducted between two parties to eliminate erroneous hypotheses. It also involves physicalist engagement with the matter of life, its particles and unfoldings.

So herewith I present you, in twenty-one theses, a minimal ethics for the era of a supposed planetary emergency:

1. The universe is constantly unfolding but it also temporarily stabilizes into entities.
2. None of the entities in the universe is preplanned or necessary.
3. Humans are one class of such entities, a class that is as accidental and transitory as any other class.
4. The differentiation between process and entity is a heuristic, but it allows us to develop a discourse about the world and about ourselves in that world.
5. The world is an imaginary name we humans give to the multitude of unfoldings of matter.
6. Transitory stabilizations of matter do matter to us humans, but they do not all matter in the same way.
7. Ethics is a historically contingent human mode of becoming in the world—and of becoming different from the world.
8. Ethics is therefore stronger than ontology: it entails becoming-something in response to there being something else, even though this "something else" is only a temporary stabilization.
9. This response is not just discursive but also affective and corporeal.
10. Ethics is necessary because it is inevitable: we humans must respond to there being other processes and other entities in the world.
11. Our response is a way of taking responsibility for the multiplicity of the world and for our relations to and with it.
12. Such responsibility can always be denied or withdrawn, but a response will have already taken place nonetheless.
13. Responsibility is not just a passive reaction to pre-existing reality: it involves actively making cuts into the ongoing unfolding of matter in order to stabilize it.
14. Material in-cisions undertaken by humans can be ethical de-cisions, even if the majority of such cuts into matter are nothing of the kind.
15. Even if ethics is inevitable, ethical events are rare.
16. Ethics requires an account of itself.
17. Ethics precedes politics but also makes a demand on the political as the

historically specific order of sometimes collaborative and sometimes competitive relations between human and nonhuman entities.

18. As a practice of material and conceptual differentiation, ethics entails violence, but it should also work toward minimizing violence.
19. There is therefore value in ethics, even if ethics itself needs no prior values.
20. Ethics is a critical mobilization of the creative principle of life in order to facilitate a good life.
21. Ethics enables the production of better modes of becoming, whose goodness is worked out by humans in the political realm, in relation with, and with regard to, nonhuman entities and entanglements.

See also: AFFECT, ANTHROPOCENE, CATASTROPHE, EVOLUTION, FICTION.

Evolution

Priscilla Wald

Debates surrounding biological theories of evolution are evident in the multiple meanings of the word itself, with its etymology haunting its subsequent meanings. The *Oxford English Dictionary* offers its etymology in the "action of unrolling a scroll," a "lapse of time," and a "tactical manoeuver to effect a change of formation."[1] Revelation and deliberation survive in its earliest incarnations. Darwin did not use *evolution* in the first edition of *On the Origin of Species*, yet as James T. Costa points out in his introduction, he gave it the last word—literally—when he concluded that edition with a poetic meditation on the "grandeur" of a "view of life" based on laws of NATURE that show how "from so simple a beginning endless forms most beautiful and most wonderful have been and are being evolved" (Darwin 2009 [1859], 490). In its past participle form, the verb edges on the nominal, as though itself evolving from process to principle. And the debates ensuing from the publication of Darwin's magnum opus gave *evolution* new life.

Darwin famously struggled with his theological training as he confronted evidence of gradual changes in a world in which he could not ultimately see the hand of God. Human destiny was not preordained. Evolution, after Darwin, named the inexorability of biological change. The insight initially troubled not only religious beliefs but also a scientific status quo that assumed the stasis of species. Myth met science in evolution, as Darwin and

1. "Evolution, n." OED Online. June 2016. Oxford University Press. http://o-www.oed.com.catalog.multcolib.org/view/Entry/65447.

his interlocutors listened to the insistent voice of the dead, speaking through fossils that inscribed stages of development and through the unique flora and fauna of distant lands. Evolution gave humanity a new creation story in which life was continuous and ever-changing, and humankind shared kinship with all living things in a delicate yet enduring web of planetary existence. The past bequeathed its secrets to the present through laws set in motion by the vagaries of chance. The FUTURE held its mysteries still.

But humanity will have its gods. Few concepts seem to be as difficult to accept as the role of chance in human destiny. Words may have betrayed Darwin when he named the mechanism of those laws "natural selection," perhaps encouraging an ineluctable attribution of agency, even design, to a godlike Nature who strives toward harmony and equilibrium: the best of all possible worlds. Noting the gradual changes in species of flora and fauna when breeders selected for desired traits, he envisioned nature "selecting" traits that confer reproductive advantage in a particular environment, leading eventually to the widespread dissemination of the traits within a population. The subtlety of a chance mutation is easy to miss when the breeder's intention becomes "natural selection."

Darwin's story was hard to tell in a culture that read the triumph of human civilization in the cultivation of flora and domestication of fauna. Landscape painting celebrated humankind's subduing of wilderness; the act of painting imposed order and meaning on even the wildest nineteenth-century scenes: nature shaped by human imagination and will. Darwin returned the brush to forces beyond human control but also brought the mysteries of creation for the first time within human reach. The new Adam may have evolved from apes, but he could still name the animals, and this Adam could even tell the story of their creation.

Ironically, the answer to the most important puzzle of Darwin's story began in the garden of a monastery with the experiments of his contemporary, Gregor Mendel. It would, however, be nearly half a century before Mendel's rediscovery led to the science of genetics and, in turn, to an understanding of how a copying error could blossom into a gradual but dramatic metamorphosis. In the wake of the first global conflagration, people eager for renewed belief in humanity, progress, and purpose embraced the idea of harnessing the forces of evolution. 4-H clubs and fit family competitions flourished, as they enthusiastically enlisted in Nature's utopian efforts of selecting for the "fittest" of humanity. Until, that is, UTOPIA placed barbed wire and armed guards around its perimeters.

Genetics survived the taint of eugenics, as science promised to turn the horrors of war into new discoveries for the betterment of humankind. But an uncertain future continued to haunt the story. As cultural commentators watching the atomic dust settle pondered the fate of humankind, genetically informed evolutionary biologists and the practitioners of the new mass genre of science fiction who followed their work gave that question a biological spin. In its capacity for abstract thought and its ability to translate thought into action, humanity might be the most "evolved" species, but, contra human hubris, evolution is neither teleological nor even assured.

As the cosmic adventurers of sci-fi encountered life-forms stretching the reach of the human imagination, they and their earthbound counterparts learned the evolutionary lessons of human finitude. Sci-fi offered readers glimpses into distant futures and interactions with more advanced species that regarded humanity as curiosities, pets, or the next meal. To the uncertain ends of human bellicosities and the more distant vagaries of evolutionary chance, a new threat soon surfaced in the fading voices of the distant past. If fossils had fueled humanity's quest to understand its relation to the dead, the EXHAUSTION of fossil fuels foretold its overconsumption of their legacy. From biologist Rachel Carson's urgent warnings (1962) about toxins coursing through the web of life, including the blood and genes of human beings, to a 1969 UN report heralding the "Crisis of Human Environment" in the imminent exhaustion of planetary resources, humanity learned of the impending consequences of its voracious appetites and its overreliance on the fantasy of a bounteous and solicitous Mother Nature. Sci-fi scenarios such as the 1973 film *Soylent Green* made the lesson gruesomely literal in the revelation that the much-touted solution to an exhausted food supply left people unwittingly eating their dead. The message: as we exhaust our fuel supply, we consume our planet, our history, ourselves.

Whither humanity indeed? What stories are we telling about evolution now, and with what consequences? Humanity occupies every geographical niche and has interbred across every distinct population. Have we, as some biologists suggest, reached the limit of natural selection? Or perhaps the limit we have reached is our definition of life itself.

For the interdisciplinary environmental scientist Eric D. Schneider and the science writer Dorion Sagan, the future of life depends on asking the question differently. Evolutionary theory does not account for life's increasing complexity; instead nonequilibrium thermodynamics, and its precept "nature abhors a gradient," suggests natural selection may be but part of a broader understanding of life as fundamentally involving the transformation of ENERGY (Schneider and Sagan 2005, 25). We have come full circle to animism, but with a twist. For Schneider and Sagan there may be a cosmic purpose to which humanity is not central but instrumental, even catalytic: "life appears not as miraculous but rather as another cycling system whose physical, material, and ultimately mundane purpose is to get rid of prior complexity in accord with the second law [of thermodynamics]" (ibid., 323). Evolution is a change in energy.

Each story returns to the insight that human beings are unique but not exceptional. So proclaimed the evolutionary biologist Julian Huxley in *Evolution: The Modern Synthesis* (1942), which gave the new genetically informed evolutionary biology its name. Humankind, he insisted, had to outgrow the need for a deus ex machina named God or Nature and, recognizing responsibility for the inordinate impact of the human presence on the planet, "become the trustee of evolution" (1942, 576).

The stories we tell have consequences. How we understand the world shapes how we inhabit it; the questions we bring to this topic—What made us? Why are we here? To what do we answer?—fashion the future of humanity as much as do the mutations that confer selective advantage, especially considering humankind's unprecedented capacity to influence both the selection and the environment. What's in a name? The answer could be life itself.

See also: ELECTRICITY, ENERGY, FICTION, FUTURE, NATURE.

Exhaust

Anna Sajecki

> As Maitland stood weakly by the roadside, waving with a feeble hand, it seemed to him that every vehicle in London had passed and re-passed him a dozen times, the drivers and passengers deliberately ignoring him in a vast spontaneous conspiracy . . . . He was now more tired and shaken than at any time since the crash. Even in the warm, exhaust-filled air he shivered irritably; he felt as if his entire nervous system was being scraped by invisible knives, his nerves drawn through their slings.
>
> —J. G. BALLARD, *Concrete Island*

The world of British science fiction writer J. G. Ballard is one of motorways and cars; highways and automobiles emblematize changing technological landscapes and emergent postmodern geographies, betokening capitalism and Americanization. At the beginning of the 1970s, when the environmental effects of automobiles came under increasing scrutiny, another aspect of the car garnered attention: exhaust. AUTOMOBILE exhaust is a secondarily produced waste resulting from energetic depletion. Think of the car as a system: gasoline in the form of fuel drives the system and is required for it to function, but this energetic imperative and the burning of fuel transforms the physical constitution of gasoline. The remnants, made up mainly of carbon dioxide, form a particular type of waste called exhaust. This system has its own symbolic potential: the intermingling of auto exhaust and human exhaustion in the Ballard passage above hints at a burgeoning awareness in the early 1970s of both the environmental impact of the car and the potentially devastating, exhaustive impact of an emerging societal logic. In *Concrete Island* (1974)—a tale about one Londoner's crash onto a cement island surrounded by motorways—the burning of fuel paradigmatically enables the expression of the LIMITS and exhaustive elements of a nascent neoliberalism, thereby generating an early index of a system that was then hardly constituted.

Concrete Island is set in 1973—a year that signaled the increasing failure and exhaustion of Keynesian economic principles and the early materialization of neoliberalism. Economic problems such as inflation had been escalating in England. British policy had curtailed investment and confidence in industry, meaning that by 1970 British industry was underequipped and productivity was poor (Spittles 1995, 28). Richard Nixon's 1968 decision to withdraw the American dollar from its fixed rate conversion into gold had created a worldwide CRISIS in monetary instability and inflation—a decision whose effects were compounded in the 1973 OPEC crisis, which caused runaway inflation in Britain, forcing the nation to ration gasoline and face the possibility of an exhausted oil supply. With prevailing instability, economists had begun to consider alternatives to the Keynesian model. In Chile, Chicago School economists were experimentally implementing the first model of neoliberal governance, which called for minimalist government intervention and privatization of industry in the service of an unrestricted economy (D. Harvey 2005, 12). The ability to thrive in society became solely a matter of individual responsibility. Although neoliberalism would not become Britain's dominant mode of economic governance until Margaret Thatcher's rise to prime minister in 1979, its logic did surface periodically in Britain in the late 1960s and early 1970s under the Labour government (Hall and Jacques 1983, 19–20).

In *Concrete Island*, the protagonist Robert Maitland crashes his car for two reasons: because he had been speeding and because he had become exhausted, "already tired after a three-day conference" (Ballard 1974, 9). As Michel Foucault states, the neoliberal individual—more specifically the Americanized neoliberal subject—is an "entrepreneur of himself" because he has become solely responsible for his economic and social success (2008, 232). The individual's competitive and efficient drive enables his prosperity within neoliberal society, and subsequently his energetic output also helps maintain the strength of neoliberalism as a system. However, the logic of entrepreneurship—and hence Foucauldian neoliberalism—presupposes the continued demand for increased success and, consequently, requires the dedication of greater ENERGY and speed by way of increased competitiveness. Just as gasoline fuels cars and thus the greater apparatus of AUTOMOBILITY (see Urry 2006), the system of neoliberalism is fueled by the input of energy; and just as exhaust becomes the waste of automobility, Ballard demonstrates that the burnt-out subject—the subject who has worked too hard and can no longer muster the energy to contribute economically—becomes the waste byproduct of the neoliberal system.

With both cars and neoliberalism, the burning of fuel supports the continued movement of the system; the combination of speed and exhaustion as reasons for Maitland's crash registers a noteworthy departure from Enda Duffy's inference that in the modern era, the speed of driving is connected with the experience of pleasure (Duffy 2009, 257). *Concrete Island* offers a more contemporary alternative vision of the speeding subject in late-modern society, where the confluence of speed and exhaustion anticipates the discourse of work stress and burnout that arose in the 1970s alongside neoliberalism (Wainwright and Calnan 2002, 25–26). As Maitland continues to investigate his crash, he realizes that "he had almost willfully devised the crash, perhaps as some bizarre kind of rationalization" (Ballard 1974, 9), a subtle suggestion that the neoliberal imperative to perform competitively had been taking its toll.

Ballard's vision of neoliberalism sees the exhausted subject as societal waste, devoid of any functionality. This impression is imprinted in the space of Maitland's crash: departing from the trope of the ocean desert-island, Ballard maroons Maitland on a concrete island in London, with constant reminders of a proximity to civilization—the smell of automobile exhaust, the noise of car engines and honking, etc. Yet Maitland is largely ignored by the motorway drivers. His existence is ethereal; he is an almost invisible presence within civilization, like the waste substance of car exhaust itself. As the title of *Concrete Island*'s third chapter, "Injury and Exhaustion," suggests, Maitland is himself a version of "exhaust"—he crashes because all his energy has gone into his job as an architect, as well as his extra-marital affair. The primarily neoliberal-economic rationale for Maitland's thinking is belied both by his self-identification with his job—Ballard first introduces us to the protagonist as *architect* Robert Maitland—as well as by supporting character Jane Sheppard's confusion and belief that Maitland is a businessman. However, after Maitland crashes, his failed economic transactions on the island with two other social outcasts suggest that his physical ejection from the highway is coupled with his disappearance within the system of economic circulation that predominates in neoliberal society. Maitland's placement at the heart of an urban tangle of concrete highways with its mass of drivers figuratively suggests that if the protagonist is an invisible gaseous presence within society, it is because he is bereft of any material subjecthood that would grant him access to the surrounding society. Ballard reveals the highway system to be a ready metaphor for economic circulation in a noninterventionist society, with the removed driver representing the economically incapable individual who has automatically lost his social standing within the system, becoming little more than "exhaust." Just as car exhaust is a wasteful byproduct, Maitland too becomes the burnt-out byproduct of a system that demands ever-increasing levels of energy.

Ultimately, however, Ballard reveals a crack in the neoliberal system, forcing a departure from the exhaust allegory, a divergence that rests with the protagonist's evolving viewpoint at the end of the novel. After the crash, Maitland aligns himself with the island, a space of waste marked by abandoned rusted cars and skeletons of collapsed Edwardian and wartime buildings. The protagonist detaches the exhaust pipe from his crashed Jaguar and uses it as his crutch to supplement his own badly injured limb, materially attaching the very concept of gasoline exhaust to his own exhausted body. He becomes nothing more than the exhausted leftovers of neoliberal society: as useless and insubstantial as the car exhaust diffused in the atmosphere. At the end of *Concrete Island*, Maitland fails to escape the island; however, a speech act—his utterance "I am the island" (71)—suggests what Michel Foucault and Judith Butler would call an instance of "radical subjectivity," in which a subjectivity created by the dominant discursive system later becomes a position of dissent: a place from which one can illuminate the limits and fault lines of that system (Foucault 1978, 101; Butler 1997, 15). Thus, just as automobile exhaust and fossil fuels in the 1970s increasingly symbolized the limits of highway culture, so too does Maitland's positioning—albeit through the lens of human subjectivity—concretize a space from where neoliberalism can be protested and delimited.

See also: ANTHROPOCENE, ENERGOPOLITICS, FICTION, IDENTITY, INFRASTRUCTURE, ROADS, RUBBER.

Exhaustion

Franco Berardi

Modern culture valorizes ENERGY, vitality, and expansive potency. In biopolitical terms, we may say that *energolatria*—the worship of energy—is another name for modernity. The defining characteristic of modern bourgeois economy is the imperative of growth, which shapes aesthetics, ethics, and political strategies. Expansion is the immanent dynamic of modern capitalism. The lack of growth, by contrast, is associated with exhaustion and all of its negative connotations.

The rise of capitalism in Europe was fueled by nationalism and its romantic cult of youthfulness and aggression. Nationalism affirms the energetic potency of young populations against senescent static civilizations. In the late nineteenth and early twentieth century, young countries (Italy, Germany, and Japan) rose against old empires (RUSSIA, Austria, and the Ottomans) that were heading toward collapse because of the decrepitude of their institutions. The quintessential cultural expression of this early-twentieth-century cult of youth, energy, and aggression was Futurism, which devalued old-fashioned styles and despised old people and women for their weakness. Emerging from the aesthetic politics of Futurism, Italian fascism presented itself as the epitome of these emergent young nations.

But this cult of energy lost ground in the late twentieth century in the face of declining economic growth and the aging of the world's population. Whereas Futurism was based on contempt for women and the elderly, late modern consumer culture does not abuse old age. It simply denies it and pretends that old people can be young if they partake of the feast of consumerism and the cultural style of never-ending energy.

An important moment in this shift was the Club of Rome's *The Limits to Growth* (Meadows et al. 1972), which sparked awareness of the scarcity of oil and the finiteness of the earth's resources. After the War of Yom Kippur in 1973, the Western world experienced its first oil shock. What is less well-understood is how this looming possibility of exhausting the earth's resources is connected to simultaneous demographic trends like the widespread decline in birth rates and increasing life expectancies, which together have created an unprecedented demographic reality: humankind is growing old. Energy is being depleted from the physical body of the Earth, as well as from the biopolitical body of humankind. Although the oil economy can be revived (as is happening in North America) through the use of new extraction technologies like FRACKING, this development constitutes an even more violent aggression toward the planet. Similarly, attempts to revive the competitive edge of aging societies disrupt the social balance and lead to violence against old people, as shown by the forced prolongation of work time and the authoritarian postponement of the retirement age.

In the 1990s, neoliberal discourse took the view that because we live longer, we have to WORK longer or else we create problems for young people. In Europe, CANADA, and the United States, there has been pressure to postpone and privatize pensions. This shift from public pensions to private retirement investment is obviously in the interest of financial capitalism but not in the interest of society in general, and certainly not in the interest of old people who are forced to work longer than expected or of young people who cannot get a job because workplaces are forcibly occupied by senior workers. For capital, however, the situation appears convenient and efficient: a senior worker requires only one salary, while a young worker must effectively be paid twice, once in wages and again in pension funds to the retiree whom the young worker has replaced. Additionally, if the old worker postpones retirement and continues to work, the unemployed young person is more likely to accept any kind of casual, precarious, short-term labor, the sort on which contemporary capitalist enterprise thrives. Amazingly, it is the old who are blamed for these social circumstances. STATISTICS about youth unemployment are often explained with the statement: "It's because people are living so long!" Young people are ideologically counter-positioned against the baby boomers: "Those communists who are living so long, who hope to live off the state—they never die. So you can't find a job!"

The cultural logic of *energolatria* underlies this fear of exhaustion and the attempts to stave it off: this logic links economic expansion with social well-being and identifies energy with the FUTURE. The prevailing cultural imagination is unable to think the future outside of the framework of growth and expansion; politics, too, is only able to follow the rules of modern *energolatria*: growth at all costs, because the only alternative seems to be depression and impoverishment. Yet far from being a natural NECESSITY, this seeming lone alternative is the consequence of a cultural prejudice that recognizes energetic economic growth as the only healthy condition of social production. Western culture is frightened by the very idea of "exhaustion" because it has been linked to the expectation of failure.

But now the contrary is true. Energy is not infinite! Our very perception of the future in relation to energy and exhaustion needs to change. We must disentangle the concept of progress from the pretense of infinite growth. By accepting the biosocial reality of senescence and the economic reality of the exhaustion of physical resources, we might open a new horizon for the political imagination.

The identification of energy with desire is a trap into which, like the Futurists, Gilles Deleuze and Félix Guattari also fell before they came to understand that desire must be dissociated from energy and that exhaustion involves its own flows of desire. While their *Anti-Oedipus* (published in 1972, just before the first oil shock) emphasized the infinite potency of energetic desire, Deleuze and Guattari reframed the problematic of desire in *What Is Philosophy?* (1991), in which they write of becoming old and of friendship as a condition for sharing knowledge. In the conclusion, "From Chaos to the Brain," they describe chaos as problem of velocity: the disjuncture between the excessive speed of the world and the relative slowness of the brain. This relationship involves suffering, the other side of desire. This dark face of desire is the result of a universe that is going too fast. In his last work, *Chaosmosis* (1995), Guattari writes of *spasme chaosmique* (the "chaosmic spasm")—the perception that one is no longer able to follow the rhythm of chaosmotic (or energetic) desire. Guattari calls for a new balance between the brain and the universe. Desire is not only energy and speed. It is also the ability to find another rhythm.

Here we can finally question the cult of energy and the rhetoric (whether Futurist or neoliberal) that insists, "I want to be active!" Are competition and ACCUMULATION the only ways to be active? Are not caring, contemplation, and dreaming all forms of activity, no less useful and no less enriching than productive labor? To be freed from the conceptual trap of infinite growth, let us try to see things from another point of view: that of relaxation, of no responsibility, of self-care.

From this point of view, exhaustion is not failure but instead a new condition of pleasure and comprehension. This shift in understanding is not an individual task; rather, it involves a broader social repositioning of desire and expectation. How do we inscribe the reality of death into the political and aesthetic agenda of a social group that is currently weak, and not only for biological reasons? The older generation of Europe should become the subject of a cultural revolution aimed at preparing Western society for a long-lasting agreement on the redistribution of wealth and resources. Such a cultural revolution can only begin with a rejection of the cult of energy: in other words, with a critique of this energetic juvenilism permeating modern culture and its ideology of unbounded growth.

See also: ENERGOPOLITICS, LIMITS, SUSTAINABILITY, WORK.

Fallout

Joseph Masco

In his 1964 film, *Red Desert*, Michelangelo Antonioni depicts a terrifying conundrum of late modernity: a world of technological marvels, whose price is local culture and the environment. The film is set in an Italian industrial town, where Monica Vitti plays the increasingly distraught wife of a petrochemical executive. The film veers from an examination of Italian industrial design—the beautiful sculptural forms enabled by PLASTICS, steel, and glass that constitute a radical break with local craft traditions grounded in organic materials—to the natural landscape destroyed by industrial production. The characters inhabit spectacular high modernist living spaces but traverse without comment the ecological ruin surrounding the chemical plant, which contaminates air, soil, and water without interruption. The opening credits—a blurry image of an industrial plant's EXHAUST towers set against an ear-splitting synthetic soundtrack—suggest that toxicity and alienation are the perverse conditions for modern subject formation in a world organized by synthetic chemicals. How can one distinguish industrial "progress" from violence?

Antonioni's first color film, *Red Desert* is an act of pure cinema, with Vitti's psychic troubles played out in visceral and highly atmospheric terms: the camera moves abruptly from objectified views of the industrial city into her inner space, a shocking technique that raises the question: Is the toxicity she experiences material or psychic, lived or imagined, or both? He achieves this effect by painting whole environments—streets, trees, even people—to index her psychic state, replacing the wide color palette of the film (an achievement in its own right) with surreal monotone sequences. In one scene, Vitti walks along a seemingly ancient cobblestoned village street with a potential lover. They encounter a fruit vender, a traditional symbol of environmental flourishing. But in Antonioni's vision, the farmer is the inverse: his body, the cart, the fruit, and the entire urban space are a uniform gray, a signal that we have moved into Vitti's interior vision, where the world is rendered flat and lifeless (see Figure 4). Vitti, a young mother with a highly successful husband, is traumatized by visions of a toxic universe that threatens everyday life through both illness and social alienation. Her psychic distance from the people she loves is mirrored by her sensitivity to the toxicity around her, a ruination that goes unnoticed by other characters.

Vitti becomes a sentinel for toxic modernity. Her character enjoys all the benefits of capitalist production—status, style, and wealth—but she can only see the ongoing destruction of industrialism, which obliterates social codes as well as Italian village landscapes. The film ends on a note of false optimism, as Vitti and her young son observe yellow poisonous smoke pouring out of their family's industrial plant. Her son asks about the safety of birds and Vitti tells him the birds are smart now and know to fly around the danger. Antonioni asks how, in the age of industrial modernity, we can avoid its toxic fallout. How can human beings navigate the atmospheres (social, material, psychic) generated by the global petrochemical economy?

Red Desert arrived on movie screens two years after Rachel Carson's *Silent Spring* (1962) alerted readers to how industrial chemicals were remaking global ecologies. The film opened shortly after the signing of the 1963 Limited Test Ban Treaty, which eliminated NUCLEAR weapons tests in the atmosphere and oceans and on land. Scientific alarm about the global effects of radioactive fallout had risen over the past decade, as nuclear tests were shown to have been planetary environmental events linking air, water, soil, and all living

FIGURE 4. Still image from *Red Desert* depicting the everydayness of air pollution. (Michelangelo Antonioni, *Red Desert*, 1964)

beings in complex circulations. Like the nuclear test ban, *Red Desert* responded to these emergent environmental dangers. The film diagnoses the psychic and material effects of industrial modernity, which transforms not only capital and class but also senses, emotions, health, and NATURE.

Antonioni's project is critical but not nostalgic. The film offers no sense that a recovery of the village life of his youth is possible; modernity is a one-way trip. Indeed, celluloid film is itself a chemical medium, explosive in mass form and produced by a global industry. Antonioni's devastating insight is that vision is both toxic and enabled by toxics. Synthetic chemicals make new worlds and destroy previous worlds. Vitti's character simply sees more than others and refuses to anesthetize herself—and thus seems increasingly unhinged. *Red Desert* remains one of the first and most exacting considerations of industrial fallout, tracking the affective and material effects of chemical airflows and signaling the need for a critical theory of unintended consequences.

Fallout remains an important concept for our age. A term of relatively recent origin, *fallout* designates an unexpected supplement to an event that causes long-term damage: aftermath, reverberation, negative side effect. Fallout is the CRISIS inherent to a process that remains unacknowledged until lived through; understood retrospectively but lived in the future anterior, fallout is a form of history made visible in negative outcomes. We live today in the age of global fallout, the various aftermaths of the twentieth century. Industrialism, militarism, and capitalism are massive fallout-generating practices, producing reverberating crises involving climate, energy, finance, and war.

Fallout comes from the verb to "fall out," which designates a social break or conflict: the fight that separates comrades, marking the end of intimacy, shared purpose, and social pleasure. Military personnel also fall out from standing at attention. Falling out thus involves individual actions and lived consequences, a postsociality lived in isolation from collective action or the war machine. As a noun, *fallout* is an invention of the nuclear age, having appeared in English soon after the US atomic bombings of Hiroshima and Nagasaki and referring to radioactive debris put into the atmosphere by a nuclear explosion. Marked as precipitation, fallout involves a gradual settling of nuclear materials and effects over a wide area. By 1960 fallout studies had demonstrated the global environmental effects of nuclear detonations, transforming the bomb's security function into one of proliferating material and psychosocial insecurities. Fallout formally links human actions, technological and human sensory capabilities, atmospheres, and ecologies in new configurations of contamination. It operates on a wide range of temporal frames; its threat to health is both immediate (radiation illness) and long-term (cancer). Importantly, fallout challenges how we understand time, space, and the nature of the "event."

Since the onset of the Cold War, fallout has been most often understood as the bomb's lesser injurious form, so that the nuclear event has been split perversely into the planned detonation and its "unexpected" atmospheric effects. A completely predictable aspect of any nuclear event, fallout was nonetheless officially coded as a side effect within US national security practice. Much as drug companies today split desired from undesired effects of molecules, the political economy of fallout construed as unanticipated side effect is a foundational misrecognition with massive consequences. Nuclear national security logics work through bizarre metrics that recognize certain forms of destruction while disregarding others.

This split vision of harm in nuclear fallout shapes many other collective practices. Whether as radioactivity or the impacts of the carbon cycle on climate, fallout produces cumulative effects that only become visible in the destabilized ecological system. How many toxic industrial processes fall into this category of the unseen but cumulatively damaging or deadly? Such processes raise urgent questions of scale and perspective, not only what to see but how to see. The atom bomb is not only an emblem of industrial modernity but also an allegory for a larger set of processes now collectively understood as petrochemical fallout, emerging anthropogenic shifts across earth systems. The toxic fallout of the twentieth century continues to shift the constitution of bodies and ecologies, psyches and economies, requiring a new politics of clean air, soil, and water. Fallout today is both material and conceptual—a way of talking about legacies and futures, toxics and natures. As a simulation of cesium fallout from the 2011 nuclear accident at Fukushima implies, fallout reveals the interconnectedness of atmospheres and industries, lives and energy economies, affects and toxins. Invisible to human senses but tracked by new remote sensing and visualization technologies in the earth sciences, fallout has become legible as a global environmental event in close to real time. There is no excuse for not responding on an equally planetary scale.

We are, however, more likely to remain in a state of toxic suspension, mirroring Vitti's character from *Red Desert*, who can feel the effects of industrial fallout all around her but does not even consider mobilizing to minimize the environmental impact of her family business. Thus, we seem destined to relearn the lessons of fallout with each new generation, even as our technological capacity to observe and measure environmental toxicity has grown exponentially over the past decades. Unlike Antonioni's more perfect expressionistic vision, contemporary fallout is not marked with a shocking yellow that warns of danger and facilitates the tracing of its longitudinal effects. No, in the twenty-first century, despite our new prosthetic senses and much greater collective understandings of the psychic and material costs of the petrochemical revolution, fallout continues unabated, expectedly unexpected—indeed, it billows and flows.

See also: AFFECT, CHINA, EMBODIMENT, EXHAUSTION, NUCLEAR, PETROREALISM, RISK.

Fiction

Graeme Macdonald

Like the first law of thermodynamics, literary fiction requires momentum. Fiction relies on propulsive devices: basic units of charge that power action, event, and consciousness, calibrated by laws of narrative motion and physical, material impressions of kinetic and potential energy transference. (These need not involve tales of actual motion or much, if any, movement—consider Beckett's minimalism or the generic predicates for entropy in naturalist writing.) Fiction requires and is measured through potential—what fuels its psychosocial dynamics, the impetus of plot and character development, and its chronotopic ability to traverse multiple times and spaces. Fiction absorbs, exudes, circulates, conserves, and converts energy, not only on the formal level of narrative or metaphor but also in its production, dissemination, and reception. (Is it churlish to note that you are reading this on once-WOOD or once-oil?)

What has gone mostly unremarked, however, is the inextricable connection between the propulsive energy *of* fiction and the attention to energy *in* fiction—the stuff and material forces that make things go and happen within literary worlds. This despite the spectacular products of primary and secondary energy conversions perceptible throughout modern literature: imagine *Anna Karenina* or *One Hundred Years of Solitude* without coal-powered locomotives! The novels of Joseph Conrad without wind or steam. Consider the vast fiction of twentieth-century suburbia—fossil-fueled worlds saturated in carbon-based products—suddenly shorn of plasticity, deprived of AUTOMOBILITY or domestic electric power, bereft of pharmaceuticals, denied the cheap food supplies of prime-moved fertilizer! Ever since *Don Quixote* registered wind power, fiction has engaged sources of heat, light, and motion. In countless novels (open one and read a random page!), energy appears and makes history. The abiding theme of transition in the history of the novel is often reflected in transformations in energy and fuel provision. Questions of capacity and power supply shape all literary worlds. Literary history is full of energetic potential.

Only recently have literary critics begun to trace the physical and aesthetic forms and variants of material energy resources that propel characters, events, and storylines throughout the history of fiction (and culture and history more generally). Energy criticism is asking questions of a sort that have long been fundamental to literary study: Does literature shift in accordance with the dominant energy of its era (Szeman 2007; Yaeger et al 2011)? Might it play a role in reproducing—perhaps inadvertently or unconsciously—a predominant energy culture (Hitchcock 2010; Barrett and Worden 2014; LeMenager 2014)? How does literature use energy and vice versa? Are literary modes—like social formations—engendered by developments in fuel or resource use to a greater extent than we have understood (Wenzel 2006, 2014b)? Can we imagine modernism outside an oil-electric context (Neuman 2012)? Realism without COAL? Romanticism without wind or water?

Beyond divining the specific fuels that drive literary plots, a critical awareness of energy in fiction can begin to discern the relations among cultural forms, energy use, and predominant modes of production (Szeman 2007). Recognizing literary fiction as a historical cultural resource suffused with energy, in form and substance, fosters a critical understanding of how historical events and political formations are created and sustained by energy resources that, in turn, create and are in part created by a specific energy culture. The task of energy criticism is not limited to works explicitly concerned with energy resources, despite the numerous texts from world literature that can be considered "energy classics." We might even dare to posit that *all* (or perhaps *any*) fiction is a reservoir for the energy-aware reader.

Energy might have remained a permanently subsumed feature of literary fiction had energy (particularly oil) not become a predominant, pervasive concern in contemporary culture. The environmental dread we are living through has sparked critical interest in excavating the furtive relationships between energy and culture. Reading for energy requires confronting an "energy unconscious" (Yaeger 2011, 306) that—particularly in the Global North—belies blinding degrees of saturation. We are up to our eyeballs "in oil," and yet we fail to register its ubiquity in social life. The effort to discern the energy in cultural production and literary fiction depends upon recognizing the potent social fictions of energy that are inhabited and reproduced across petroculture globally. These social, economic, and corporate fictions assume (and thereby entrench) the NECESSITY of current ENERGY REGIMES dependent upon exhaustible resources, as in the (oil-based) "fictive capital" derivatives of financial modeling systems. These energy fictions occlude transnational sites of production and extraction and prop up globally uneven consumption patterns and living standards reproduced by the inexorable ACCUMULATION of "natural" energy resources, which are imagined to be inexhaustible, often unseen and unhandled by the majority of consumers. The fiction extends to the conversion of "bad" energy into "good" energy by corporate greenwashing: a prime example of the need for robust energy criticism.

Energy criticism has an important role to play in articulating how literary fiction has both enabled and challenged these social fictions of energy. The fantasy of unhindered and waste-free energy flow—always-already a degraded desire in a systemic culture of nonrenewables—unconsciously pervades most, if not all, cultural production from the coal age onward, where the accelerated mobility and compressions of space and time enabled by carbon-driven capitalism and petrotechnology have altered the shape and geography of literary plot, not to mention the available global constituencies of character, custom, and style. Despite being the engine-room of narrative, energy tends to surface as an explicit matter of concern in fiction only periodically, in times of resource angst. But we must seek to do more than merely revealing energy's "hidden" ubiquity and our bedevilment by it. Once we discern the 500 MW reactor in the corner of the parlor or the derrick in the drawing room, what then?

Beyond tracing the role of energy in literary history, what does energy criticism have to offer the future? We might consider the rather different ways in which literary fiction imagines worlds without oil. Since the oil shocks of the early 1970s, a substantial body of fiction about energy has emerged—concerned not only with LIMITS but also with exploration and (over)production, capacity and consumption, and conversion, distribution, and commodification. In both serious and popular literature and film, "post-oil" narratives are mostly depletion-anxiety, carbon-fretful speculations on the eventuality of a world without freely flowing oil—gasoline in particular. This future projection of less offers lessons about modernity's past and present: the uneven access, distribution, and consumption of energy as a capitalized resource, which may partly explain why imagined futures rarely posit less energy as automatically good. Although generally set in the future, speculative fiction's energy scenarios foster critical (re)cognition of the present. Such fictions, working out the irrepressible logic of contemporary petro-finance and ongoing carbonization, are at once a shape of things to come and a glimpse of things as they are for the billions of fuel-poor on the planet.

Oil endures and we endure it. The fictional nature of this endurance, and its incompatibility with petroleum's radical finitude, is the ultimate challenge facing our superenergized world. Perhaps our criticism, like our technology and terminology, is insufficiently refined. We need to reconsider our historical sedimentations of genre and period. *Petrofiction* certainly means stories about platforms, drill bits, and combustion transport, resource colonialism, deadly SPILLS, and exploration rights. But it is also about the relation between the oblique and surface world of fuel: an everyday world reliant on oil consumption but far removed from its backstage processes of extraction, refining, and delivery. Is it trite or redundant to claim, given the global cultural reach of a fossil ENERGY SYSTEM, that all contemporary fiction is petrofiction? Is fuel that fundamental to culture?

The promise of alternative modes of energy should be the terrain of cultural critics, trained in speculative imagining, interpreting dreams, and busting myths. But new extraction techniques mean that fossil fuels are resurgent and remain controversially abundant.

Might unconventional energy's deferral of peak culture hinder the development of new narrative conventions and literary genres? Whither hydrofiction? Windpoetics? Nuclear drama? What will a nonhydrocarbon imaginary look like after the reign of oil?

If a future world of postprime moving is inevitable, bringing with it diminishment or undesirable energy types, then perhaps literary texts that predate the era of oil can be re-energized for their glimpses of worlds outside the carbon web, fueled by other energy sources. Emma Woodhouse gazes admiringly at "the abundance of timber in rows and avenues" on the estate of her future husband in Jane Austen's *Emma* (1995 [1816], 323). Konstantin Levin considers a sea of human labor scything crops in *Anna Karenina* (Tolstoy 2002 [1873–1877] 3.XII.275). The transformative power of a water wheel running a nail factory commences Stendhal's *Le Rouge et le Noir* (2003 [1830]). Can such pre-oil energy memories be useful for a post-oil world?

Did people really walk "sixty miles each way" on errands and business, as Mr. Earnshaw does matter-of-factly in Emily Brontë's *Wuthering Heights* (1847)? Will literature after oil become more pedestrian? Certainly post-automobilic narratives of on-foot struggle, such as Cormac McCarthy's *The Road* (2007 [2006]) or Joshua Ferris's *The Unnamed* (2010) invite us to reattune ourselves to an embodied ambulatory aesthetic with a rich literary history, from Rousseau to Baudelaire, Beckett, and Sebald. Stendhal's famous aphorism from *Le Rouge et le Noir*, that "a novel is a mirror on a highway walked," was eclipsed in an era when the mirror was more likely to reflect a highway burned up by an SUV. But the image comes back into focus in this post-oil framework as a peculiarly speculative form of pedestrian realism. Reading fiction in this light offers chronological backflips, with refueled scenes from literary history opening a window on a possible future energy imaginary. Reading these scenes of life before oil while contemplating a life after oil demonstrates how fiction might offer a particular kind of renewable energy resource that can help to fuel the future.

See also: AUTOMOBILITY, ELECTRICITY, ENERGY, ENERGY REGIMES, ENERGY SYSTEMS, EXHAUSTION, PETROREALISM.

Fracking

Imre Szeman

As conventional sources of oil and gas become depleted, nations have turned increasingly to "unconventional" (i.e., expensive and difficult to access) forms of ENERGY. Shale gas—natural gas trapped in the rock of black shale—has quickly become one of most important of these. The process known as hydraulic fracturing, or "fracking," combined with an increased capacity to undertake horizontal (as opposed to vertical) drilling, has transformed a handful of countries into surprise energy superpowers in the past decade. It cannot help but appear as a cruel historical irony that these same countries launched the hydrocarbon era: instead of paying the price for their wanton fuel consumption, the original COAL power, the United Kingdom, and the land of the oil baron, the United States, are now flush with natural gas and (in the United States) shale oil. One newly discovered gas field in northwest England promises to fuel the country for more than six decades (Chazan 2011), while 2013 marked the year when the United States transitioned from oil penury back into its beloved caloric overabundance (Hussain 2013b). As with history, so too with energy: first time as tragedy, second time as farce.

The benefits of new sources of domestic gas are obvious. Countries once dependent on foreign sources of oil and gas can now fulfill domestic DEMAND and add to the national bottom line by becoming energy exporters. Proponents of fracking also point out that gas is better for the environment than oil or coal, as it generates fewer emissions of carbon dioxide and nitrogen oxide. And, of course, fracking appears to mitigate, if not eliminate, one of the defining geopolitical dramas of the twentieth century, in which liberal democracies were (again, from the view of proponents) held hostage by oil oligarchies: the West had to mess endlessly about in the politics of the MIDDLE EAST with huge repercussions for all involved (though disproportionately greater for those living in the region). But beyond such economic, environmental, and political practicalities, fracking brings even more important ideological benefits. The energy future it promises affirms a view of history in which things always work out in the end for the good guys—a geological confirmation of the deeply sedimented liberal belief in the inevitable expansion of rights, freedoms, and the wallets of the middle class. This view of fracking reinforces a sometimes shaky techno-utopianism, the belief that for all manner of problems, the nature of historical development is such that "technological solutions arrive just in time and never fail to come" (Szeman 2007, 814). Finally, because fracking promises to generate enormous amounts of oil and gas, it seems in a flash to have dissipated worries about fossil fuel depletion. According to the World Energy Council, as a result of fracking and other new extraction technologies, fossil fuels will be the most important form of global energy until and well beyond 2050 (Hussain 2013a). Fracking means business as usual, with the West in the driver's seat and the environmental consequences of fossil fuels pushed off into the distant FUTURE.

What has hampered fracking's affirmation of global capitalism's necessity and inevitability, however, are those same technologies whose miracle appearance seems to have rescued the West from its own worst practices. Fracking condenses the environmental consequences of modernity into one complex figure, pitting the future of capitalism against the future not just of this or that social formation but of the continued existence of life on the planet. There are innumerable other signals of the implications of contemporary energy use, including the overabundance of atmospheric CO_2 and its repercussions for global temperatures, sea levels, crop growth, and so on. But if this predicament has largely eluded attempts to name and explain it in such a way as to generate adequate social and political change, fracking has produced a response quite different than what we have come to expect from those modes of contemporary subjectivity mapped by Slavoj Žižek and Lauren Berlant as "cynical reason" and "cruel optimism," respectively (Žižek 2012, 312; Berlant 2006, 23). The highly technical and specialized process of fracking has seized public attention because it requires water—huge amounts of it, between one and three million gallons for each well drilled. It is not only the amount of water used that is alarming but also what is added to it to break up shale rock and release the precious petrocarbons locked within. Fracking fluid contains toxins and known carcinogens, including methanol, ethylene glycol (a substance in antifreeze), isopropyl alcohol (a solvent used in a wide-range of industrial liquids, including cleaning solutions), and a host of other chemicals, most of which remain secret in the name of resource competition and are never divulged to the public. The worry is that these chemicals will find their way from fracking water into groundwater used for drinking, bathing, and growing crops. Once there, it is difficult to know how or if these chemicals can be removed from the environment—except, of course, by making their way into the bodies of plants and animals, and not only those near fracking sites but living bodies located anywhere and everywhere.

Gas versus water; the steady-state operations of capitalism versus the continued existence of life itself: these stark oppositions are at the center of debates about fracking across the planet but especially in the developed world. The economic and political importance of oil and gas obtained through fracking has generated intense and protracted dispute in the United Kingdom, which has pitted the government of David Cameron and industry against NGOs, community groups, and locals in affected areas. The Royal Society for the Protection of Birds launched the first formal objection to fracking in the United Kingdom, citing its impact on endangered wildlife and the lack of adequate government assessment of an industrial process that will result in as many as 100,000 wells (Harrabin 2013). In the face of Chancellor George Osborne's intent to generate policy that would make the United Kingdom "the most generous tax regime in the world" for the gas industry, Water UK, the representative of water suppliers in the country, worried that fracking might contaminate the water supply (Gosden 2013). Opposition has been forceful and intense from those who live in and around the sites where the largest shale gas company, Cuadrilla Resources, plans to drill (Waldie 2013).

In the United States, the oil and gas industry has long been granted exemptions from the Clean Water Act (1972) and the Safe Drinking Water Act (1974), and fracking received a specific exemption from the latter in the 2005 energy bill, a move long planned by Vice President Dick Cheney's closed-door energy task force. In California, Pennsylvania, and other fracking sites, state and local governments perform little (if any) oversight regarding the composition of fracking fluid or the disposal of wastewater, which has generated mounting concerns among affected communities about the devil's bargain of allowing companies to drill (Levine and Sears 2013; Wilber 2012). In TEXAS, water use for fracking has already led to drought in thirty communities, turning many residents from supporters to opponents of the oil and gas industry (Goldenberg 2013b). The importance for industry of maintaining control of the public message about fracking was made brutally evident by the ban imposed in 2011 on two Pennsylvania children; under the terms of a settlement with Range Resources, they will not be able to speak about fracking or the Marcellus Shale for the rest of their lives (Goldenberg 2013a). Like many public disputes, fracking involves a struggle over what constitutes legitimate knowledge, with government and industry claiming the authority of science to dismiss what they view as unsubstantiated and irrational fears of those directly affected by the process. As with similar disputes, the appearance of what Cymene Howe terms "anthropocentric ecoauthority" matters more than facts on the ground: the consequences of fracking on the environment have (quite deliberately) been little studied (Howe 2014; Soraghan 2011).

In an interview about his book, *Disassembly Required: A Field Guide to Actually Existing Capitalism*, Geoff Mann commented that "'freedom' is meaningless without adequate food and water: I believe the main reason to be anti-capitalist is that capital deprives most of the world of precisely this stuff of life" (Mann quoted in Kilian 2013). In the IMAGE of tap water set ablaze with a lighter—an image central to documentaries on fracking like Josh Fox's *Gasland* and Cameron Esler and Tadzio Richards's *Burning Water*—the abstractions and uncertainties of resource extraction take a terrifyingly concrete form. Given the stakes involved, we can expect to see ever greater conflict between money and earth, industry and the public, gas and water, such as the traumatic and violent confrontation over fracking between the Royal Canadian Mounted Police and the members of the Elsipogtog First Nation in New Brunswick, CANADA, in October 2013 (Lukacs 2013).

Might fracking, an abstruse and secretive process with a bizarre, unlovely name, be what finally forces us to reconsider our commitments to and desires for CH_4 versus those for H_2O: what we need for the economy as opposed to what we need for life?

See also: CHINA, COMMUNITY.

Future

Todd Dufresne

1. The future looms over the present, provoking and haunting us, not as a postdated check, even less as a blank check, but as a reality check.

2. Capitalism's success can be measured by how effectively it has erased the myriad possibilities of the future, how well its version of the future is accepted as natural, given, inevitable. In extremis, the capitalist narrative blots out all alternative futures. De facto, it must contend with alternative futures proposed by social and political opposition, science fiction, and utopic literature. However, the de facto experiences of toothless alternatives only work to reinforce the ideal of capitalism in extremis. The result? A defeatism that declares that reality itself has been conquered, subject to principle, efficiency, instrumental reason.

From within the bubble of capitalism, one is inclined to believe the future has already arrived, realized as an eternal present. In the words of Margaret Thatcher's conservative slogan, "There Is No Alternative" (TINA), no possibility for thinking a future beyond or outside neoliberalism and capitalism.

3. Every economic CRISIS since the fall of the Berlin Wall has belied the fiction of TINA. But the world financial crisis of 2008 made it official: the capitalist future is over, dead. The years from 1989 to 2008 were the death throes of capitalism, a period of convulsive boom and bust—and, not incidentally, the end of easy oil (T. Mitchell 2011). In Marxian language: the dynamism of history, lulled to sleep by a cold war détente between capitalism

and communism, has picked up once again, revealing the revolutionary forces of a quite different future.

And there's the rub. The futurity of the future has come rushing back. The 2008 crisis was not just about the death of the economic theory and practice called capitalism, the death of the neoliberal ideology that rationalized globalization, and the end of petropower. It was actually a global existential crisis—an identity crisis that defines the winners and losers of oil capitalism.

Today the future has become what it used to be: an onslaught of alternative narratives, hopeful and fearful anxieties. The future is dangerous—but also alive. Not since the time of Marx, perhaps, has the future been *present*, not as the fatalistic completion of some grand narrative of *Geist* but as possibility, openness, plurality. That is to say: as futures.

4. *Justice, truth, and capitalism*. Under capitalism, the future is managed and quantified as human and financial investment, risk management, insurance payments, compound interest, amortized debts, defined benefits, and tax/retirement/estate planning. As for justice, one is reminded of Cephalus, who argues in Plato's *Republic* that justice is repaying one's debts (Plato 2008, 142–43). Subtext: justice is the advantage of the rich. When the great sophist Thrasymachus takes up the debate, he merely radicalizes this advantage in universal terms, namely, as the advantage of the stronger (ibid., 153). Simply put, justice is never having to repay one's debts. Justice is sheer caprice. Might makes right.

Plato's target is decadent, cosmopolitan Athens. His response to Thrasymachus is to rigorously imagine a better kind of justice. No more Socratic ignorance in the face of immorality and injustice. No more suspension of judgment. No more bullshit (a sentiment he shares with Thrasymachus). Instead, Plato offers a righteous counterblast of absolutist idealism: justice is the enlightened certainty wielded by the One-Who-Knows against the nihilism of the sophists. Plato's republic is a world turned right side up, purged and sanctified by philosophy and the philosopher king. In its essence, the birth of philosophy is reactive, reactionary, and even revolutionary—a program for CHANGE.

What is justice today? According to the regulators, politicians, and bankers, justice is never having to repay their debts—the advantage of the stronger—even as the poor are swamped in impossible indebtedness. Austerity as community, organized theft as virtue, environmental degradation as full employment, and oligarchy as freedom and democracy—in short, might once again makes right. The mighty pass finance, trade, tax, and environmental laws that suit their own narrow self-interest.

And truth? It is spun by MEDIA hacks, sold by an army of lobbyists, willfully ignored by corrupt politicians, and buried under the obfuscation of just-in-time science and its laughable rhetoric of objectivity. We witness today a return of the sophists on a global, institutional scale. Truth is treated with ideological dispersant, the discursive equivalent to Corexit, that calamitous chemical compound sprayed over the calamitous oil SPILLS of Alaska and the Gulf of Mexico (*The Big Fix* 2012). When oil power and oil democracy are at issue, one calamity always seems to follow another. But the first calamities are always justice and truth.

The legacy of such spectacular bad faith is not just the destruction of the economy and of the natural environment. It is the demise of capitalism altogether—in its claim on the future and its right to exist. Another legacy is the rise of righteous counterblasts, from comedian Russell Brand's (2013) editorial denunciation of capitalism to environmentalist David Suzuki's "Carbon Manifesto" (2013) and Pope Francis's (2013) lengthy first *Apostolic Exhortation*. Jeremiads are the new black.

Obviously the cultural context has changed since ancient Greece. No one trusts absolutism and essentialism anymore, not after the horrors of the twentieth century. But, like Plato, we find ourselves at the end of one future and before the subsumption of others. We once again need reactive, reactionary, and revolutionary thinking. We need a new philosophy, one appropriate to a world enhanced and diminished by the ideology of global capitalism, a world remade as the "ANTHROPOCENE."

5. A crisis or perceived outrage can drive a mob to plunder what it can from local stores—food, alcohol, cigarettes, diapers, baby food, electronics, guns—and burn the rest. Today this behavior perfectly describes the activities of the ruling elite, who have been busy plundering the environment, deregulating policies, exacerbating inequalities, and hoarding tremendous wealth. Looting, stockpiling, and fortifying. Naomi Klein (2007) rightly calls it "disaster capitalism" (355–408), a last burst of activity, the better to arm and secure the elite mob inside a protective bubble or "green zone" of their own making. Still, pity the elites this much: their future has been destroyed by their own ceaseless activity.

6. *Doom capitalism*. We hear much about an apocalyptic, dystopic, and catastrophic future. But understand that it is primarily the elite, with almost everything to lose, who see doom on the horizon—not just superstorms, lost oceanfront properties, and diminishing oil but an existential doom borne of these very conditions. Pitchforks, nooses, pikes, and guillotines: the revenge of the low, the very people they have impoverished (and killed) over decades of neoliberal rationalization and globalization. Consequently, the elite have doubled down, strip-mining every facet of social welfare. Of course, their brazen thievery of the public sphere, condoned if not facilitated by government stooges, is rational only as endgame, as survivalist desperation. To repeat: their behavior is possible only after the end of capitalism. For disaster capitalism is really the reductio ad absurdum of the promise of capitalism.

We all know that the future is full of hazards. But let us not be fooled: the first and greatest hazard is the elite, not just because of their riotous behavior but because of their dark prognostications on the future. Certainly we cannot allow the elite to dictate our imaginations yet again and, to that extent, steal the future. Their future, we already know, is the futureless future of "permanent austerity" (Weisenthal 2013). Their future is a theft of future, a doom capitalism in the wake of oil capitalism. Our future, by contrast, must bend to what is beautiful and just, not what is ugly, hateful, vengeful, wicked, and inhumane. Our future must remain a human future, one responsive to the reality of the natural environment.

7. *Silver lining*? When climate increases two degrees and possibly more; when water is scarce or fouled; when oil is gone and alternative fuels are insufficient to maintain con-

sumption levels in the West; when food supplies cannot meet need or DEMAND; when superstorms ravage hinterland, heartland, and coast; then we will have fully met the disastrous future of late-twentieth-century capitalism. The debt generated will be paid for in mass migration, limited prosperity, hardship, and war. It will be paid over generations.

A common hope is that science and technology will protect (some of) us from the worst effects of this future (Szeman 2014). Perhaps. But in truth, the futurity of the future lies elsewhere: in a humane and collectivist social and political future; in the social justice that lies in thinking the future of economics differently; and in the dreams and flights of imagination that realize a future filled with promise, life, and laughter. It is trite, but let us say it anyway: this future of alternatives is now struggling to become our present.

See also: CATASTROPHE, CHANGE, DISASTER, EXHAUSTION, INNERVATION, UTOPIA.

Gender

Sheena Wilson

In January 2014, Tenelle Star—a 13-year-old student from Balcarres, Saskatchewan, and a member of the Star Blanket Cree Nation in Treaty Four territory—made headlines in Canada when school officials ordered her to remove her hoodie and, later, to wear it inside out (Subdhan 2014). On the front, the hoodie read, "Got Land?" and on the back, "Thank an Indian."

That same month, MEDIA controversy also surrounded hearings about the proposed Enbridge Northern Gateway Pipeline, which would carry tar sands oil from Alberta to ports in British Columbia. Tom Isaac, lawyer and Indigenous rights consultant for Enbridge, declared on CBC Calgary's *The 180 with Jim Brown* (2014), "Certainly from my vantage point I don't see an inherent weakness in [the pipeline review] process. . . . What the courts have actually said . . . is that the balancing act between societal interests, on the one hand, and Indigenous interests, on the other, are to be decided by government." Isaac's binary—societal versus Indigenous interests—raises questions about how he (or the courts, governments, and industry) defines "society." Where does society end, and where do Indigenous interests begin? Where does Tenelle Star fit in Issac's definition of *society*? Or women in general? And where do race and class intersect with these discourses of power, particularly in relation to natural resources, oil, and the interests that purport to develop them?

These questions became urgent in light of the Canadian government's 2012 labeling of certain environmental and First Nations groups as radical extremists and its creation of a counterterrorism unit to protect Alberta's natural resources and INFRASTRUCTURE, at the same time as it funded a forty-million-dollar ad campaign promoting the oil sands and Canadian resource development (Tait 2012; Canadian Press 2013b; Cheadle 2013).

As they consolidate neocolonial and neoliberal petro-agendas, these political and media discourses shape public opinion and foreign policy. Built upon restrictive conceptions of women and Western feminism, they have ramifications for women and for ethnocultural communities and other marginalized groups in Canada and other petrocultures around the world. Such strategies sustain Big Oil's environmental destruction and obfuscate activists' efforts in gendered ways by silencing and rhetorically banishing women leaders and by depicting men of color as "petro-terrorist-gang-members"—a process identified by Heather M. Turcotte in US media representations of Niger Delta resistance (2011a, 216). Class is also used as a rhetorical management strategy to neutralize resistance. Tropes of model citizen-consumer and/or model citizen-entrepreneur offer access to (middle) class identity—a seemingly positive escape from the negative feedback of racialized and gendered discourses, but one that nevertheless nullifies petro-resistance in the short term and, in the long term, maintains existing inequities.

Take, for example, Canadian mainstream media representations of the Idle No More movement started by four female activists in November 2012, and Chief Theresa Spence's forty-four-day hunger strike that began in December 2012. Media coverage of these protests reinforced traditional IDENTITY tropes that perpetuate inequity and injustice. Spence and Idle No More were protesting omnibus bills C-38 and C-45 and aspects of Section 35 of the Canadian Constitution, which weaken environmental protection and Indigenous land rights, allow the expansion of large-scale oil production, and pave the way for associated projects like PIPELINES. However, Spence's demands were never explained by the mainstream media, which focused instead on her health and whether her actions actually constituted a hunger strike, since she consumed tea and fish broth (Campion-Smith 2013; Javed 2013; Hopper 2013; Kay 2013). These accounts depoliticized her intentions and her struggle. The slippage from tactics to semantics brought her ethics into question; she even faced allegations of financial corruption, with media reports perpetuating racist, sexist, and classist stereotypes (Levant 2013b; Press 2013; Taylor 2013). Media accounts publicized exhortations that she refocus her energies on her family and COMMUNITY; they rhetorically banished her to the private, domestic sphere by reinvoking her status as mother-grandmother, while providing space for male Indigenous leaders to declare, "It's time for the men to step up" (Campion-Smith 2013). Such practices rationalize women "as unpolitical and external to the political economy" (Turcotte 2011a, 208). They reinforce extreme notions of conventional gender roles and elide the leadership of Indigenous women activists. At the same time, Indigenous men and their leadership strategies were described as "tactical," "aggressive," "extreme," and "angry" (D. Ross 2013).

Media coverage also undermined the Idle No More movement by linking it to violence—both violence initiated by "aggressive elements within the existing [A]boriginal leadership structure" and that enacted against Indigenous people yet depicted in a way that blames the victims for eliciting racist responses (D. Ross 2013; Barrera 2013). The gang rape of an Indigenous woman in Thunder Bay, Ontario, for instance, was alleged to have been retaliation against Idle No More. These reports insinuate that violence is endemic to resistance efforts; they replicate the intention of the aggressors, who aim to discourage Indigenous youth and their communities from activism.

This rhetorical strategy spectacularizes and markets violence as a reaction to Idle No More rather than as a persistent legacy of the colonial logic, human rights abuses, and gender-sexual violence upon which petro-states are founded. Furthermore, these racialized, gendered representations of Indigenous resistance contrast starkly with sanctioned forms of consumerist resistance marketed predominantly to middle- and upper-class (and generally white) women. In mainstream popular media, women's relationships to environmentalism tend to be reduced to trivial issues and fashion advice (see Planet Green 2014).

Mainstream media reports on Tenelle Star's hoodie protest repeat many of these patterns. Indigenous press sources described Star and her mother as conscientious protestors and members of a larger community movement concerned with protecting land and treaty rights, which pose obstacles to ongoing industrial oil projects. In the mainstream press, Star became a damsel in distress through repeated mentions of the pink hue of her sweatshirt and her "confused" response to the controversy. She was chastised as being "rude" and "cheeky" ("First Nation Teen" 2014)—things a young woman should never be. Just as Theresa Spence's body and diet, the validity of her hunger strike, and her status as an older female figure were invoked to depoliticize her, references to Star's gender and youth undermine her as figure of resistance.

Star vanished from the mainstream mediasphere in late January 2014; some of the last reports about her noted that her family was advised by police to deactivate her Facebook account because she had become the victim of racist cyber attacks. Vigilante reaction and the state's paternalist concern converged to silence Star and cut short her protest. The mainstream media predictably conflates race, gender, and issues of security: Star is figured as a victim of violence who can be best protected by withdrawing from petro-protest. The authorities and the media fail to link this instance of racially motivated violence to a history of violence against Indigenous women endemic to the Canadian petro-nation's colonial legacy.

From the hallways of high schools to the runways of high fashion,[1] in popular culture and late monopoly capitalism, women's images—and women as a concept—are systemically co-opted to serve national and international petro-politics. Consider the fictions of Ezra Levant's *Ethical Oil*, which contrast rigid notions of female gender norms in CANADA with other petro-states in order to justify exploitation of Canadian oil reserves. Consider also the ironies of BP's early-millennial "Beyond Oil" campaign, which uses women's bodies and female identity tropes steeped in a history of patriarchy to rebrand the company as an ENERGY innovator (see Wilson 2014). These gendered constructions intended to undermine women and their political agency are intricately connected to other marginalizing discourses at the intersections of race, class, and petro-politics.

1. For example, the "Water & Oil" cover shoot in *Vogue Italia* (Meisel 2010).

Whether the discussion is of pipelines or wind turbines, understanding how women's resistance to petro-politics is strategically managed, subverted, and neutralized by mainstream power discourses provides insight into how the inequalities of our current "society" (as understood by Enbridge consultant Tom Isaac) are sustained. After all, it is not oil that places "society" and its needs in opposition to Indigenous or women's interests or the interests of the un- or underemployed. It is not oil that designates certain people as model neoliberal entrepreneurs or consumer-citizens in binary opposition to enemy Others: environmentalists and petro-terrorists. It is not oil that co-opts middle- and upper-class consumers through falsely progressive discourses that sell the promise of alternatives. Oil merely fuels those interests. Critical petro-intersectionality provides a lens through which to trace and detangle the webs of relations and to expose how the inequities of race, class, and gender are not only perpetuated in our current petroculture but also actively deployed as rhetorical strategies to literally and figuratively buoy and sustain existing power sources: oil and the neoliberal petro-state.

See also: ABORIGINAL, AMERICA, CANADA, EMBODIMENT, IDENTITY, MEDIA, PETRO-VIOLENCE.

Green

Toby Miller

Green can signify displeasure, even disgust. For example, "he turned green" or "it is indefensible to have green lawns in Los Angeles." But the term is more complex than that. It is simultaneously serene, beneficial, disturbing, corrupted, radical, and conservative: green consumption, green certification, new (green) deal, and greenwashing. When I typed *green* into a partially coal-fired search engine, the first entry to pop up was Greenpeace's website.

In the late 1960s and early 1970s, the word *pollution* was in vogue to explain environmental hazards. It was everywhere, yet localizable. The problems it described occurred when particular waterways, neighborhoods, or fields suffered negative externalities from mining, farming, and manufacturing. The issue was how to restore these places to their prior state: pristine, unspoiled, enduring. Pollution was about corporate malfeasance, governmental neglect, and public ignorance, and how to remedy their malign impact. It could be cleaned up if governments compelled companies to do so—and would soon be over, once those involved understood the problem.

But when greenhouse gases, environmental racism, global warming, and environmental imperialism emerged on the agenda, pollution reached beyond national boundaries and became ontological, threatening the very earth that sustains life, in demographically unequal ways. A new word came to describe the values and forms of life encompassing a planetary consciousness that might counter this DISASTER. Signifying new possibilities and a greater, more global sense of urgency, *green* emerged to displace the more negative and limited term *pollution*.

This beguilingly simple term quickly transformed into a complex polysemic mélange. Today, *green* can equally refer to local, devolved, noncorporate empowerment or to international consciousness and institutional action. The term is invoked by conservatives, who emphasize maintaining the world for future generations, and by radicals, who stress anticapitalist, postcolonial, feminist perspectives. *Green* may highlight the disadvantages of technology, as a primary cause of environmental difficulties, or regard such innovations as future saviors, via devices and processes yet to be invented that will alleviate global warming. It can favor state and international regulation or be skeptical of public policy. It may encourage individual consumer responsibility or question localism by contrast with collective action. It can reflect left-right axes of politics or argue that they should be transcended, because neither statism nor individualism can fix the dangers we confront.

This massive, conflictual expansion in meaning has generated a wide array of instrumental uses. So green environments are promoted as exercise incentives (Gladwell et al. 2013), encouragements for consumers to use quick-response codes (Atkinson 2013), ways of studying whether plants communicate through music (Gagliano 2012), attempts to push criminology toward interrogating planetary harm (Lynch et al. 2013), gimmicks for recruiting desirable employees (Renwick, Redman, and Maguire 2013), and techniques for increasing labor productivity (Woo et al. 2014).

Most important, big polluters make cynical use of other institutions to improve their public image, seeking "a social license to operate" through links with allegedly benign entities (art and sport) that seem far removed from their core business. This surprisingly overt term has been adopted with relish by corporations to describe their diplomacy with local, national, and international communities, undertaken by sponsoring truth and beauty (Thomson and Boutilier 2011), as if that were all we need to know about their operations.

A classic example is BP and the ideological work it does at Britain's Science Museum, where schoolchildren are encouraged, in the words of the corporation's magazine, "to explore and understand how energy powers every aspect of their lives and to question how to meet the planet's growing demands in the FUTURE" (Viney 2010). A "partnership" between the two institutions was necessary because of "a shared concern over the public lack of awareness of energy-related issues." The initiative features "an interactive game where visitors play the energy minister and have to efficiently power a make-believe country by balancing economic, environmental and political concerns before the prime minister fires them" (ibid.).

This alliance poses a clear challenge to environmental science rather than an invitation to dialogue; it positions BP as a benign intermediary between present and future, science and childhood, and truth and innovation rather than as one of the worst polluters in human history. The game presents BP and the Science Museum as reasonable people in a world of extremes, capable of measured, fair-minded engagement with the issues (unlike hotheaded, green-gaseous environmentalists). For those on the left, this is a prime instance of greenwashing, a cynical means of deceiving the public. Vibrant social movements oppose such activities (e.g., the "Greenwash Guerrillas") by pointing out the contradictions of cultural institutions that espouse green values they conveniently disregard if sponsorship from Big Oil awaits ("Activistas y artistas" 2010).

When we ponder such uses of green spectacle, it is easy to fall into either a critical camp or a celebratory one. The critical camp would say that rationality must be appealed to in discussions of climate change and competition for emotion will ultimately fail. Why? The silent majority does not like direct action, corporations outspend activists, such occasions preach to a light-skinned, middle-class eco-choir, MEDIA coverage is inevitably partial and hostile, and crucial decisions are made by elites, not in streets.

Conversely, the celebratory camp would argue that a Cartesian distinction between hearts and minds is not sustainable, a sense of humor is crucial in order to avoid the IMAGE of environmentalists as finger-wagging scolds, corporate capital must be opposed in public, the media's need for vibrant textuality can be twinned with serious discussion as a means of involving people who are not conventional activists, and a wave of anti-elite sentiment is cresting.

Absent external evaluation of the social composition of activists, the nature of old, middle-aged, and new media coverage, and subsequent shifts in public opinion and reactions from lawmakers, it is difficult to be sure about the impact of radically green spectacles. I am generally inclined toward the skeptic's view of populist activism—that it is mildly amusing and disruptive but is basically pranks without proof. But I do not feel that way in these instances, because I think the lugubrious hyper-rationality associated with environmentalism needs leavening through sophisticated, entertaining, participatory spectacle. A blend of dark irony, sarcasm, and cartoonish stereotypes can mock the pretensions of high art's dalliance with high polluters. (Consider, for example, the ruckus and Liberate Tate culture-jamming regarding BP's sponsorship of the Tate Modern museum that ultimately led BP [denials notwithstanding] to withdraw its support.)

Regardless of how one evaluates the tendencies it describes, *green* has become maddeningly over-present in our lives. It does such contradictory work that motor-racing mavens and messianic eco-martyrs alike can invoke the term with equal credibility. This reflects both the CRISIS to which the word refers and the variety of responses it connotes. I just hope the folks who shop via QR codes, or visit exhibitions underwritten by BP, take a critical view of *green* that is alert to its co-optation as well as its value. And that radical, ludic users of the concept are as alive to the need for research to test the efficacy of their play as they are to their desire to be naughty adults acting out in public.

One hopes that the most abiding legacy of green politics and theory will be the development and installation of the precautionary principle into everyday life and policymaking. Opposed to conventional cost-benefit analysis, which looks at the pluses and minuses of consumer satisfaction versus safety, the precautionary principle places the burden of proof onto proponents of industrial processes to show they are environmentally safe. The idea is to avoid harm rather than deal with risks once they are already in motion: prevention, not cure.

This principle can lead us to a new kind of citizenship. The last two hundred years of modernity have produced three zones of citizenship, with partially overlapping but also distinct historicities and implications for the environment. These zones are the political (conferring the right to reside and vote), the economic (the right to work and prosper), and the cultural (the right to know and speak). They correspond to the French Revolutionary cry "liberté, égalité, fraternité [liberty, equality, solidarity]" and the Argentine left's contemporary version "ser ciudadano, tener trabajo, y ser alfabetizado [citizenship, employment, and literacy]." The first category concerns political rights; the second, material interests; and the third, cultural representation.

Today, we need green citizenship, blending these three forms, so that we operate from an ecological perspective across political, economic, and cultural life. That would make green less hideously white (and middle class) and would force mandarins to address spectacle and pranksters to address bureaucracy.

See also: CORPORATION, RENEWABLE, UTOPIA.

Grids

Cymene Howe

Channeling a million megawatts of current through millions of miles of wire, the electric grid in CANADA and the United States has been called the world's biggest machine (Achenbach 2010). It enables an abundance of electric life—not just gadgetry but economies, industry, social space, medicine, and perhaps the stabilization of modernity itself. Circulating a force indispensable to daily life in much of the world, grids are conduits for systems of social organization. But grids are frail machines, prone to breakdown and blacking out. Rummaging in the dark, we realize that we know grids best by their failure. Grids disappear into the background; we become unconscious of them. In this sense, electric grids are exemplary INFRASTRUCTURE: embedded, largely invisible technical arrangements conveying services to human populations. However, grids are more than transmissional tools, more than a "thing" (Appadurai 1986; Bateson 1972; Latour 1993). Grids are the working relationship between humanity and the electron. The more deeply we read the grid, the more we see its potency and status as a vital entity. It is worth considering not only how grids supply certain, valued modes of life but also how the grid itself is interpreted in sociobiotic terms, as a machine with a life of its own.

Gridlife

When Thomas Edison flipped the switch in the Manhattan offices of J. P. Morgan in 1882, he launched a life-altering phenomenon: ELECTRICITY for commerce and home, the illumination of outdoor space, and the birth of the grid. Accustomed to gaslight and oil, some early customers worried they might find electricity dripping from sockets onto the floor (D. E. Nye 1990, 152). The grid signified more than power rushing behind the walls. As both a substitute for labor and a form of capital, plentiful, portable electricity would overturn the previous industrial order (ibid., 233–34). During the Great Depression, public works projects extended the US electricity network to the country's hinterlands. Farms began to glimmer. But grid expansion also instituted "bioterritoriality" and ensured state management of RURAL outposts: good old-fashioned governmentality by the kilowatt hour (Alatout and Schelly 2010).

The original grid-world of wires, meters, towers, substations, and switches is not radically different today (Schewe 2007). The three grids that make up the US system—Eastern, Western, and Texan—are monitored by engineers whose computers collect data from hundreds of thousands of response points on the system. Engineers communicate with the grid to monitor DEMAND and control production so that the thermal conditions of the wires are kept in equilibrium. Too much power in the wires risks prostration; the grid is vulnerable to overheating, expansion, and dangerous drooping of its lines. The grid has a METABOLISM.

Blacking Out

In summer 2012, approximately one-tenth of the world's population suddenly found itself electronless when India blacked out. As much as two thousand miles of territory and 670 million people went without, though many Indians reported that they are used to it. Local power outages are commonplace (Yardley and Harris 2012). But an outage this vast was unprecedented, the largest in world history. To say that the gears of industry ground to a halt is too mechanical a metaphor; electronic productivity ceased to exist. The grid was temporarily braindead. Worries about looting and fear of the dark loomed in accounts of the event. Fear was also an affective consequence in 2003 when the US Eastern Seaboard was de-electrified. Wall Street trading ceased. Flights were grounded, trains halted, and sweaty office workers trudged home because of shuttered subways. What would have been a local loss of power in Ohio cascaded across the Eastern grid, from Detroit to Brooklyn to Ontario. The largest power outage ever in North America was brought on by hot weather, cranked air conditioners, computer failure, and unpruned trees. Unlike the rolling black- and brownouts imposed by Enron's energy tyranny in California, these power failures were unplanned, a sign of debility and human and mechanical errors. They are missed messages with recursive impact.

Whether engineered or accidental, the temporary death of the grid surfaces its promissory life. In California, as in India or Zanzibar, the grid's existence and operation is equated with economic health and prosperity (Winther 2011). The grid supplies the promise of progress wherever it is installed; it is supposed to deliver redemption from poverty and underdevelopment. From a technocratic point of view, "electricity by itself is not a panacea for economic and social problems [but is] nevertheless believed to be a necessary requirement for economic and social development" (Wolde-Rufael 2006, 1106). The grid is so enmeshed in the logic of progress that governments subsidize electricity production and distribution, at a loss, in Mexico, Morocco, and many other countries. But as blackouts and aging infrastructures show, while the grid may offer the magic of "modernity," it is also a machine with an almost biological vulnerability to harm. Policy makers and security planners warn that "critical infrastructure" like the electric grid is susceptible to attack. "Vital systems" are also "vulnerable systems" (Collier and Lakoff 2008).

Smart Grids

The 2003 blackout, along with increasing concern about climate change, has created an ultimatum for electric grids: they have to get smart. The US grid, while among the best in the world, has been described as a "kludge." *Kludge* is an onomatopoetic term, a homely word for an ugly device. It describes a technical instrument engineered to perfection that, by growing organically and additively, becomes ungainly: functional but technically inelegant. As we continue to gorge on electricity, we might rightly be described as "creatures of the grid." But one could also say that the grid is our creature. A "monster" that, as Bruno Latour (2006) might say, we have failed to adequately love. Grids in the United States, Canada, and Europe mostly function, most of the time. But they are overdue for CHANGE. The meters that measure household consumption are 1920s technology. Experts predict that the US grid will undergo more renovation in the next ten years than it has in the last 100.

The grid was originally designed to meet demand rather than to moderate consumption, but the smart grid will change that. Industry experts explain that the smart grid will be self-healing, able to feel, report, and remedy its own injuries and sagging flows of electricity. More acute nerve endings will allow technicians to easily locate damage and outages. The smart grid will have a biofeedback loop that signals where consumption demands are high, and it will recalibrate accordingly. It will finally remedy the plague of RENEWABLE energy's intermittency because an interconnected grid will allow constant access to backup power sources.

A smarter grid promises to make us "smarter" too. Consumers will be able to program appliances to run outside peak hours and reduce electric bills or to ensure their electricity is all GREEN all the time. The smart grid will sharpen our electronic sensibilities as we become conscious managers of our electric life. If the old "dumb" grid facilitated governmentality in urban and rural contexts, the smart grid will create new horizons of neo-governmentality predicated on energic rationalities and technologies. Our skills as "calculative selves" (Paterson and Stripple 2010, 359) will be accelerated and tested by smart grids that inform us, in ever more minuscule metrics, of our quotidian consumption. For better or worse, we will likely be transformed too, becoming reflexive self-regulators, panoptic participants in our electrical worlds.

Ungrids

Earth at Night is an assemblage of satellite photographs of Earth by night. Assembled by NASA, the IMAGE is at once beautiful and indicative, mapping both lumens and economic development through the powers of the grid. Brilliant meshworks cover most of Europe and the east and west coasts of North and South America. Cities are legible by their luminescence alone. The jagged edges of South Asia and CHINA's urban conglomerations shimmer; Japan is a beacon in night waters. Other places are as black as pitch. The interior of the African continent still appears, from space, like Joseph Conrad's abyss.

So what of the ungridded? According to the World Bank, 1.1 billion people, about one-sixth of the world's population, is gridless: without regular, reliable access to electricity (World Bank 2016). Policy makers call this "energy poverty." Where infrastructures do not tread, however, other forms of electric life are beginning to propagate. The British company Eight19 (the minutes and seconds it takes for sunlight to reach the earth) is equipping some East African households with small SOLAR generators, networked by cell phones and paid for with scratch-off cards. The company's founder calls this arrangement the "un-grid" (Guay 2012). Cell phones, radios, light bulbs, and televisions rhizomatically emerge in new domains. The un-grid appears nonhierarchical and multiple, allowing things to exist in a place where, as Deleuze puts it, "light scours the shadows" (1986, 52).

Continental philosophy and micro power stations may seem an improbable union. However, German lawmaker and social philosopher Hermann Scheer sees considerable social, political, and cultural potential in recreating ENERGY distribution at a smaller scale. Shorter energy supply chains, he argues, would change the world. For Scheer, localized production—as opposed to dependence on globally distributed fossil fuels controlled by states and corporations—would foster more egalitarian forms of social life and "new political, economic and cultural freedom" (2004, 67). If the grid is our working relationship with the electron, then inviting power in and making it more proximate may be the premise for new imaginaries of that relationship. A more intimate grid may be a more egalitarian grid. Or, perhaps more accurately, the un-grid may hold the promise of a different sort of electric social life.

See also: ABORIGINAL, CHARCOAL, NETWORKS, NUCLEAR, OFF-GRID, RENEWABLE.

Guilt

Noah Toly

Right collides with right!

—AESCHYLUS, *The Libation Bearers*

"Old Forecast of Famine May Yet Come True" read the headline of a 2014 *New York Times* article by Eduardo Porter about the possibility of FUTURE food shortages caused by climate destabilization. The opening question, "Might Thomas Malthus be vindicated in the end?," invoked Malthus's *An Essay on the Principle of Population*, a treatise so bleak that it earned economics the nickname "the dismal science" (Porter 2014; Malthus 1993 [1798]; see also T. Carlyle 1889; Groenwegen 2001; Marglin 2010). Malthus's premise was simple: because population increases exponentially, DEMAND for nourishment would outpace the supply of food, which grows only arithmetically. Eventually, food shortages would cause conflict, famine, and disease, checking runaway population growth as "sickly seasons, epidemics, pestilence, and plague advance in terrific array, and sweep off their thousands and ten thousands" (Malthus 1993, 61).

As Porter rightly noted, Malthus's gloomy forecasts were "soon buried under an avalanche of progress" (2014). The fate Malthus supposed unavoidable "even for a single century" has been evaded for more than two (Malthus 1993, 61). Since Malthus's time, population has boomed. So has food production. Though some communities suffer hunger and malnourishment, food shortages have not caused global population collapse. While demographic projections for the twenty-first century suggest the need to increase food production and improve distribution for a still-growing world population, technical solutions seem to be at hand, as they have been in the past.

Developments in agriculture, INFRASTRUCTURE, and transportation have allowed us to produce and distribute enough food to meet the needs of a rapidly growing population while consuming more food per capita than Malthus could have imagined. Technics—described by Lewis Mumford as "that part of human activity wherein, by an *energetic* organization of the process of work, man controls and directs the forces of nature for his own purposes"—seem to have overcome what was for Malthus an unassailable logical force (Mumford 2000, 15, emphasis added).

But if technical solutions have delivered society from the grim calculus of population implosion, why do some say that Malthusian crises may soon be upon us? Because climate change threatens to affect agriculture, infrastructure, and transportation in ways that destabilize the global food supply. However, while most depict the link between climate change and Malthusian crises as ad hoc—as if Malthus may finally be vindicated by some tangential CATASTROPHE—their connection is actually far more systematic and necessary.

Consider Malthus's motivation, to which his original title alludes: *An Essay on the Principle of Population, as it Affects the Future Improvement of Society, with Remarks on the Speculations of Mr. Godwin, M. Condorcet, and Other Writers*. Malthus wrote to prove that society is not perfectible. Faced with the scenario he outlined, human beings would have to forgo one or more goods to secure others. When one of his interlocutors, Thomas Godwin, maintained that people would adjust to population pressures by giving up sex, Malthus noted that sex was a real good and to insist upon giving it up was to make his point. This tradeoff was too much: If the cost of avoiding food shortages was to forgo sex, then society was clearly not perfectible.

By highlighting tradeoffs, Malthus underscored what theologian Edward Farley describes as the tragic structure of the human condition. Farley defines "the tragic" as "a situation in which the conditions of well-being require and are interdependent with situations of limitation, frustration, challenge, and suffering" (1990, 29). The tragic demands "effortful existence" in the face of LIMITS. In the case of Malthusian limits, our effortful existence in the face of dismal forecasts relies upon an "*energetic* organization of the process of work." Our agricultural, infrastructural, and transportation technologies require a hydrocarbon energy regime to power our resistance to Malthusian logics. But deliverance through this regime has brought its own costs. The conditions of well-being have been entangled with limitation, frustration, challenge, and suffering. Mining and drilling have destabilized fragile ecosystems and vulnerable communities. Air, water, and soil pollution undermine life and livelihood. Climate change threatens planetary distress and renewed prospects for starvation, war, and disease. We have traded one catastrophe for another. The cost of avoiding Malthus's dismal future has not been forgoing sex; it has been destabilizing the climate. For the past two centuries we faced the tragic choice—only at first unwittingly—of *either food or a stable climate*, but for the next two centuries we may face the prospect of *neither food nor a stable climate* as the possibility of food shortages returns.

Every option for resolving this climate crisis involves further tradeoffs and costs, deepening our entanglement with the tragic. On one hand, assuming an unchanged ENERGY REGIME, a return to emissions levels that would stabilize climate within the next two centuries would entail a return to consumption levels of the early nineteenth century (Posner and Weisbach 2010). That would involve forgoing many nontrivial goods. On the other hand, hydroelectric, NUCLEAR, SOLAR, wind, and geothermal generation are viable alternatives to oil, gas, and COAL; they may be the only path to climate stability without massive economic and social disruption. But transitioning to new sources of ENERGY has its own costs. Investing in cleaner energy sources means abandoning dirtier, but cheaper, sources. Funds invested in an energy transition might otherwise be invested in other goods, such as education, public health, and nutrition. Whatever option we choose, we incur costs. We forgo some goods—real, nontrivial goods—to possess others.

When every choice, including inaction, has its costs, we must face the question, "Who will bear burdens?" In most cases, we have two options: we can act in ways that displace frustration, limitation, and suffering onto others—future generations, vulnerable populations, or nonhuman NATURE—to secure the conditions of well-being for ourselves, or we can embrace limitation ourselves so that others can secure well-being. Farley calls for the latter, for a response to the tragic that "tak[es] up the aims and goods of [local] particularity into agendas oriented to the well-being of broader environments" (1990, 270). Would that it were always so simple. It is often impossible even for the most magnanimous to bear costs themselves. Sometimes any decision incurs costs for others. Even the most admirably other-oriented among us sometimes face choices among future generations, vulnerable populations, and nonhuman nature.

Thus our experience of the current energy regime involves a perpetual conflict between values. No matter how we seek to resolve problems at the intersection of energy, environment, and society, we give up some good, incur some cost. We must forgo, limit, or destroy one good to possess another. This predicament, for Paul Ricoeur, is characteristic of the tragic: "That a value cannot be realized without the destruction of another value, equally positive—there, again, is the tragic. It is perhaps at its height when it seems that the furthering of a value requires the destruction of its bearer" (1986, 323). Recognition of these "non-dialectical contradictions" (323)—for which no resolution can embody the good of both alternatives—is often accompanied by what Ricoeur describes as a sense of "ineluctable guilt" (313), which can evoke despondency. Faced with the unrelenting rigor of the tragic, "the very order of the world becomes a temptation to despair" (323).

However, recognizing the truth of ineluctable guilt can be a source of creativity, freedom, and hope in the face of energy crises. By disabusing us of illusions of self-justification, it expands our range of potential responses to energy. There may be no one self-justifying response that we must choose but instead a plurality of potentially legitimate responses; this plurality can be a source of creativity at the energy-environment-society nexus. Lacking the possibility of a perfect society, we may recognize a variety of potentially legitimate ways to organize social life. Weakening the grip of necessity on energy politics may increase both our sense of freedom to choose among alternative futures and our sense of responsibility for the outcome—there can be no hiding behind NECESSITY. Finally, a sense of ineluctable guilt opens up the possibility of an "unverifiable faith" (Ricoeur 1986, 313), hope in a resolution that we cannot see or achieve ourselves. The price of this creativity, hope, and freedom is recognizing the truth of ineluctable guilt, a truth that "will set you free, but not until it is finished with you" (D. F. Wallace 1996, 389).

See also: DEMAND, FRACKING, FUTURE, UNOBTAINIUM.

Identity

Geo Takach

In the industrialized world, petroleum fuels the INFRASTRUCTURE of our societies and the logistics of our lives. Yet its ubiquity and power transcend gas pumps, foodstuffs, and countless other delights of contemporary existence. Sure, oil fills state and private coffers, builds Brobdingnagian beacons like Dubai, and incites the odd bloodbath. But it can also color the soul of its sites of production by defining expressions of local values and representations of that place to the world. Take my home province of Alberta, CANADA—to which many ecologically concerned global citizens would hastily add, "please." Now playing ball with behemoths like Saudi Arabia and VENEZUELA (behind which Alberta ranks third in recognized petroleum reserves), the province boasts what has been called the largest industrial project on Earth, the bituminous ("tar" or "oil") sands (Leahy 2006). Alberta illustrates the power of petroleum to fuel the culture of a place as much as it powers the growing fleet of tank cars clogging its primary paved artery, renamed the Queen Elizabeth II Highway in a grand if unwittingly ironic nod to a more overt icon of colonialism. Oil has shaped the province's sense and representation of self—its identity—with significant repercussions.

For ten millennia, what is now Alberta was a site of ingenious, indigenous buffalo hunting, and after an outbreak of white party-crashing in the late eighteenth century, it became a vast fur farm lining European hats and investment ledgers. In the late nineteenth century, private ranching opportunities gave way to a nation-building project by the Canadian government and the Canadian Pacific Railway Company. These institutions recruited settlers to serve as a human land bridge to another British colony, British Columbia, while thwarting the threat of US incursion and promoting Rocky Mountain tourism. Thus did Alberta assume the twin mantles of national breadbasket and international eye candy, with limitless wheat fields, azure skies, and soaring peaks, accented by the faded but brilliantly mythologized and marketed ethos of the cowboy on the open range. When settlers realized that farming made a far better living for railways and middlemen than for themselves, they turned to prairie radicalism, uniting to form grain pools and start populist, progressive protest parties of national consequence. But previous spending sprees on infrastructure and the Great Depression helped Alberta become the only Canadian province to go bankrupt. Then, miraculously, oil changed everything.

A major wildcat strike in 1947 propelled Alberta from an indebted, largely agrarian byway into modernity as "Canada's energy province." Aided by American know-how and cash, the province drilled enough physical holes to dig out of its fiscal one and to build hospitals, ROADS, and schools from oil and gas revenues. The spudding ceremony at the fabled Leduc No. 1 oil well was carefully choreographed (complete with dramatic gas flare and gusher), and oil superseded wheat in both the provincial economy and the popular imagination. Transposing Alberta's sodbusting ethos, extraction became valorized as the heroic taming of the frontier with physical determination and awesome technology (Davidson and Gismondi 2011). Today, the industry conflates the province with oil, as in its "Alberta Is Energy" campaign (Canadian Association of Petroleum Producers 2010a). And why not? Alberta depends on ENERGY (mostly oil) for about 22 percent of its GDP, 30 percent of its budget, and 75 percent of its exports (Government of Alberta 2013a, 2013b). No wonder that any criticism of the accelerated, unbridled extraction of bitumen is dubbed "un-Albertan" (Canadian Press 2012).

If we accept that identity is created relationally through discourse to "maintain a sense of consistency in the social order constructed through the discourse" (Tann 2010, 165), then representing place-identity—a rhetorical expression of a unique sense of a place and its social and symbolic capital (Campelo, Aitken, and Gnoth 2011)—becomes significant. With globalization, the competition for human, physical, and financial resources requires place-identities to be defined and marketed more aggressively (Dittmer 2010). Because place-identity involves the power to set local social norms, it can be used to maintain political and economic hegemony (van Ham 2010). Conversely, as place-identity is multifaceted, positional, and contested (J. Anderson 2004), it may also help engage citizens in public issues (Dahlgren 2009) like environmental concerns.

This explains why an immense public-relations battle has emerged, situating Alberta as the epicenter for a conflict between demands for continuous economic development fueled by nonrenewable resource extraction and concerns about the unsustainable ecological costs of development. Mounting international alarm over Alberta's (mis)management of the bituminous sands is evident in controversies like the proposed Keystone XL pipeline to transport bitumen through the United States and the European Fuel Quality Directive labeling the product "dirty oil" (D. Bennett 2013b) and in critical, high-profile articles (Kunzig 2009), books (W. Marsden 2008; T. Clarke 2008; Nikiforuk 2010), and independent documentary films and advocacy videos (*Dirty Oil* 2009; *H_2Oil* 2009; *Tipping Point* 2011), which collectively depict Alberta as a greedy, shortsighted environmental pariah. Alberta's oil industry and its ardent cheerleaders, the provincial and federal governments, have responded with advocacy campaigns depicting Alberta as a friendly, secure, principled, and ecologically responsible source of oil (Canadian Association of Petroleum Producers 2010b; Government of Alberta 2010; D. Bennett 2013a).

With massive sums at stake—such as $2 trillion in new investment and more than $623 billion in royalties for the province forecasted by 2035 (Honarvar et al. 2011)—the bitumen-fired battle for Alberta's reputation and identity threatens casualties beyond the ecological, social, or economic, colossal as those may be. These casualties are cultural, touching on core aspects of Alberta's place-identity.

The first casualty is language, as proponents of intensifying extraction seek to naturalize the practice with agrarian allusions (a petroleum-processing bastion becomes "Alberta's industrial heartland" [Alberta's Industrial Heartland Association n.d.]). Proponents also argue that bitumen—long called "unconventional" oil because it is massively more difficult and costly (and resource- and emission-intensive) to extract than conventional oil—is becoming "conventional" (Alberta Energy Regulator 2013) or even "GREEN" (Koring 2013). They urge publics to eschew the longstanding colloquial (if incorrect) use of "tar sands" and "oil sands," taking up the Alberta government's decision to naturalize the latter term by branding the former "the language of environmental extremists" (Arab 2011), while creating a new, naturalizing compound word, "oilsands."

Another casualty is history. Oil has consigned the province's formative, collectivist impulses and populist radicalism to the attic dust of cultural consciousness, banished by neoliberal imperatives preaching the primacy of market forces in governing human affairs, situating citizens as consumers, and denying contrary voices. The province has flouted the academy's historical purpose by moving to tie the mandates of Alberta's postsecondary institutions to the government's economic priorities (CBC News 2013)—that is, to oil extraction—and by including oil interests as "partners" in redesigning its public-school curriculum (Government of Alberta 2014).

Oil has made Alberta Canada's wealthiest province (Government of Canada 2013), but the quality of its democracy—the third cultural casualty noted here—may be the poorest. Alberta's petroculture illustrates Friedman's (2006) "First Law of Petropolitics," which links the strength of democracy inversely to levels of oil production and specifically to governments' dependency on oil-related revenue rather than taxation, the latter inviting more engagement by and accountability to citizens. Alberta boasts the lowest taxes in Canada (Government of Alberta 2013b). Its fiscal dependency on oil exposes Albertans' employment opportunities and access to essential public services like environmental protection, health, and education to the vagaries and volatility of oil prices and global politics, making their public finances the most unstable and unpredictable in Canada (Landon and Smith 2010). Democracy? Since 1947, there have been only two changes in provincial government (in 1971 and 2015), scant opposition (government members have held about

80 percent of the seats in the legislature [Elections Alberta 2012]), and the lowest electoral turnouts in Canada (CBC News 2008). Before the 2015 upset, the provincial government fostered a climate of conformity (Takach 2010).

The conflation of Alberta with oil extraction offers a cautionary tale of the power of petroleum to fuel societies and create immense wealth (for some) and to transform the identity of a place. Oil has catapulted Albertans from agrarian, radical populists to turbo-capitalists literally betting the farm on bitumen extraction—the more and the faster, the better. Alberta's pre-eminent filmmaker, Tom Radford, says, "We are no longer building a culture as a place, an identity, but as an economic opportunity."[1] Albertans, or at least their economic and political masters, may think they have made oil their own. But given the current trajectory of the stewardship of the bituminous sands in the hands of the latest in a long line of colonialists, it looks like the other way around. Is this how you would want to be remembered?

See also: COMMUNITY, ENERGOPOLITICS, PIPELINES, RESOURCE CURSE, TEXAS.

1. Tom Radford, interview with the author, Edmonton, Alberta, April 4, 2012.

Image

Ed Kashi

The Niger Delta is where Nigeria's plagues of political gangsterism, corruption, and poverty converge. What is happening in the Niger Delta is nothing short of a militarized insurgent struggle against the violent machinery of state security forces with a terrible reputation. At the same time, there is something distinctive about this violence; the militants' struggle is a backlash against a long history of exploitation, the presence of transnational oil corporations, a style of politics dependent upon violence, and myriad groups, gangs, and cults with no leadership as such.

Under such dynamics, sustainable development for the local environment and population is inconceivable. The damage done by oil extraction and its attendant forms of violence to the livelihoods and environment of the Niger Delta, Africa's second largest wetlands, will take generations to repair. More disconcerting still is the lack of viable INFRASTRUCTURE or a vision for how to remedy the damage done over the past half-century since oil and gas were discovered in this land traditionally inhabited by small farmers and fishermen.

My work as a photographer has led me to many places, and it is plainly apparent that our planet is under tremendous stress. The vital relationship we humans hold with the earth has become a core theme of my work. With a growing inequality in living standards, the voices of those who live without are growing in number, clamoring to be heard as the precious resources upon which their existence depends are depleted beyond repair at a rate that cannot be sustained. Nowhere are the effects of this dysfunctional relationship more evident than in Madagascar. My work responds to the global cry to stop and take responsibility, seen through images of the people of southeast Madagascar, in a cross-examination of the intricate ties that bind them to the earth.

My work on the Niger Delta led me to explore the violence in the ongoing battle over oil among humans and between humans and Earth, which could not be a more striking contrast to the quiet struggle for survival I witnessed in Madagascar. This relative silence has allowed the world to forget this beautiful yet fragile island. Drawing parallels between these very different sites, I am led to the unmistakable conclusion that the way we are living on this earth is not sustainable. In the Niger Delta, it is a classic tale of multinational corporations plundering natural resources, in cahoots with corrupt local and national officials, to create a degraded environment, a diminution of local skills, livelihoods, and opportunities and a decrepit infrastructure that is polluted and crumbling. In seemingly pristine Madagascar, a different tale of unsustainability is unfolding. Local people, who live subsistence lifestyles, cannot continue their ways because of the destruction of forests, the lack of water, and the inability to produce enough food for themselves anymore. This predicament is not the result of corporate or government greed; the way they have lived for thousands of years is simply not working anymore—a sobering reality to comprehend.

For the Malagasy, everything necessary to sustain human life comes from their island earth. It is their food, their shelter, and often their only source of income. Holding it all together is an astoundingly depleted forest system. Systematically plundered by both external forces and local inhabitants sustaining their way of life, what remains of Madagascar's forests are mere small fragments, home to hundreds of endemic species, many yet to be named or even discovered. These forest fragments provide livelihoods to the majority of the island's human population. Despite recent forest conservation laws, the people of Madagascar, caught in a cycle of grinding poverty that eradicates the luxury of free will, continue to risk fines and imprisonment in pursuit of forest-based survival. Through the millennia, the people of Madagascar have relied on forests to make CHARCOAL for cooking, for WOOD to heat their homes, for materials to build dwellings, fences, boats, hunting tools, and weapons. They have cleared some forested areas for agricultural plots to grow food.

Recently the situation has become critical. Another year of insufficient rain means failing crops across the southeast. In search of fertile land, farmers engage in the illegal practice of slash and burn agriculture, or *tavy*, in which entire forest fragments are razed to the ground in return for one year's soil fertility (see Figure 5). At the end of the year when the soil is unusable, the farmers move to the next patch of forest and repeat the cycle, sowing the seeds of desertification that sees sand dunes encroaching on previously arable land, oceans receding, and communities migrating to already overcrowded towns to seek a livelihood the land can no longer provide. Never has the cycle of poverty and resource depletion been more evident than in the farmers I met walking down a dry riverbed toward the ocean. With no more arable land, they were on their way to join other newly conscripted farmers-turned-fishermen, too great in number to operate in balance with the ocean's limited supply. When they caught pregnant lobsters, they'd remove the egg sacs

FIGURE 5. Tavy, or burning of the forest to clear for planting, is illegal, but local farmers continue to do this despite the massive reductions of their forests in Analavinaky, Madagascar, on Jan. 12, 2010. (Ed Kashi/VII)

FIGURE 6. These young girls, ages 11–13, use a paste made from the tsiambara plant's roots to beautify their skin in Belay, Madagascar, on Jan. 14, 2010. They usually leave it on for 5 days and keep on redoing the process except on market days. The meaning of this process is called "I don't want to show you." (Ed Kashi/VII)

FIGURE 7. In Afiesere, near the troubled oil city of Warri, local Urohobo people bake "krokpo-garri," or tapioca cakes, in the heat of a gas flare on July 28, 2004. Since 1961, when Shell Petroleum Development Company first opened this flow station, residents of the local community have worked in this way. Life span is short for these people, as pollutants from the flare cause serious health problems. (Ed Kashi/VII)

FIGURE 8. Smoke billows from the Trans-Amadi Slaughterhouse, the main abattoir of Port Harcourt, Nigeria on May 31, 2006. The conditions are very bad, with a lack of infrastructure and hygiene. The animals are killed in the open, their blood spilt into the waterways below and their skin is burned by the flames of old tires, which is what causes the thick clouds of black smoke that hang over the scene. (Ed Kashi/VII)

to make them appear saleable on international markets. First the earth, then the sea—the people of Madagascar are running out of options.

Within Fort Dauphin, the urban center of the southeast, the population has exploded in this devastating process of unwilling urbanization, while infrastructure has not kept up. The mining giant Rio Tinto has arrived in this beautiful town, luring unemployed people with the hope of better things to come. Sadly, with the initial phase of the mine's construction over, mine jobs are not accessible to the majority of the Malagasy. The only infrastructure to have benefitted from their presence are ROADS to the mine and campsite, an eerie replication of Western suburbia inhabited by a floating population of expats and migrant workers.

Isolated from the country's capital, the southeast is strangled not so much by a lack of political will as by a lack of political capacity; with current fears over political instability meaning that the majority of international assistance has long ceased, the survival of the people of southeast Madagascar rests on the charities, NGOs, and other organizations working alongside the people in helping some of the poorest, most vulnerable communities lift themselves out of poverty sustainably, while protecting the environment from further harm.

My work has always been predicated on preserving the dignity of my subjects. Within this selection of photographs, you will see not only the problems that Madagascar faces but also violence perpetrated in the Niger Delta by forces all too familiar. No longer in photojournalism is it sufficient to document the challenges; rather, we need to empower each other through images like these to show how powerful collective action can be. This remarkable development in the field of visual journalism works through two modes. In many cases, I now work in collaboration with NGOs, foundations, nonprofits, or other advocacy organizations using my skills as a visual storyteller and my connections to the MEDIA to advocate for a particular issue or cause. That was the case with my work in Madagascar. Another effective path is creating images and narratives through my own efforts and in collaboration with media outlets, which then become useful to advocacy organizations and outreach campaigns. My work in the Niger Delta illustrates that approach. In fact, Oxfam America used my Niger Delta work to create a traveling exhibition for universities around the United States as part of a successful campaign to get Congress to pass a law requiring more transparency in the extractive industries. This kind of success represents a new way to use photojournalism to create positive CHANGE in our world.

See also: NIGERIA, PETRO-VIOLENCE, PHOTOGRAPHY, RESOURCE CURSE, STATISTICS.

Infrastructure

Jeff Diamanti

The safest place for rich people to dump their money during a recession is usually the one most likely to receive government stimulus—historically, public works projects aimed at shoring up both employment rates and the physical carrying capacity of an economic setting. Whether new, refurbished, or expanded, infrastructure has always functioned as a kickstarter for macroeconomic recovery, leading (so this fantasy goes) to long-term growth. The idea is somewhat paradoxical: the current configuration of a given economy (call it economy A) has either 1) reached a natural limit or 2) never really worked but works so badly now that its contradictory core has become inoperative; thus economy A needs to become A+ (the same but bigger, faster, and with the capacity to become much larger than the original A). Hence, an "infrastructure BOOM" after proverbial decline; more of the same but different.

Even in times of relative stability, infrastructure (more, different, better) occupies a unique position in the economic imaginary of growth: it seems necessary and appears, almost in spite of social constraints, to itself produce growth and is thus indistinguishable from it. Infrastructure thus resembles bonds in the medium-view of business cycles. Neither is a commodity in the strict sense (although both are increasingly treated as such), but a deficit in one or the other immediately spells dire consequences for growth. With bad (or no) ROADS, electrical GRIDS, or waste management systems (whether through negligence

With many thanks to Marija Cetinić and Mason White for their invaluable feedback.

or strike), all three moments in the cycle of ACCUMULATION—production, circulation, and consumption—dry up. This is why the Organization for Economic Co-operation and Development (OECD) is devoting so much attention to long-term policy forecasts regarding infrastructure in developing and developed economies alike: needing roughly 3.5 percent of global GDP to keep apace with capital's growth needs, infrastructure is becoming a major economic sector in its own right (a roughly 2.5 trillion–dollar industry in 2012) (OECD 2007). Unlike bonds, the stuff beneath both house and factory is always immediately social and political—increasingly so as investors find new ways to turn the infrastructure of social and economic life into a cash grab. Bonds, however, are mediation par excellence. If you burn down the treasury, you still have to round up the accountants; but if you take down the power plant, the workday comes to an end.

Nobody Works in a Blackout

Only very old materialists would assign such political primacy to roads, sewage systems, and power plants. Newer materialists would say such systems have a life of their own. In any case, infrastructure fits neither within the art history taken up by ARCHITECTURE nor in the naturalism implied by architecture's external constraint, much less the economic structure that gives value to land and buildings. How then, given the sensitivity with which investors have recently turned their attention to infrastructures (see Torrance 2009), to develop a political and epistemological relationship to infrastructure? Access to basic services such as ELECTRICITY and water increasingly defines the process of proletarianization at a global scale (see Endnotes 2010) and, as Paul Virilio has shown regarding electricity, also facilitates the "absolute power" of state militaries (Virilio quoted in Jakob 2001); infrastructure today has become something like the naked medium of political economic friction. This claim is not far from Angela Mitropoulos's—which is also Hannah Arendt's—that politics, after a definite moment in the history of economics (around 1945) "is premised not on a subject . . . but on the *infra*, the unassimilable plurality of that which lies between" (Mitropoulos 2013, 115). And once the social gets electrified and the primary source of ENERGY is not COAL but oil (also not too long after 1945), *infra*structure is the place where (economic) value and (social) energy share a provisional identity.

At the level of economic growth (or, in an older idiom, the forces of production), infrastructure regulates the value-time of a given economic setting. Any regional economy will have a limit beyond which production, circulation, and output are system-clogging (A1 in the above formulation, where a natural limit is expressed in traffic jams, power outages, and slow delivery times), which is why upgrading electrical grids, transportation NETWORKS, and bandwidth availability is synonymous in policy-speak with economic stimulus. President Obama's plan to open an "infrastructure bank," which would hold in reserve billions of dollars for states most desperate for new hardware, literalizes the CRISIS of public coffers in the United States (which have suffered hand in hand with municipal public works departments) and shows how much *infrastructure* is a code word for economic growth that tells us something about the internal logic of growth itself (Baker and Schwartz 2013). An infrastructure bank would fundamentally alter the financing of urban construction. It would remove tax barriers to public and private pension funds investing in infrastructure bonds, a 1990s financial innovation designed to open the market to private investment. Notwithstanding the fact that this model stinks too much of what Republicans would call "nanny-state economics," the idea is to treat infrastructure as a commodity like any other in order to transfer the costs of upkeep onto users or publics and the risks onto pensioners. Energy spent outside the factory creates as much value as inside and should thus (so the logic goes) be brought to market.

At the level of political relations (or, in an older idiom, the relations of production), the watts, water, and waste flowing across today's social bedrock constitute the immediate conditions of social reproducibility as such, without which—say, in a postapocalyptic wake—we would have nothing but its symbolic remains. If that is true—if infrastructure is the moment and place where labor power as a value-creating commodity and a concrete social relation is fueled (with watts, water, and waste removal)—then what it is "between" are two antinomical moments of the value form itself: the moment between concrete and abstract and that between labor in itself and capital in itself. If the "hidden abode of production" is for Marx where all the secrets of the commodity are to be found (1977, 279), then it is in infrastructure—the material bedrock upon which social, political, and economic life now depends for its energy—where the secret of labor's metamorphosis into labor-for-capital occurs or its *re*production into a (indeed, the only) source of exchange value. Which is why nobody works in a blackout.[1]

Everyone Is a Homeworker During a Blackout

In a long-term blackout, everyday habits become life-threatening: food turns sour; weather is unmediated; water becomes undrinkable; use-value trumps exchange-value. Variegated moments of production, circulation, and consumption are isolated, emptied of value-time, leaving only social time determined by reproductive (not productive) needs. Stored fuel becomes a source of heat, streets a place to find one another. A unique materialism emerges, capable of isolating moments of what is otherwise the unity of labor and labor power (the worker on a labor market). If infrastructure is the medium where politics and economics become one (i.e., autonomous in appearance during the work day), then its breakdown is also the separation of social energy and energy as value. Although most of us cannot help but wait for the return of power, the city-wide blackout generates a series of important disarticulations (Luke 2009). Without functioning infrastructure, system and subject sublate

1. Of course, it is not quite true that *nobody* works during a blackout. The economic function of those who do work—municipal workers and caregivers—is rendered visible as the panic of a day without profit and a day with life-threatening shortages become identical; such labor is increasingly considered "essential services," rendered exceptional before the law.

one another; the aesthetic genre of the electrified city mutates temporarily into a postwork naturalism. Labor power is returned to the worker as collective labor, indistinguishable from the reproductive labor otherwise keeping the whole thing afloat. Objects become either useful for labor or not—a tyrannical and ruthless materialism inverting the class tyranny under capitalism.

Circuit breakers and aqueducts are more likely found in engineering histories than the art historical canon of architecture studies. The beaux-arts inheritance of buildings makes them primary sites for an archive of cultural historicity, not unlike painting or poetry. Theirs is a humanities standpoint, given their universal function: to house people. Yet infrastructure is what houses all the stuff in between: it gives time to moments between production, circulation, and consumption—indeed it makes time the unitary medium across which all three become instances of one another. Infrastructure is the most immediately historical of any medium for describing the proximity of politics and economics or culture and WORK. Though there is as yet no critical theory of infrastructure, its rapidly growing status as an object of capitalist desire, and our increasing reliance on its vital, cultural, and political function, will demand more of our collective attention.

See also: CRISIS, DAMS, DISASTER, ENERGOPOLITICS, LIMITS, WORK.

Innervation

Robert Ryder

No imagination without innervation.
—WALTER BENJAMIN, "One-Way Street"

Innervation is nowadays a predominantly neurophysiological concept, generally used to refer to an anatomical detail: "the route of the nerve on its way to a given organ" (Laplanche and Pontalis 1988, 213). But in the field of psychology, the term has a more turbulent history. Often thought in tandem with kinesthetics, innervation can be found in early psychological studies of the articular, tendinous, and muscular complexes (Baldwin 1960, 549). Since its inception, innervation has generally been regarded as a mode of energy transfer or conversion—and therefore a process of stimulation, in stark distinction to its current neurophysiological definition as part of the anatomy. But exactly what kind of ENERGY is being converted, in which direction, and to what ends have been hotly debated since the days of psychologists Wilhelm Wundt (1832–1920) and William James (1842–1910). The history of innervation in psychology thus acts as a barometer for gauging the history of the psychophysiological problem (i.e., whether psychology can or should be done with or without physiology).[1]

The term was also taken up by several film and cultural theorists of the early twentieth century, most notably Walter Benjamin (1892–1940). Innervation in MEDIA and cultural studies today often takes its point of departure from Benjamin's peculiar use of innervation

1. Freud also makes limited use of the term *innervation*. See Laplanche and Pontalis's entry (1988, 213), as well as Hansen's comparison of Freud's and Benjamin's interpretations of the term in *Cinema and Experience* (2012, 135–37).

and his theory of film in particular. Miriam Hansen's analysis, both in "Benjamin and Cinema: Not a One-Way Street" (1999) and in her final book, *Cinema and Experience: Siegfried Kracauer, Walter Benjamin, and Theodor W. Adorno* (2012), is the most incisive reading of Benjamin's use of the term. With a nod to Hansen,[2] I turn first to the origin of the term in psychology following Wundt and James and then to Benjamin's reframing of it from a psychophysiological into a technocultural sense with social and political implications. These formative interpretations of innervation illustrate the extent to which energy—as that which is being rerouted—itself fluctuates among electrophysical impulses, human consciousness, and experimental metaphor.

Although energy transference between physical and affective impulses was a topic under discussion throughout the nineteenth century, the terms *innervation sensation* (*Innervationsempfindung*) and *innervation feeling* (*Innervationsgefühl*) were coined by the father of experimental psychology, Wilhelm Wundt, around 1863. With these terms, Wundt meant specifically efferent or "out-flow" impulses, that is sensations that "derive uniquely from the central innervation of the motor organs, and their origin would therefore be central rather than peripheral" (1904, 182). He developed this notion to counter the term *muscle sense*, which, while employed by a number of French, German, and English psychologists since the beginning of the nineteenth century, became restricted by the second half of the century to refer to afferent or "in-flow" impulses from the "peripheral" muscles toward an inner "feeling of effort"—the will. While for Wundt "feelings of innervation" are not the only sensations contributing to the feeling of effort—there are at least three (Scheerer 1989, 41)—he was nevertheless sharply criticized by William James for entertaining their very existence. James dismissed its efferent direction as specified by Wundt; he instead maintained "that the feeling of *muscular* energy put forth is a complex *afferent* sensation coming from the tense muscles, the strained ligaments, squeezed joints . . . etc." (James 1880, 4). As Hansen writes, James was not arguing against the concept of innervation as such but "against the idealist notion of innervation being coupled with, and depending on, a sentient, conscious 'feeling' rather than the physiological fact of 'discharge into the motor nerves'" (Hansen 2012, 324). At the crux of the debate are two differing views of energy: was it a physiological "feeling of muscular energy" being transposed into nervous impulses that eventually lead to emotions (James), or did it derive from conscious "feelings" then expressed via the motor organs (Wundt)? Whether the origins of energy are interpreted here as concrete and literal or affective and metaphorical determines how one reads innervation.

While it remains unclear whether Walter Benjamin borrowed the term from Wundt, Freud, or others, Miriam Hansen and Susan Buck-Morss agree that innervation for Benjamin is less a defense mechanism than "a mimetic reception of the external world, one that is empowering" (Buck-Morss quoted in Hansen 2012, 137). Hansen makes the further claim "that Benjamin . . . understood innervation as a *two-way* process or transfer, that

2. See also Anne Rutherford's entry on mimetic innervation in the recent publication of *The Routledge Encyclopedia of Film Theory* (2014).

is, not only a conversion of mental, affective energy into somatic, motoric form but also the possibility of reconverting, and recovering, split-off psychic energy through motoric stimulation" (Hansen 2012, 137). While Benjamin's use of the term is idiosyncratic, one element it shares with previous formulations is that innervation always signals some relation to a body or body part. But when reading Benjamin, *body* should never be assumed to be a *human* body or to function as a complete whole or univocal container of the self. Benjamin associates innervation with functions or parts of bodies and their singular roles in memory and habit formation, a fascination that he maintained up until his last published essay, "On Some Motifs in Baudelaire" (2003 [1939]). Innervation as a concept lends itself to Benjamin's inclination to address bodily experience not traditionally as a container that delimits one object from another but as a surface, opening, or volume that includes both outside and inside, both "peripheral" motor and "inner" psychic energies. If innervation for Benjamin is indeed "a *two-way* process or transfer" (Hansen 2012, 137), this makes the debate about whether innervation is efferent or afferent moot and requires us to think that the energy in question must derive from motor, psychic, and mnemonic sources, none of which can be clearly demarcated. In other words, an energy's source or ontology is not as much at issue for Benjamin as its malleability or transferability.

This includes energy's transferability into the metaphorical realm. Since for Benjamin *body* never simply means the human body, innervation and the energy it transfers are also no longer limited to the individual physical body but can be expanded to include the collective. Benjamin introduces what might best be called a politics of innervation when he writes at the end of his essay "Surrealism" (1929), "The collective is also bodily [*leiblich*]. . . . Only when in technology body [*Leib*] and image-space so interpenetrate that all revolutionary tension becomes bodily collective innervation, and all the bodily innervations of the collective become revolutionary discharge, will have reality surpassed itself to the extent demanded by the *Communist Manifesto*" (Benjamin 2001, 217–18). Once again, technology sets the stage. But how? Its role in "bodily collective innervation" becomes somewhat clearer when, in his unfinished *Arcades Project*, Benjamin refers to "the idea of revolution as innervation of the *technical organs* of the collective (analogy with the child who learns to grasp by trying to get hold of the moon)" (Benjamin 2002, 631, emphasis added). Benjamin imagines here a collective body whose organs are not technical devices per se but social organs that develop in direct association with technology, which includes a mode of appropriating that technology. The development of such sociotechnical organs, however, is necessarily a slow learning process, one that, like the child who "learns to grasp by trying to get hold of the moon," will include gestures of motor-perceptual miscognition. But as Hansen writes, "for Benjamin . . . this miscognition fuels creative and transformative energies, anticipating an alternative organization of perception that would be equal to the technologically charged environment" (Hansen 2012, 143). We have strayed far from James's physiological rendition of innervation when one such transformative energy for Benjamin is revolutionary action.

Benjamin injects collective receptivity and technological transformativity into an otherwise psychophysical term, rerouting *innervation* into a critical term for cultural studies. In doing so, he opens up pathways for thinking about energy as both physiological and metaphorical, with less emphasis on its source than on its transferability qua innervation. Put another way, with Benjamin's version of innervation, energy becomes not simply metaphorical but allegorical: it is a nexus through which physiological and psychic stimulation as well as technological reception and revolutionary action innervate each other, thereby fueling imaginative ways to interpret our developing social organs as they interact with and appropriate different technologies.

See also: AFFECT, COMMUNITY, ELECTRICITY, EMBODIMENT, METABOLISM.

Kerosene

Mark Simpson

> All the yard-arms were tipped with a pallid fire; and touched at each tri-pointed lightning-rod-end with three tapering white flames, each of the three tall masts was silently burning in the sulphurous air, like three gigantic wax tapers before an altar . . . . "Aye, aye, men," cried Ahab, "Look up at it; mark it well; the white flame but lights the way to the White Whale! . . . Oh! thou clear spirit of clear fire . . . I now know thee, clear spirit, and I now know that thy right worship is defiance . . . . Oh, thou clear spirit, of thy fire thou madest me, and like a true child of fire, I breathe it back to thee."
>
> —HERMAN MELVILLE, *Moby-Dick*

The "Candles" chapter in *Moby-Dick*, which initiates the narrative's final drive toward CATASTROPHE, asks us to confront and reckon the metaphysics of fuel. The *Pequod* is battered by a typhoon—a result, thinks Starbuck, of Ahab's reckless and relentless chase after Moby-Dick—when Saint Elmo's fire illuminates the lightning rods atop its masts and the "keen steel barb" of Ahab's harpoon. Where Stubb strives to find in these "'corpusants . . . a sign of good luck,'" portending a hold filled "'with sperm-oil,'" Starbuck sees only DISASTER: electric evidence of divine opposition to the ship's "'ill voyage.'" For Ahab, though, the corpusant flame illuminates the leviathan's pursuit—and so fires his monomaniacal fixation. Hence his ominous declaration that the "'right worship'" of the "'clear spirit of clear fire . . . is defiance'"—a defiance immediately enacted when, seizing his "burning harpoon" like a torch, "with one blast of his breath he extinguished the flame." Meant to galvanize his crew in their quest for the whale, Ahab's defiance succeeds only in causing "many of the mariners [to] run from him in a terror of dismay" (2007, 442–44).

Critics tend, with good reason, to take this episode to announce the novel's climactic disaster in the play of light and dark: extinguishing the uncanny, divine flame of the corpusants, Ahab guarantees the *Pequod*'s descent into the ocean's dark "vortex" (ibid., 449). But what meanings might emerge if we refract the material (rather than metaphysical) poetics of fuel? Approached this way, catastrophe looms less in the plunge from light into dark than in the intensification of light itself: an intensification—"clear spirit of clear fire"—already underway when Melville published his novel in 1851 and spurred during the ensuing decade by the chemical and financial typhoon named kerosene. At stake is the premonitory, impossible knowledge of another sort of emergency altogether—what, following Patricia Yaeger, we might term the novel's energy or fuel unconscious (2011, 306): its capacity to intuit the incipience of kerosene, the clear spirit already present in the ENERGY REGIME of the early 1850s that would, by decade's end, effectively displace whale oil as illuminant and industrial lubricant and thereby help to obliterate WHALING as an industry.[1] In a narrative inaugurated by loomings and saturated with apocalypse, one of the all-but-impalpable loomings involves the imminence of the kerosene era (and conceivably the immanence of a kerosene imaginary?) as it would contribute to the apocalypse of whaling.

"Kerosene" is at once the trademark name—secured by Nova Scotian geologist and entrepreneur Albert Gessner in several patents in the early 1850s—and the common term for fuel (whether gas or oil) distilled and refined from a range of bituminous hydrocarbons. The nominal slippage between the proprietary and the generic suggests the conceptual and cultural volatility of kerosene as fuel. As Paul Lucier explains in his cultural history of COAL and oil consulting in nineteenth-century America, kerosene triumphed over its market rivals—whale oil but also linseed oil, olive oil, rapeseed oil, lard oil, turpentine, and camphene—"because it offered three advantages . . . : safety, brightness, and cost" (2008, 153). Yet the very stability of kerosene's chemical properties seems, in retrospect, inextricable from its explosiveness as commercial property: kerosene was the object of an intensive BOOM, craze, or fever through the mid- to late 1850s, with major manufacturers increasing from three to as many as seventy-five between late 1857 and early 1860. And it became just as quickly the casualty of a precipitous bust, triggered in 1860 by a "coal oil glut" alongside "rising costs of cannel coal," that provoked widespread bankruptcy (ibid., 161). "In the boom-and-bust frenzy," observes Lucier, "Kerosene was . . . transformed from a patented trademark into a household product. In America, kerosene (little k) became the generic description for all mineral-based lamp oils, including those manufactured from a recently discovered raw material called petroleum" (ibid., 161).

Arguably, those most able to profit in the longer term from the volatile career of kerosene in this era were scientific experts, repeatedly called upon to consult in court cases over control of manufacturing science and procedure, a conjuncture that helped to establish the hegemony of scientific expertise itself in regulating, indeed producing, ENERGY as industry (Lucier 2008, 162–85). Kerosene culture here offers another kind of incipience: that which anticipates the modes of expert knowledge crucial to the consolidation and occlusion of political power in the petro-fuel economy (T. Mitchell 2011, 109–43). While we might associate kerosene's "clear spirit of clear fire" with an emerging narrative realism around 1850—one devoted to sharp outlines and illuminated interiors—the more

1. As Lance E. Davis, Robert E. Gallman, and Karin Gleiter observe, "by 1860 the whale fishery, which had been far and away the most important supplier of illuminants and lubricants in 1850, retained only tiny fractions of those markets, both at home and abroad. The fishery was almost as large as it had ever been, but the overall market was very much larger than it had ever been. Furthermore, the use of coal to produce gas and oil [that is, kerosene] presaged the extinction of the whale fishery, given the American reserves of cheap coal" (1997, 368). See also Dolin (2007, 335–41).

pressing correlate involves speculation and credit: the power, in an ideology and aesthetic of expertise, to prise apart apprehension from knowledge and understanding, so as to capitalize on resulting investments of unknowing trust and faith. Channeling the obscurity of expertise, perhaps kerosene fever involved its own descent into darkness after all.[2]

In the 1850s, kerosene was residual and emergent simultaneously: a rediscovery of centuries-old methods of bitumen use (especially for illumination) yet also an innovative refinement and reapplication of those methods within the social and economic configurations of industrial modernity (see Yergin 1991, 22–25). Thus understood, kerosene seems like an interstitial resource—a threshold fuel crucial to, yet superseded in, the passage from whale oil and coal to petroleum and ELECTRICITY. But even if outmoded by subsequent developments in fuel and energy, kerosene nevertheless persists deep into the petrol era,[3] burning on today as a staple source of heat and light for millions globally who lack access to basic electricity and as a key component in the jet fuel that fires the modern aviation industry. This jarring dichotomy organizes kerosene's obdurate afterlife while also illuminating the violent asymmetry of the distribution, among those many neglected and those few privileged within the global order, of resources in energy, mobility, and prosperity—a violent asymmetry looming within the hydrocarbon era from the outset.

See also: ANIMAL, CORPORATION, ENERGY REGIMES, FICTION, PETROREALISM, WHALING.

2. Published on the cusp of the kerosene craze, Melville's 1857 novel *The Confidence-Man* viciously examines the emerging ethos and aesthetic of credit in AMERICA in this moment. As Lucier shows, "the so-called Coal-Oil Man" was a recognizable type of con man: "gullibility and respect produced the same effect, namely, the acceptance of authority, regardless of its authenticity. Confidence men merely exploited the public's predisposition to defer to those who professed a knowledge of technological science" (Lucier 2008, 160).

3. Yaeger observes, "looking at the 'ages' of energy will never be a tidy endeavor, since fuel sources interact" (2011, 308).

Lebenskraft

Alice Kuzniar

In considering contemporary cultural discourse about ENERGY, we should not forget a different discursive usage in late-eighteenth- and early-nineteenth-century Germany, when the concept of *Kraft* (power, energy, force) was current in the life sciences, referring to organic generative and regenerative processes. *Kraft* and *Lebenskraft*—a vital life force—accrued a range of meanings whose ultimate demise coincided with the technologization and instrumentalization of so-called natural resources—COAL and oil—used in powering life. It was as if the formerly inner, organic propulsion of life became exteriorized in the shift to fossil fuels: energy was generated from outside rather than from within oneself. With this exteriorization, earlier understandings of the connection between life and energy drained away.

Coined by physician and botanist Friedrich Casimir Medicus in 1774, the term *Lebenskraft* quickly gained widespread use, although it was preceded in 1772 by Paul-Joseph Barthez's concept of *principe de vie*. Johann Gottfried Herder extolled the powers (*Kräfte*) that underlay human and ANIMAL life. "Physiology," he writes, "of the human or any animal body [is] nothing but a *realm of living forces* . . . . Everything that we call matter is therefore more or less in itself enlivened; it is a realm of active forces that according to their nature and their relationships form a complete whole [*ein Ganzes bilden*]" (1994, 774).[1]

Lebenskraft was debated among physiologists, naturalists, philosophers, and physicists—from Johann Friedrich Blumenbach and Gottfried Reinhold Treviranus to Johann

1. All translations are by the author.

Wilhelm Ritter. But *Lebenskraft* raised more questions than it answered. Was vitalism to be explained as irritability (demonstrated in the muscles) or sensibility (demonstrated in the nerves), as medical specialists after Albrecht von Haller were to investigate? Blumenbach and Treviranus investigated infusoria and zoophytes under the microscope in order to trace what was invisible to the naked eye and yet underlay processes of generation, which Blumenbach would term *Bildungstrieb* (1781) or *nisus formativus* (1787). Not restricted to generative powers, Ritter (1798) and Alexander von Humboldt (1798) studied vital life forces in the guise of electrical impulses in order to determine whether they were chemical or magnetic in nature. And how were these forms of energy related to human physiology? By 1801, the discovery of a galvanistic fluid was being linked to *Lebenskraft*; even as late as 1820, animal magnetism formed the basis of the physiology of Georg Prochaska.

For all these writers, *Lebenskraft* was a postulate to be confirmed by observational inquiry. Although the belief in vitalism was premised on hiddenness in nature, it invited experimental intervention so as to make its unseen process evident. Because *Lebenskraft* could not be pinpointed, scientists formulated rules that governed it, as if these would then provide the key to NATURE. *Lebenskraft* represented the EMBODIMENT of nature, yet it was an abstraction from nature. It was an undefinable term that explored, as Joan Steigerwald notes, the "border zones of life" between plants and animals and between the lifeless and living (2014, 106). Despite this uncertainty, however, through *Lebenskraft* the entire organic and inorganic world shared the same substantiality. Nature was seen as a whole, and one life force permeated all of it. As Herder writes, "we do not have the senses to examine the innermost being of things; we stand on the outside and must observe. The more clear-sighted and still our gaze, the more the living harmony of nature reveals itself to us" (1994, 778).

The most famous medical proponent of *Lebenskraft* was Christoph Wilhelm Hufeland, physician to Goethe. Hufeland's most significant work was *Macrobiotics, or the Art of Prolonging Life*, published in 1794 and still in print today. Hufeland enumerated eleven characteristics of *Lebenskraft*, including the body's capacities to react to environmental stimuli and to fight against destructive influences such as decay and frostbite, as well as the presence of *Lebenskraft* within every part of the body. To guarantee longevity, *Lebenskraft* had to be regulated, not subjected to excessive physical or mental stimulus. In line with eighteenth-century anthropological medicine, regulating *Lebenskraft* involved for Hufeland both moral and physical realms. He understood it as an objective reality yet paradoxically beyond human sight and comprehension. He compared *Lebenskraft* to other physical powers (gravity, electricity, magnetism), "which, at bottom, signify nothing more than the letter that expresses the unknown quantity in algebra. We must, however, have expression for things whose existence is undeniable though their agency be incomprehensible" (1979, 25).

Yet Hufeland would soon qualify his claim about the undeniable existence of *Lebenskraft*. In an article, "My Concept of the Vital Force," Hufeland (1798) posits *Lebenskraft* as merely a cipher or heuristic term: it designated the inner font of vitality and regeneration in the body, but it was ultimately unknowable. Others began to express their skepticism as well. In his essay on *Lebenskraft* in 1795, the physician Johann Christian Reil argues that such forces have no final, absolute grounding in experience (*Erfahrung*); hence we cannot ascribe material existence to them (1910, 2). *Lebenskraft* is a subjective concept, the form in which we imagine the connection between cause and effect (ibid., 23). Kant, too, in the third *Critique*, acknowledges the necessity of positing a "self-formative force [*in sich bildende Kraft*] which cannot be explained solely by the capacity for movement (mechanism)" (1983, 486). But he also calls this force that synthesizes the parts of an organism an "inscrutable property" (ibid.); it serves merely as analogous to life (ibid., 487). It must remain a regulatory concept, not constitutive of being (ibid., 487). Kant acknowledges that an organized being is not merely a machine, but we have to rely as much as possible on mechanistic explanations of how bodies work, otherwise we engage in lazy reasoning.

Such warnings, however, went largely unheeded for half a century. For instance, Arthur Schopenhauer in *On the Will in Nature* (1836) searched for the term *Wille* in Haller, Treviranus, and writings on mesmerism and plant physiology, because he equated it with a life force: "life is the appearance of the will [*Erscheinung des Willens*]" (1962, 407). Because *Lebenskraft* still had such currency, by the late 1830s the prominent physician and landscape artist Carl Gustav Carus wrote exasperatedly that physiology was burdened by abstruse conceptions about life (1838–1840, xi). He deemed problematic the understanding of *Kraft* as something objective rather than an operational abstraction (ibid.). Even the late-nineteenth-century German zoologist Otto Bütschli used the terms *Lebensenergie* and *vis vitalis* (1901).

Today the sole afterlife of the concept *Lebenskraft* arguably lies in its use in homeopathy. Prior to the third edition of his major work, *The Organon of the Healing Art*, the founder of homeopathy, Samuel Hahnemann, did not use the term *Lebenskraft* extensively. But, to emphasize nature's independent capacity for agency and self-healing, by 1829 he rephrased *Leben* and *Natur* as *Lebenskraft*, *Lebensprincip*, *Lebens-Energie*, and *Lebens-Erhaltungs-Kraft*. The vital force can be retuned with the assistance of homeopathy. Hahnemann's understanding of vitalism goes beyond that of the physicians of his time, who restricted their discourses to human health, and approximates the position of the physicists such as Ritter, who saw *Lebenskraft* as permeating all matter. Hahnemann trusted in the spirit-like animating force in the remedy that had the capacity to heal. Hahnemann successively diluted tinctures in order to exponentiate this dynamic action. More than any other physician of the time, Hahnemann thus brought the vitalism of the botanical realm (later in life he was to include the mineral) to bear on the medicinal cure. Hahnemann writes: "everything in nature lives and is force [*Kraft*]; we must only know how to bring it to life and to develop its power" (2001, 726).

By ending with homeopathy, my point is not to advocate an anachronistic return to vitalistic notions. Homeopathy needs to be seen as a cultural product of its time. But it is worthwhile to note the resurgence of Romantic tenets, whether in the form of homeopathy's popularity today or in academic discourse on a Deleuzian vibrancy inherent in matter. These tendencies indicate a desire for thought and practice that pose an alternative to the hegemony of "unnatural" relationships to nature, including external energy sources not capable of the renewal previously ascribed to *Lebenskraft*.

See also: AFFECT, ELECTRICITY, ENERGY, GRIDS, INNERVATION.

Limits

John Soluri

When Chilean Army Captain Arturo Fuentes Rabé arrived in Tierra del Fuego on official business in 1918, he rented a car—"a small, old Ford"—for a "rather steep" price (Fuentes Rabé 1923). Within a decade of the first Model T rolling off an assembly line in Detroit, automobiles had reached the southernmost part of the Americas, introducing a defining element of twentieth-century petrocultures to a place that a generation earlier had been the territory of an indigenous foraging/hunting society. Automobiles appear frequently in Fuentes Rabé's account of his travels in Tierra del Fuego, but unlike other recent studies that view cars as a kinetic, transformative force in urban Latin America, Fuentes Rabé's narrative underscores the improvisational character of AUTOMOBILITY in a region where ROADS were largely absent (Errázuriz 2010; Giucci 2012; D. Miller 2001; Wolfe 2010). Roadways are not unique to petrocultures, but petrocultures may be uniquely dependent upon them, a point so obvious that it can be easily overlooked until one ventures "off-road."

When Fuentes Rabé arrived in Tierra del Fuego, sheep ranches (*estancias*) dominated the landscape and economy. Livestock far outnumbered the region's human population, and both sheep and people tended to migrate seasonally. Fuentes Rabé's tour included visits to several estancias. On his first such visit, he traveled in a horse-drawn coach to meet a Chilean rancher named Quintana. Nearing the ranch, Fuentes Rabé heard a loud noise and spied the approach of a "strange vehicle" (1923, 1:121). The coach driver explained cryptically, "It's the 'airplane' of Mr. Quintana." Fuentes Rabé soon learned that the "airplane" had been a "little luxury car" that wore out because of the rough conditions. Quintana fabricated his own replacement parts "adapted to the climate and terrain." After describing the modifications in detail, Quintana concluded with pride, "it is a work entirely of my own" (quoted in Fuentes Rabé 1923, 1:122). In this anecdote, the car's exoticism is signaled by its initial denomination as an "airplane," but the story is about neither speed nor alien technology's transformative power. Instead, the car as material object undergoes a process of creolization in response to Tierra del Fuego's (virtually roadless) environment, transforming the vehicle into something neither indigenous nor foreign.

Adaptability and RESILIENCE are central themes in Fuentes Rabé's account of his travel in the rented auto. The trip begins inauspiciously when the vehicle stalls while crossing a river. The experienced chauffeur pulls the vehicle from the river with an improvised pulley and the journey continues only to be impeded shortly thereafter by a very steep ridge. The passengers disembark and make an arduous hike to the summit; the auto eventually reaches the top by the force of pulleys and motor (Fuentes Rabé 1923, 2:15). As the journey continues, Fuentes Rabé extols the virtues of the "valiant Ford," particularly its agility: "The road narrows and enters a pass; innumerable streamlets fall from the hillsides and zigzag their way across the road. The agile Ford avoids these constant obstacles, its four wheels sinking from time to time into small bogs. What a well-adapted machine for these remote regions!" (ibid., 2:15). Significantly, the auto is almost always identified with its US manufacturer, in sharp contrast to the never-named chauffeur whose presence diminishes over time as he is discursively displaced by repeated references to "the machine" (*la maquina*).

Fuentes Rabé does not reveal any anxieties over the machine's presence in the "garden" of Tierra del Fuego. For example, when attempting to cross a long stretch of sandy beach, he portrays the Ford as a diminutive yet resilient machine that yields, but does not succumb, to the forces of NATURE:

> The Ford had to work like the devil to free itself of the sand. The waves are enormous and surge frighteningly fast; many times, the little machine had to yield before the massive force, superior to its power . . . . But, the Ford has never been lost . . . recovering its strength, it sheds the sand that covers it and continues its march undaunted. (Fuentes Rabé 1923, 2:132)

But not all automobiles enjoyed such good fortune in Tierra del Fuego. Passing through a particularly treacherous stretch of swampy ground, Fuentes Rabé happens upon a "mute witness" of the dangers confronting drivers: the remains of a strange-looking chassis mounted on the wheels of a Ford (ibid., 2:95). He later learns from local inhabitants that a German mechanic had built the vehicle, dubbed the "Chug-Chug" by his bemused neighbors. On its first trip, the vehicle ran out of fuel; undaunted, the mechanic declared, "Give me some gas and I'll make it over the ridge this time" (ibid., 2:96). However, on his second journey, he lost his way and drove into a swamp. A few days later, the mechanic was spied carrying the motor in his arms.

Taken together, these stories demonstrate the uncertainty of automobility in the absence of roads. They offer little sense of speed or even steady movement. Far from annihilating space, automobile travel makes Fuentes Rabé acutely sensitive to changes in terrain.

His account is filled with instances when landscape features—including bogs, rivers, hills, and beaches—inhibited the Ford's movement. If not for the periodic interventions of the anonymous chauffeur and the passengers, the machine would have utterly failed. Indeed, much of the praise showered on the "excellent Ford" referred to the ease with which it could be lifted or pulled when the motor gave out. In the case of the Chug-Chug, the final IMAGE of the mechanic carrying his engine inverts the very idea of a motorcar facilitating human transit.

These stories about early automobiles in Tierra del Fuego are simultaneously stories about roads (or the lack thereof). Even the tale of the Chug-Chug was not intended as a commentary on the mechanic's poor engineering skills but rather the deceptively treacherous road conditions. In fact, the abandoned remains of the auto that had piqued Fuentes Rabé's curiosity transformed into a signpost of sorts (*objeto demarcador*; Fuentes Rabé 1923, 2:96) signaling a transition in road conditions. The first time Fuentes Rabé and his entourage came upon a "surprisingly" good road maintained by a large sheep company, he wrote at length about the "general clamor" over the lack of maintained roads and bridges. He drew sharp contrasts between the "wide," "concrete-like" roads maintained by sheep companies and the narrow, bumpy ones neglected by state authorities. He urged the Chilean government to invest in road construction and maintenance in order to improve conditions for migrant workers and to create economic opportunities for small landowners who could not afford to build and maintain their own roads (ibid., 2:17–18).

Given the historical and geographical contexts, Fuentes Rabé's experiences with automobility are not terribly surprising. His admiration (bordering on wonder) for the Ford was in keeping with Latin Americans' fascination with Henry Ford in the early twentieth century (Giucci 2012, 1–50). In Santiago, Chile's largest and wealthiest city, the number of automobiles was rising sharply (Errázuriz 2010, 374). Fuentes Rabé's call for public investment in road building was not unique either. In the United States, state governments were just beginning to develop tax schemes to finance the construction of highways in the 1910s; the federal government would not get involved until even later (Sabin 2005). In Tierra del Fuego, the rapid expansion of the wool trade created both the wealth and trade NETWORKS necessary to import a small number of automobiles, but it did not give rise to the INFRASTRUCTURE necessary to turn gasoline-powered automobiles into a routine form of transit.

In contemporary petrocultures, roadways are ubiquitous, forming an essential infrastructure that has become so naturalized as to draw attention primarily in their absence or deterioration. Planes, trains, and automobiles do not so much annihilate space or time as they reconfigure them in ways that are uneven, inconsistent, and highly dependent on elaborate infrastructures that constantly deteriorate from use and environmental forces. Indeed, today, nearly a century after the Model T reached Tierra del Fuego, one cannot drive from Motor City to the End of the World, suggesting that there are political, social, and ecological forces that continue to limit the immense transformative power of petroleum in Latin America and beyond.

See also: AMERICA, AUTOMOBILE, EXHAUST, INFRASTRUCTURE, RESILIENCE, ROADS.

Media

Lisa Gitelman

W. J. T. Mitchell and Mark B. N. Hansen's *Critical Terms for Media Studies* (2010) includes entries for *new media* and *mass media* but not for plain old *media*, while, in an essay published the same year, John Guillory outlines "The Genesis of the Media Concept," noting the several centuries across which the concept of media was "absent but *wanted*," until the new communication technologies of the nineteenth century may be said to have arrived to beg the question (Guillory 2010, 321; Durham Peters 2001). Not until the twentieth century did the term *media* emerge fully in what Raymond Williams calls its "technical" sense, designating that parade of technological forms—the telegraph, telephone, phonograph, cinema, radio, TV—whose intensive capitalization and widespread familiarity would eventually help to institutionalize "the Media" in common parlance and prompt the study of media within the academy (R. Williams 1983, 203–4). Media, it would seem, have forever been tricky to apprehend concisely as such, while developments in that middle distance of the later nineteenth century seem to have been crucial. I first encountered the parade-of-technological-forms version of media when, fresh out of graduate school, I joined the staff of the Thomas A. Edison Papers Project. Telegraph, telephone, phonograph: that is how the Papers Project kept track of the Edison subject matter we published, but the same index terms kept coming up in our sources too, among them magazines like *The Electrician* (1878), *The Electrical Review* (1882), and *The Electrical World* (1883), this last punningly self-described as "a weekly review of current progress." Even though the earliest phonographs were mechanical rather than electrical, the parade-of-forms version of media seems to have arrived coincident with a self-consciously electrical age.

Before *media* earned its new technical sense, the term referred generally to any in-between agency or substance. Early US letters patent, for instance, specifies things like filtering media for rendering solutions less particulate. Likewise, patents mention nutrient media, insulating media, conductive media, grinding media for lenses, and the actuating mechanical media of levers and gearwork, as well as light-sensitive media for photographic processes. The jump from *media* in this old, general sense to the new world of electrical communications would involve some ingenious component devices and new NETWORKS of connection, but it further required the addition of content—meaning, messages, information, signal, and noise—absent in any appeal to filters, lenses, gears, or the like. Today, under the influence of computing and its "new media," one might call this crucial addition "the content layer," but to nineteenth-century ears that figure would have made little sense. Better instead to think of a charge, a jolt, an electrical impulse, a live wire. The new world of mediated communications was powerfully wired. By the 1880s, so many telegraph lines connected to the New York Stock Exchange that one observer noted, "No bird could fly through their network, [and] a man could almost walk upon them." Darkening the street below, this thicket of wires confirmed at a glance that Wall Street was "the focus to which all currents of American progress and energy converge" (Stedman 1905, quoted in Hochfelder 2013, 101). The devastating blizzard of 1888 would help to teach New York City the prudence of burying wires rather then stringing them atop poles, but wiring hardly slowed.

In 1884, *The Electrical World* declared "The Age of Wire," acknowledging its indispensability to electronic communications amid the sudden proliferation of other uses, everything from barbed wire fencing to the new suspension bridges, from elevator and streetcar cables to mattress springs and clockwork ("The Age of Wire" 1884, 106). So powerfully did live wires and electric cables take hold of the popular imagination as a feature of modernity that the possibility of instantaneous communication without them seemed all the more remarkable, whether encountered as the question of spiritualist communication or as the seeming miracle of wireless telegraphy. In 1897 Guglielmo Marconi called it "Signaling Through Space Without Wires." Over the next decades, wirelessness would help to point anew at the wire-fullness of modernity, so much so that *The Wireless Age* (1913), one might say, became the *Wired* (1993) magazine of its day.

See also: ELECTRICITY, GRIDS, INFRASTRUCTURE, SPIRITUAL.

Mediashock

Richard Grusin

More than a decade after 9/11, the networked world remains in an acute state of "mediashock." At the first sign of meteorological turmoil, social unrest, financial turbulence, or natural cataclysm, news media shift into 24–7 crisis mode, generating on-the-ground reports, live updates, multiple commentaries, and breaking news. CNN pioneered this mode in global cable news as far back as the 1980s, but the media's obsession with remediating disaster and premediating shock has intensified in the twenty-first century, jump-started by the events of 9/11 but escalating since then. With the exception of regularly scheduled events like the Olympics or World Cup, national elections, or award ceremonies, the most intensive media coverage occurs in response to crises or disasters that operate according to their own temporality—whether dramatic falls on the world's financial markets; hurricanes, earthquakes, volcanoes, or tsunamis; geotechnical accidents like mine explosions, oil SPILLS, floods, or NUCLEAR meltdowns; violent terrorist acts like suicide bombings, assassinations, or hijackings; or political upheavals like strikes, demonstrations, occupations, or riots. The immersive impact of mediashocks is felt even more quickly in the age of social media; now it is not only the traditional news outlets that are spurred to action in a crisis but also Twitter streams, Facebook feeds, YouTube, e-mail, blogs, and instant messages that multiply exponentially in (or as) the aftermath of such events.

Contemporary mediashock has four main aspects: the preoccupation with CRISIS and DISASTER; the physical shock that MEDIA exerts on the human bio-organism; the effect of media on public norms or collective affective formations; and the geophysical, geopolitical, and geoaffective shock to the global assemblage of humans and nonhumans, of social, technical, and natural actants. In each case, the affectivity and technicity of media themselves are related to the affectivity of the crises they mediate. These geotechnical media events have a hybrid ontology: not natural, social, or technological, they emerge as complex assemblages, new kinds of actants that are related, but not finally reducible, to the explosion of new information and media technologies in the past few decades.

Consider, for example, the physical impact on our mediatic system of a catastrophic event like the 2011 Japanese earthquake, tsunami, and nuclear meltdown. Not only did the Tōhoku earthquake produce numerous aftershocks in the earth's crust, generate several tsunamis, and cause serious damage to the Fukushima Daiichi nuclear plant; it also generated affective shocks in the mediatic system. These affective shocks fueled media effects that we might understand as mediatic aftershocks or tsunamis that reverberated through global print, televisual, and socially networked news media. Bruno Latour's (1986) conception of the transmission of "immutable mobiles" can help to explain how the Tōhoku earthquake generated media aftershocks through a direct chain of inscription and translation. The seismic waves produced by the quake were registered by seismometers connected to seismographic instruments that translated these seismic waves into inscriptions that could be read by scientists, compared with other seismographs, and transmitted through the networked communication infrastructure to academic, governmental, and media institutions. Such media aftershocks were then incorporated into news reports in print, televisual, and networked media, fueling geoaffective shocks that were connected back through a chain of "immutable mobiles" to the shaking of the earth produced by the subduction of the Pacific Plate.

This sequence of transmission is not mechanistic. Mediashock emerges both from individuals who tweeted, texted, e-mailed, Flickred, or updated their Facebook status and from the agency of the news reports, tweets, texts, e-mails, photos, or status updates themselves. Mediashock is not identical with this chain of immutable mobiles but rather names the agential intensity or event that comes into existence through this mediatic agency. Like all human and nonhuman agency, mediatic agency is distributed and heterogeneous, interwoven among humans and nonhumans, organic and technical actants and NETWORKS. Our mediatic system is a complex assemblage of state, corporate, and informal media; of technical infrastructure, finance capital, and affective labor; of bodies, screens, and devices.

The links between the Tōhoku quake's seismic aftershocks and its media aftershocks demonstrate how the media system itself is physically and affectively affected by crises and how media function as active translators and mediators of physical shocks. Media transform and remediate the shock of geophysical disaster into new media formations that themselves work to modulate individual and collective affectivities of people and things across a widely networked and hypermediated world that overwhelms us with the force and scope of its flows. In the disastrous aftermath of the Japanese quake and tsunami, these flows worked very much like media aftershocks caused by the initial quake. And of course the earthquake generated numerous other aftershocks, producing additional smaller quakes and generating additional shocks in the realms of economic, social, cultural, technical, and ecological systems.

Why should the mediation of disaster in contemporary media take the form of mediashock? Perhaps because we find ourselves subject to global assemblages that move, change, and grow according to logics, trajectories, or forces that exceed the control or mastery of humans or nation-states. Throughout history, these forces have been primarily natural: storms, earthquakes, floods, tornadoes, hurricanes, drought, avalanches. Such events are known in the legalese of insurance contracts as "acts of God." But today such catastrophic events are increasingly generated by human and technical agency or, as Naomi Klein (2007) might say, by the forces of disaster capitalism. Fostered by the complex sociotechnical assemblages of twenty-first-century finance capitalism, phenomena and events like global climate change, financial collapses, nuclear meltdowns, oil spills, and famine now take on the force of "acts of God," in the sense that they are not controllable or stoppable by individual or collective human agency but can only be responded to, modulated, adapted to, or endured.

The nuclear reactions in the fuel rods at the Fukushima Daiichi plant are a perfect example of this kind of event. They could not be stopped by human or technical agency because they operate according to their own nonhuman time-scale having to do with the half-life of uranium rather than the periodicity of the mediatic or other human systems. They might be minimized, modulated, or redirected not by acting upon them directly as if they were inert, passive matter but rather by accepting the fact that their agency, trajectory, and development operate according to their own laws, their own temporality, their own scale. All such events generate a sense that we can do little more than watch, navigate, or negotiate large, complex forces already in action.

Instead of being separate from, external to, or secondary to disasters like the Tōhoku quake and its aftereffects, mediation is immanent to them. Crisis or CATASTROPHE does not exist without mediation as such; these conceptual and ontological entities are not prior to or independent of mediation but instead constituted as forms or categories of mediation. The same multinational conglomerates that own media outlets invariably also manufacture products or provide services that cause the events that are covered by the media: these corporate entities help to cause the disasters or crisis situations that in turn generate the phenomenon of mediashock. General Electric, until recently the parent company of NBC, designed and helped to manufacture the nuclear reactors whose cooling system failed at Fukushima. Thus when NBC News reported on the radiological and affective FALLOUT from the reactor core meltdown at Fukushima Daiichi, its mediation of the event could not be separated from the event itself, as General Electric was involved in both the disaster and the mediashock that it generated. If multinational capital operates according to a shock doctrine of disaster capitalism that fosters and takes advantage of technical, geopolitical, and natural disasters in order to take control of markets and governments across the world, then it is not a stretch to suggest that mediashock names the media regime of disaster capitalism, a form of disaster mediation that functions in the twenty-first century to fuel the spread and distribution of state, corporate, and informal media across the globe.

See also: AFFECT, ANTHROPOCENE, CATASTROPHE, CORPORATION, EMBODIMENT, GRIDS, NETWORKS.

Metabolism

Adam Dickinson

In all living things, ENERGY is consumed in specific physical and chemical reactions. Respiration, reproduction, digestion, blood circulation, and waste removal are functions of these metabolic processes. One driver of metabolism is the endocrine system, which regulates the flow of hormones through the body. Hormones act as a complex exchange of messages among different states and locations. The thyroid gland secretes instructions that increase the rate at which cells utilize energy from food. Human reproduction is similarly governed by sensitivity to chemical messengers like estrogen and testosterone. But the conventional energy sources that power economies and give rise to myriad consumer products also interfere with the power systems and communication networks intrinsic to human metabolism. Petrochemicals are rewriting the metabolic processes of our bodies. Synthetic chemicals have become part of the hormonal conversation and what it means to be human.

We wear the energy interests of multinational companies in our flesh. Most North Americans have Monsanto in their blood and fat, in the form of polychlorinated biphenyls (PCBs). Similarly ubiquitous are polybrominated diphenyl ethers (PBDEs), or brominated flame retardants. Faced with a declining market at the end of the leaded-gasoline era, the bromine industry found a new outlet in flame retardant additives to commercial PLASTICS. PBDEs do not break down easily in the environment; they travel up the food chain and accumulate in fatty tissues of animals and humans. PBDEs are endocrine disrupters: they can be mistaken for hormones and lead to developmental disorders or illnesses. PBDEs have been linked to neurotoxicity, estrogenic effects, and reduced thyroid hormone function (Smith and Lourie 2009, 111). These chemicals produce their toxic effects only over time, so that regulation or production bans require protracted struggles with industry. Rick Smith and Bruce Lourie argue that "hormone disruption, like climate change, is a spin-off from society's ADDICTION to fossil fuels. The damaging effects of hormone-disrupting chemicals on fertility, the brain and behaviour quite possibly make them a more imminent threat to humankind than climate change" (ibid., ix).

In their rewriting of the biochemical messages in our bodies, endocrine disruptors reveal the importance of semiotic processes in biological systems. The emerging discipline of biosemiotics contends that communication must be understood beyond its conventional associations with culture. Drawing upon linguistics and biology, biosemiotics sees all living things within worlds of signification, a semiosphere where the production and interpretation of signs are fundamental to life. Communicative dynamics are central to the interaction of organisms within their environments and to internal endocrinological and immunological dynamics, where plentiful membranous surfaces interpret and respond to hormonal messages. Jesper Hoffmeyer asserts that "cultural sign processes must be regarded as special instances of a more general and extensive biosemiosis that continuously unfolds and acts in the biosphere" (2008, 4). Human language is an extension of the biosemiotic activity of a female dove, for example, who coos not only at the male but also at her own ovaries to stimulate the release of eggs (ibid., 118). If biosemiotics, and its interest in the evolution of semiotic complexity, encourages us to think more broadly about communication, Jane Bennett provokes us to consider material environments in similarly affective terms. Bennett argues that we have been too quick to divide the world into living beings and dead, inert matter. We "ignore the vitality *of* matter and the lively powers *of* material formations, such as the way omega-3 fatty acids can alter human moods or the way our trash is not 'away' in landfills but generating lively streams of chemicals and volatile winds of methane as we speak" (J. Bennett 2010, vii).

Given its implications for communication, how might contemporary writers respond to the predicament of chemical pollution and its potentially toxic effects on human metabolism? How do endocrine-disrupting chemicals shape literary forms and genres? Who is better positioned to think about the consequences of these extreme forms of writing than poets who work at the limits of writing? Petrochemical pollution is what Timothy Morton (2013) calls a hyperobject, entities and forces (living and nonliving) so massively distributed in time and space that they change our understanding of objects. Morton's examples include global warming, flu pandemics, nuclear radiation, oil, and chemical pollution. These phenomena defy measurement: "Because they so massively outscale us, hyperobjects have magnified this weirdness of things for our inspection: things are themselves, but we can't point to them directly" (T. Morton 2013, 12). What artistic response is adequate to this predicament? "Are there any hyper-art-objects," Morton wonders, "that perform the terrifying ooze of oil?" (ibid., 180).

Contemporary writing that is concerned with what I will call metabolic poetics might be seen as an artistic response to the hyperobject of chemical pollution. Metabolic poetics involves poetry (or acts of reading and writing) derived from or responding to biological mediums, especially in the context of homeostatic states and homeorhetic trajectories affected by interference from prevailing powers in industry or politics. Juliana Spahr's *The Transformation* (2007) offers an ethnobotanical exploration of cultural infection related to the complex politics of Hawai'i and 9/11. The book chronicles "prickly new cells" that have entered the blood, changing how the speaker and her partners think of analogy, grammar, and government (39). *The Perfume Recordist* by Stacy Doris and Lisa Robertson "consider[s] the meeting of perfume, hormones and degeneration" (2012, 236). Other examples include Evelyn Reilly's *Styrofoam* (2009) (plastic pollution as hyperobject);[1] Jenny Sampirisi's *Croak* (2011), a frog-and-girl opera that explores a human-frog relationship where the language of chemical pollution has rewritten and deformed the bodies of frogs; and Craig Dworkin's (2005) conceptual poem "Fact," a list of the chemical ingredients in the sheet of paper the poem is printed on.

My own method of inquiry concerns the metabolic poetics of oil-derived and oil-related chemicals. I am conducting biomonitoring and microbiome testing on my body to consider how the outside writes the inside in ways both necessary (certain kinds of bacteria) and harmful (chemical pollution). I am charting a biosemiotic map of the toxicological and symbiotic circumstances of my body, which I will shape into a species of writing (a chemical/microbial autobiography) that explores the subject as an assemblage of objects, an intimate, "trans-corporeal" expression (Alaimo 2010, 2) of the hyperobject of chemical pollution. To generate this "exposome," or account of environmental exposures over a lifetime (Wild 2005, 1848),[2] I am having my blood and urine tested for phthalates, PCBs, PFCs (perfluorinated chemicals), OCPs (organochlorine pesticides), OPIMs (organophosphate insecticide metabolites), PAHs (polycyclic aromatic hydrocarbons), HBCDs/PBDEs (flame retardants), triclosan (antibacterial additive), parabens, BPA (bisphenol A), and thirty-one heavy metals. I am also having my microbiome sequenced in order to discover some of the viruses, microbial eukaryotes, and fungi that inhabit my body. I am developing poetic compositional methods that extend out of the biological predicament of my body and reflect the constraints and processes I experience as a being composed of other beings and "volatile" materials (J. Bennett 2010). Preliminary tests are starting to come in, and it is amazing to consider how my body wears industrial, agricultural, and military history whether I like it or not (PCBs, organochlorine insecticides, uranium). Following Jussi Parikka's (2010) contention in *Insect Media* that organisms and other geophysical elements constitute forms of MEDIA through "various modes of transmission and coupling with their environment" (xiv), I want to see my body, the chemicals in my blood, as forms of media expressing both the biology of petroculture and my strange intimacy with the energy sources of our historical moment.

1. For another poetic response to the plastic pollution, see Adam Dickinson, *The Polymers* (2013).
2. Wild advocates that individual environmental exposures (exposomes) be mapped with the same precision as genomes (2005, 1848).

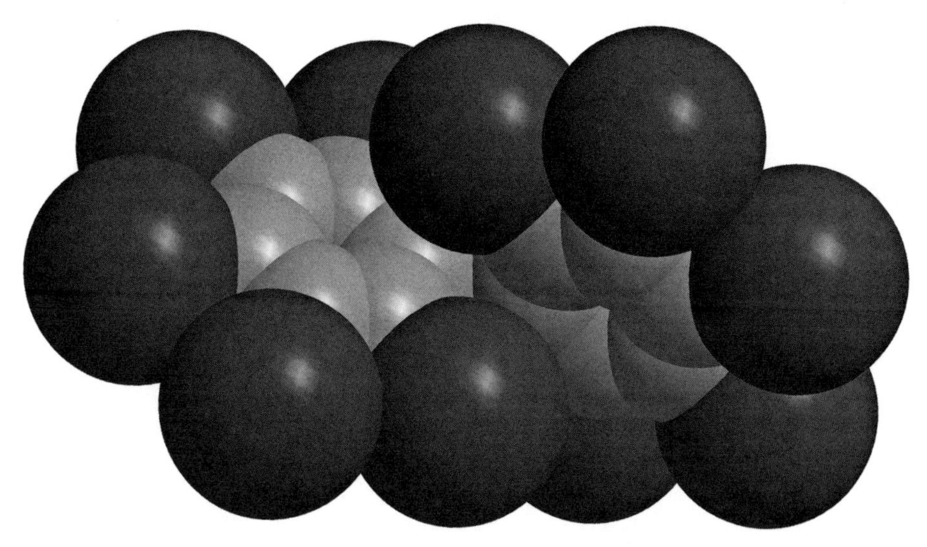

FIGURE 9. Polybrominated diphenyl ether. (Image generated by Adam Dickinson)

Below is a response to flame retardants found in my blood. PBDEs leach out of consumer products like carpets, TVs, furniture, window blinds, mattresses, and other common household items. The greatest source of PBDE contamination for humans is believed to be dust.

A BROMIDE

Polybrominated diphenyl ether, IUPAC # 47	*Plasma*	*0.04 ug/L*
Polybrominated diphenyl ether, IUPAC # 153	*Plasma*	*0.03 ug/L*

The umbrella is the starting point for a larger obfuscation. A constant mist of tiny particles rains upward, like neck hair at the cicada sex of a smoke alarm. Children outgrow the behaviors of cats, but for many years they are derelicts of skin flakes, stair runners, and upholstery. The average carpet smokes three packs a day. The glassy bits scratching your throat are leftover deterrents to predators. Dust is a conversation happening just out of earshot, it's the street talk of the Endocrine and Alderaan systems, a vector for the invectives of misdirection. Dust is a bunch of nickels your uncle gives you to get him another drink. My thoughts, like every other coagulation cascade, are made of melted lint and move around with the chirality of lost oven mitts. In the dusty barns of Michigan, the wrong bag of pale grit was mixed into cow feed. Nine million people ate FireMaster. My limbs tingle just out of broadcast range. Here come the industry standards to burn down the roofs of our mouths.

See also: AFFECT, DETRITUS, EMBODIMENT, UNOBTAINIUM.

Middle East

Juan Cole

SOLAR power could be a game changer in the Israel-Palestine conflict. Big Carbon is nowhere more problematic than in Israel and Palestine. These areas have little COAL or oil, and their richest natural gas reserves are offshore in the Mediterranean Sea. That location makes gas expensive to extract at a time when the American technique of hydraulic FRACKING is driving down global prices. Ownership of the fields is contested among Israelis, Palestinians, and the Lebanese, promising more conflict.

Moreover, carbon dioxide emissions, causing disruptive climate change, have led to increasing water shortages in the region, which are implicated in the upheavals of 2011, including the Syrian Civil War. Few regions of the world will suffer more from global warming than the Arid Zone stretching from Morocco to the Gobi Desert. The looming water CRISIS in the Middle East can likely only be resolved affordably by solar-powered hydrolysis.

In the past, petroleum has driven conflict in the region. Indeed, some American geopolitical thinkers have seen Fortress Israel as a forward operating base, guaranteeing that the West could grab petroleum fields in the Middle East if access to them were ever threatened (Thomas 2007). One impetus for Israel's repeated attempts to seize the Sinai Peninsula from Egypt was a desire to exploit petroleum fields there. Oil states like Iraq have previously used hydrocarbon revenues to fund guerilla actions against Israel, and Iraq's own wars on Iran and Kuwait were partly driven by an attempt to acquire their vast oil fields.

The occupied Gaza Strip, home to 1.7 million Palestinians, is deprived by Israel of an airport or seaport and has limited access to the outside world through Israeli checkpoints. Since 2007, allegedly to punish them for supporting the Hamas party-militia, Israel has prevented the export of most of what Gaza Palestinians produce and has limited imports, including fuel (Gaita 2010). The Geneva Convention of 1949 makes blockades on civilian populations by occupying powers illegal. Egypt is a junior partner in the blockade, since its elites also fear Hamas. The consequences of the blockade are dire: fuel shortages, a lack of import-export INFRASTRUCTURE, and dire poverty. These factors means that electricity in Gaza can be knocked out for as much as twelve hours a day, including to vital facilities such as hospitals, which endangers lives. Premature babies in incubators are at special risk from power outages. In what is perhaps a sign of the times, al-Nasr Gaza's Children's Hospital installed photovoltaic solar panels on its roof in 2014 to avert such tragedies, and crowd-funding has been launched to put solar panels on three additional hospitals (Balousha 2014; Ahmed 2015).

The cost of photovoltaic cells plummeted after 2012, making them abruptly affordable for some villagers of the Global South. For the vast swaths of the world where RURAL electrification is still rare, the panels allow access to ELECTRICITY in the absence of a grid. In conflict situations such as Gaza, where energy politics are deployed in a long-running struggle, the panels allow a potential end-run around fuel blockades.

For this reason, the photovoltaic panels themselves have become targets for sabotage. In 2012, Israel declared panels installed by Palestinians in the occupied West Bank near al-Khalil (Hebron), and funded by the European Union, to be "illegal" (Neslen 2012). In March of 2014, Israeli squatters in Mitzpe Yair (viewed as an illegal settlement by the Israeli government itself) attacked and broke solar panels owned by a Palestinian COMMUNITY in the south Hebron hills ("Settlers Destroy" 2014). Nevertheless, solar panels and wind turbines (also increasingly efficient and inexpensive) have the potential to change the lives of Palestinians under occupation. Access to electricity also allows charging of smartphones and thus aids in documentation of human rights abuses by authorities. It gives access to computers and the Web. It is a potential game changer for the stateless Palestinians.

Israel itself ought to be a pioneer in solar energy. The amount of ENERGY it could realize from solar panels in this mostly sunny clime is enormous. It has a robust scientific and engineering establishment. Israel is forced to import petroleum, and in the Middle East region that need implies importing (even if indirectly) from the Muslim oil states. In past decades, this DEMAND for oil produced somewhat embarrassing arrangements, as with the 1980s deal to provision the Islamic Republic of Iran with spare parts for American weapons in return for continued supply of Iranian petroleum. (This arrangement was in some ways the background of the Reagan administration's own outreach to Iran and thence the Iran-Contra scandal [Parsi 2007].) Even thereafter, Israel ended up importing indirectly from the Arab and Muslim worlds and therefore being dependent on them. If the Israeli political elite were really devoted to self-reliance, they would be more eager to get off petroleum.

In fact, only 2 percent of Israel's electricity was generated by photovoltaic cells as of 2015 (Wainer and Hirtenstein 2015). The country's right-wing Likud government has pledged that 10 percent of energy will be from renewables by 2020, whereas most European countries have plans to get to 20 percent renewables by then, and Germany, Spain, and Portugal have already surpassed that threshold. Israel had been a pioneer in using solar for water heating (80 percent of homeowners use this technology), but that momentum has been lost (Sterman 2009).

While government policy has not helped, Israeli entrepreneurs, scientists, and engineers are working on the problems. In order to meet its goal of 10 percent RENEWABLE energy by 2020, the country will construct the fifth largest solar power station in the world. Arava Power is building a large solar farm outside the southern city of Eilat, which will power one-third of that Red Sea resort's buildings (Gross 2015). But solar firms speak of having to fight the government every inch of the way.

One has a dark suspicion that elements in the ruling Likud government are tied to natural gas exploitation or otherwise have pecuniary motives in fighting the solar industry. As with the Likud's irrational obsession with colonizing and annexing the Palestinian West Bank, its odd romance with Big Oil and Big Gas endangers Israel's long-term security rather than guaranteeing it.

One obstacle solar panels face in the desert is that dust and sand tend to cover them over time, reducing efficiency. One Israeli kibbutz put in a solar panel installation and is using small roaming robots to keep them clean of dust (Shamah 2015), a promising technology with implications for the rest of the region.

Solar power is potentially a path to peace in the Middle East. Whereas Israelis and Palestinians and the Lebanese could fight over natural gas, there is plenty of sunshine to go around. Hydrocarbon energy is a zero-sum game (the supply is limited and the source can be owned by some, excluding others). In contrast, solar energy does not produce competition for resources. Given the rapid decline in the price per kilowatt-hour of solar electricity generation, and given the political implications of and obstacles to hydrocarbon fuel imports and distribution, solar power could well contribute to a reduction of tensions in the region.

See also: ENERGOPOLITICS, GRIDS, OFF-GRID, PETRO-VIOLENCE, SOLAR.

Nature

Louise Green

Emphasizing the social nature of language, Mikhail Bakhtin writes that when we speak, we do not find words in a dictionary but rather take them from other people's mouths (1994, 77). To use a word is always to reproduce, adapt, or resist the meanings it acquires as it circulates through other people's conversations. This is true for all words but particularly so for *nature*, which is like a worn, somewhat blurred coin that has passed among speakers for centuries.

At some point in the twentieth century a shift occurred in the kinds of conversations in which *nature* circulated. Nature no longer figured as independent, powerful, and changeable only in the *longue durée* of archaeological or geologic time. In these new conversations, nature is spoken about in terms of vulnerability, loss, and, perhaps most important, instability. Signal events that mark this shift include the atomic bombs dropped on Hiroshima and Nagasaki, the discovery of the hole in the ozone layer caused by CFCs used in refrigerators and aerosols, growing awareness of the effects of carbon emissions on weather patterns, the documentation of species extinctions, and predictions about resource scarcity and population increase.

In his book on money, Georg Simmel comments that we deem valuable things that resist our desire to possess them (2011, 64). In our time, nature becomes valuable precisely because of its scarcity—consider, for example, the ubiquity of "nature" and "natural" in advertising. Myriad consumer products acquire value through their purported association with nature—shoes, food, skin care products, as well as holiday destinations, to mention only a few. This practice conceptualizes nature as something unquestionably positive. To describe a location as natural is to imply a kind of habitability, as if all terrains are equally hospitable to human and ANIMAL forms of life. In nature documentaries, an animal's natural habitat is the terrain in which its species is understood to feel most comfortable, most at home. In this usage—emerging out of the tradition of Rousseau and the Romantics—*nature* signifies the coincidence of a physical and a metaphysical order.

Nature imagined as an abstract totality implies the conceptual ability to encompass and make intelligible radical diversity. Yet the term also implies immediacy and singularity, and in its materiality it enters into the very substance of everyday life. Perhaps part of the word's appeal and power is its ability to link disparate elements. Nature, as Neil Smith points out, acts as a pivot linking what is external, the material world, to what is internal, human nature (1984, 11). It intimates a relationship between the world of things, of which it is the sum, and the subjective experience of those things.

Thus, ecological crises disrupt the material world and introduce a conceptual disquiet. When weather patterns change dramatically, the material world, previously imagined to be the stage for human activity, becomes significant in a new way. As a result, Michel Serres argues, "Global history enters nature; global nature enters history: this is something utterly new in philosophy" (1995b, 4).

In ecology, talk about nature tends to produce a universal "we," as in the familiar lament "we are destroying the planet." This "we" hides the differential degree to which each of us contributes toward this destruction. But this "we" might be imagined in other way. Serres suggests that the pronoun "we" takes on a new meaning when neither individual humans nor even the collectivities of nation and COMMUNITY are the most significant agents. Instead, he suggests, human assemblages—such as dense megalopolises that can be seen from space—have assumed the power of a tectonic plate. In contrast to the peasant who is almost invisible to the world, a "fragile reed," human assemblages have a geophysical existence, an impact similar to that of an ocean or desert.

These human assemblages turn culture, in the sense of shared daily patterns of movement, routines of WORK, and practices of consumption, into a force of nature. The prospective renaming of the current geological epoch as the Anthropocene indicates the extent to which global history and global nature are understood as intertwined. The nature/culture binary is coming undone in an entirely new way.

Extending Serres's idea, we might consider the collective made up of mobile individuals as effecting the geophysical processes that make the earth habitable for humans and other mammals. Mobility and speed seem to be the key elements in considering this collective that, unlike the sprawling cities, is widely dispersed across space. The mobility of things is also a key characteristic of global culture. The oceans crowded with ships carrying goods, the freeways crowded with cars, the sky crowded with airplanes, all create a network of mobility operating like an ocean current, raising temperatures, altering the atmosphere, and perhaps having other currently incalculable effects. It is useful to think of this assemblage not as destroying the planet but as radically affecting the habitability of different terrains, both directly, through infrastructure and extraction industries, and indirectly, through climate change.

Timothy Mitchell describes how the material differences between COAL and oil meant that their extractive processes produced very different social interactions and political configurations. Coal is heavy and inert; oil flows along pipelines. Coal extraction requires a large labor force and produces the forms of sociality associated with shared work (2009, 403–4). Oil is driven to the surface by underground pressure (rather than human labor) and so permits the enclaved extractive operations like those in the Niger Delta (ibid., 407). What new forms of sociality might be produced by wind farms, SOLAR power, and biofuels?

For some theorists, such as Timothy Morton (2009), the word *nature* is too compromised to invoke in the project of mobilizing ecological consciousness. Morton's concern is that *nature* is a general concept that subsumes and sometimes conceals the particularities of its usages. The concern with nature evinced by some affluent environmentalisms makes certain sites invisible, not recognizable as nature. The Niger Delta, for instance, cannot hold the same emotional appeal as the Serengeti National Park. As Raymond Williams observes, industrialization divides nature into resources for production and "untouched wilderness" that itself ironically becomes a resource to be consumed (1980, 81).

However, nature still seems to hold some utopian possibility. Despite its blurred and worn aspect, *nature* still indicates something compelling: what cannot be reduced to private property. While ownership of resources derived from nature, such as oil, wildlife, or even game reserves is obviously possible, *nature* refers to something that exceeds this logic. Nature implies something held in common, a different order of engagement with the material world. Conceptualized not as economic resource or untouched wilderness but rather as an intangible commons, nature opens the possibility of a different kind of collectivity. Michael Hardt claims that "the value of the commons defies measurement" (2010, 270), but perhaps what nature can still suggest is the possibility of an alternative mode of value. Nature is that which exceeds the private and the individual and so offers a kind of friction or traction for reconceptualizing public life, a collectivity that is at once abstract and concrete—limited, shaped, brought into being by the unthought possibilities of scarcity.

See also: ACCUMULATION, ANTHROPOCENE, AUTOMOBILITY, ENERGY REGIME, GREEN, NETWORKS, NIGERIA, UTOPIA.

Necessity

Timothy Kaposy

By now the protagonists of the world's ENERGY economy are widely recognizable. Over the last two centuries, the energy industry has produced iconic figures whose biographies are studied by entrepreneurs, whose opinions about the market shape economies beyond the energy sector, and whose decisions channel a violent flow of petrodollars from extraction sites to private firms. From Standard Oil's John D. Rockefeller to ExxonMobil's Lee Raymond and Yukos's former CEO Mikhail Khodorkovsky, owners of the energy sector have consolidated corporate power to the detriment of an incalculable number of the earth's inhabitants (human and otherwise).[1] The iconic One versus the unknown Many: Is this how the grand narrative of energy's necessity is structured today?

Necessity is a crucial concept for how energy is understood and narrated. For philosophers dating back to Aristotle, the concept of historical necessity offered a way to interpret time and distinguish between possible and impossible events. Book nine of *Poetics* claims that necessity—what cannot *not* happen—indexes the difference between possibility and impossibility in the plot of a poetic work. The poet rationalizes the necessary movement of time to formally anticipate its outcome. Since the Renaissance, Aristotle's argument has validated poetic thought's "universal" relevance in contrast to history's record of particular events.

1. Conventional protagonist-driven energy narratives include: Daniel Yergin, *The Prize: The Epic Quest for Oil, Money, and Power* (1991); Steve Coll, *Private Empire: ExxonMobil and American Power* (2012); and *Khodorkovsky*, the 2011 documentary film by Cyril Tuschi.

For Hegel, the aim of rationalizing temporality is to gain accurate historical consciousness by giving contingency a form of necessity rather than vice versa. Causal explanations of time are shown to be questionable but not invalid. By critiquing the alignment of concepts and substance, what he calls "external necessity" (1975, 26), Hegel examines historical necessity on a metaconceptual plane, questioning the adequacy of concepts to rationalize time. Thought, for Hegel, tends to be normative, impelled by desires other than cognition, and rife with unintended consequences. Necessity, therefore, is an invaluable concept because narratives of causation raise questions concerning the historicity of rationality and the role of thought in producing temporality. Aristotle and Hegel show that although they trouble the thinker, philosophies of necessity are a first step toward repudiating the mystifications of religion, where everything happens because of God, and in dispelling the Epicurean tradition, where everything happens by chance. When philosophies of causation make explicit a temporality of necessity, narratives of demystification come into view.

Energy industries are unlikely settings for the study of history or protracted character development. One reason is that understanding energy's causality is difficult. Knowledge of the effects of energy use has required a decades-long collective effort based in painstaking scientific study. Though many nonexperts are capable of outlining the causal effects of energy use, the idea that one may know the *necessary* conditions of possibility for its extraction, refinement, production, circulation, and consumption is a different matter. One signal difficulty is the scale of energy. Narrators of energy must contend with the geological and atomic size of the phenomena addressed. Additionally, to describe energy with any precision requires attunement to its immanent intensity. For these conceptual, aesthetic, and metaphysical reasons, the temporalities that swirl around energy are too unstable and complex for its iconic captains of industry to be adequate protagonists. What, then, might be alternative foci of energy narratives? Rethinking how stories of energy subjectivization are told is crucial for making sense of our shared dependence on energy's tenuous industrial future.

In the opening of *Heat: How to Stop the Planet from Burning*, journalist George Monbiot explains how his initial assumptions about climate and energy were challenged. "My instincts are almost always wrong," he writes:

> I have made the mistake of confusing what is aesthetically pleasing with what is environmentally sound. For instance, I have always assumed that candles are more environmentally friendly than electric lighting, for no better reason than that I like them and that they produce less light. In his excellent textbook on energy systems, Godfrey Boyle points out that in terms of the light given off per watt of expended power, a candle is 71 times less efficient than an old fashioned incandescent bulb, and 357 times worse than a compact florescent bulb. (Monbiot 2007, xxvi)

Monbiot's point is instructive. No matter how much one desires knowledge of energy's function, its actuality can be counterintuitive. Even the most basic laws of physics are elusive without long-term investigation of their genesis and structure. Not only do the deceptive minutiae of energy undercut causal narratives, but the various scales of ecology should give one pause before identifying the human as an assured locus of knowledge.

Timothy Morton reanimates Monbiot's skepticism by explaining how one's immanent relation to energy informs the ability to gain perspective on its causes and effects. Morton's concept of the "hyperobject" names the viscosity, nonlocality, temporal undulation, invisibility, and interobjective expression of "massively distributed" objects (2013). While Monbiot is skeptical of the ability to perceive energy's function, Morton insists that energy cannot be grasped "as such" since "humans are caught in intersecting phases of time" (ibid., 68). For instance, a medical patient may worry about or even refuse radiation exposure for the diagnosis or treatment of an ailment, but radiation from decades of sun exposure or living nearby overhead hydroelectric wires tends to elicit less awareness or concern. To the extent that the perceived certainty behind knowledge of energy's infrastructure and effects subsumes the divergent temporalities borne of daily use and their perceived hazards, Morton furthers the case for dismantling the metaphysical basis of quotidian certainty.

Necessity is an enduring concept that knots these terms—causality, scale, energy—together in a bind that is almost impossible to unwind. One cannot simply abandon necessity because of the subject's limited conceptual and aesthetic capacity to know the daily function and effects of energy. After Aristotle and Hegel, perhaps the most innovative thinker on the question of necessity was Karl Marx, who translated historical necessity from a temporal concept into a spatial one, which he called "a realm of necessity." In the final volume of *Capital*, Marx frames the interpretation of historical causes in a familiar space of need and labor. If Aristotle and Hegel dwell in the ontology of necessity, Marx occupies the ontic realm of need:

> The realm of freedom really begins only where labour determined by necessity and external expediency ends; it lies by its very nature beyond the sphere of material production proper . . . . Freedom, in this sphere, can only consist in this, that socialized man, the associated producers, govern the human metabolism with Nature in a rational way, bringing it under their collective control, instead of being ruled by it as a blind power; accomplishing it with the least expenditure of energy and in conditions most worthy and appropriate of their human nature. But this nonetheless remains a realm of necessity. The true realm of freedom, the development of human powers as an end in itself, begins beyond it, though it can only flourish with this realm of necessity as its basis. The reduction of the working day is the basic prerequisite. (Marx 1991, 958–59)

By transforming necessity from a discrete philosophical concept into a readily intelligible unit of time (i.e., the working day), Marx roots historical causes in the reproductive "METABOLISM" of human labor, nutrition, sleep, housing, and health. Wresting necessity from Aristotle's metaphysics and negating Hegel's conceptual dynamism, Marx renders a spatial "realm" from history's temporal axiom. He situates historical time within ineluctable human needs to show that freedom is inseparable from the struggle to survive, thrive, and exceed basic metabolic processes that ground the species.

Since needs are adjudicated by political modifications of capital, it is difficult to root energy's use and abuse in a fully determinate notion of human need. By extension, Marx underscores freedom's relation to "the expenditure of energy" precisely because societies have built infrastructures to manage needs in the hope of freeing up space to flourish in other ways. Perhaps, then, the energy narratives we seek should dramatize a greater proximity to this collective realm of knowledge production and daily life instead of parsing once again the psyches of its unreliable capitalists, whose freedom to act and inquire is based simply on the exploitation of others and on undermining inquiry into energy's industrial effects within the planet's ecology.

See also: CORPORATION, ENERGY, METABOLISM, RUBBER, STATISTICS, TEXAS.

Networks

Lisa Parks

Every keyboard button you push, every screen you view, every ringtone you hear requires electrical ENERGY. If you are reading on an e-reader, smartphone, or computer interface right now, consider how these words arrived before your eyes, how packets of data hopped through network nodes to become digitally rendered pages for your perusal. Data, whether text, image, or sound, moves so rapidly and transparently that we rarely consider the energizing of networks, the fueling of cultures. Building on work in environmental media studies (see Cubbit 2005; Maxwell and Miller 2012; Bozak 2012), I offer three energy-media network scenarios involving water, lithium, and the sun in order to provoke critical and phenomenological thinking about the interdependencies of natural and cultural resources, the layering and coordination of networked infrastructures (Marvin and Graham 2001; Parks and Starosielski 2015), and the subjectivities formed in their interstices.

Water

The first scenario involves television, tea, and hydroelectricity in the United Kingdom, as depicted in a short video, "How the National Grid Responds to Demand" (2008). A UK national balancing engineer examines a computerized map of the national energy grid. He is managing a "TV pickup"—a surge in DEMAND for ELECTRICITY when the credits of popular TV dramas (such as *EastEnders*) begin to roll and an estimated 1.5 million people simultaneously turn on their electric kettles to make tea. The engineer scans the grid map, analyzes power availability, and draws extra load from two remote hydroelectric power plants. Once he brings the hydroelectric plants online, thousands of tons of water rush down hills, generating huge torrents of power to provide enough electricity for citizens to sip a cup of post-TV tea.

This situation reveals the layering and coordination of networked infrastructures as well as the significance of water resources. As TV networks broadcast programming, they also set schedules for fluctuations in the water supply and power grid. The end of demand in one network creates increased demand in another. Water is pumped through plumbing systems to fill kettles and it is heated by electricity generated by the gushing of water through DAMS. Water sustains the nation's electrical, television, and social networks. Similar pressures on networks happen elsewhere in the world, such as in the United States on Super Bowl Sunday. Civil engineers have reported increased demand upon water and sewer systems during commercial breaks when viewers visit the loo and flush their toilets en masse. Consuming MEDIA should be understood not only as viewing what appears on screen but also as using resources to capacitate that screen and the lifeworlds that take shape around it.

Lithium

While water resources energize television and tea drinking in the United Kingdom, lithium-ion batteries fuel the network connectivity of billions of people using portable electronic devices and a growing number of electric cars. But where does lithium come from? The world's largest lithium reserves are in Bolivia, Chile, and Argentina, and deposits have been recently found in Mexico ("Lithium" 2013). Bolivians describe their country as "the Saudi Arabia of lithium"; wary of histories of exploitation by foreign developers, they have taken a protectionist stance as a lithium BOOM unfolds. US and Chinese firms such as LG Chem and BYD have maneuvered to tap lithium reserves, recently described as "one of the planet's most strategic commodities" (Wright 2010; Colomar-Garcia and Zhao 2011; McDougall 2009).

Once extracted, lithium is transported to East Asia where assembly-line workers make rechargeable batteries for mobile devices. Mobile connectivity draws not only on lithium reserves but also on power GRIDS around the world that rely on a patchwork of fossil fuels, SOLAR, wind, NUCLEAR, and hydropower, as well as human labor. If, as Tiziana Terranova (2004) suggests, the Internet is sustained by the "immaterial labor" of its users, then mobile connectivity is also supported by the time, attention, and capital required for battery recharging.

While recharging devices is relatively easy for those who are already highly mobile, it can be a daily challenge for the disenfranchised. Sites in airports once hosting public payphones have become free charging stations for business class elites, yet mobile device users in the developing world often lack electrical access and must pay fees to charge their devices at public markets or Internet cafés, or poach grid access from a friend, family member, or employer. Mobile connectivity requires lithium and human resources and is interwoven with differential access to electrical and social power.

Sun

While water and lithium resources fuel network infrastructures and media cultures, so does solar energy. In recent years, massive solar harvesting facilities have sprouted up in sun-soaked places ranging from Abu Dhabi to Arizona, channeling energy into national or regional power grids. The Shams 1 facility near Abu Dhabi is among the largest concentrated solar power plants in the world. Its one-square-mile array of parabolic mirrors has a 100-megawatt capacity, enough to power twenty thousand houses.

A video celebrating the Shams 1 plant's inauguration applauds the United Arab Emirates' commitment to RENEWABLE energy and sustaining its citizens' high quality of life; it features a cameo of Sheikh Khalifa bin Zayed Al Nahyan, who supported the project. The video opens with shots of the cutting-edge ARCHITECTURE of Abu Dhabi's skyline and transitions to sci-fi looking scenes of the Shams 1 solar plant in the desert where uniformed technicians inspect equipment and giant orange trucks crawl through the site with automated arms extended to buff the shiny mirrors. Two young children, presumably next generation energy consumers, walk through the facility and then venture into the desert as the sun beams down on the sand around them. They ascend a massive dune and the boy uses a small mirror to reflect sunlight back to the plant. Images of engineers in action are rapidly intercut with computer interfaces, a switch is flipped, and an aerial view spotlights the mirrors rotating in unison: the solar plant goes online for the first time. In the final frame, the words, "Reflecting the Nation's Commitment" flash on-screen, followed by the Shams Power Company logo ("SHAMS 1" 2013).

While this video promotes the national dream of turning the earth's surface into fields of solar and economic power, other solar configurations are less audacious. For decades, small-scale solar systems have powered facilities and consumer electronics in OFF-GRID, RURAL communities. From Zambia to Mongolia, the sun fuels Internet access, computer and mobile phone use, satellite TV reception, and radio listening, linking remote locales to far-flung networks. Just as lithium-ion batteries catalyze cultures of device charging, solar power generates local cultures of panel installation, tinkering, and repair and intensifies interest in weather patterns.

Whether imagined as national development or rural empowerment, the physical properties of solar power systems have the potential to create a sense of insular self-sufficiency at the very moment of (and perhaps as a response to) the intensifying interdependencies and flows known as globalization. Rather than turn inward, it is vital that citizen-consumers seek relational, site-specific understandings of network infrastructures, energy sources, and global/national/local cultures so that they can participate in debates and decisions about resource futures.

By adopting a materialist approach to the study of networks and analyzing the resource requirements of television viewing, mobile connectivity, and Internet access, citizens might resist interpellation as transfixed button-pushers, particularly in the age of the smartphone, and instead re-socialize networks. This would mean treating each moment at a network interface as an opportunity to investigate what undergirds it, whether the geopolitics of extraction, emergent forms of labor, infrastructural designs and literacies, regulatory frameworks, or systems of differentiation and distinction. In the end, understanding networks in the digital age involves much more than tweeting, texting, or liking; it also involves taking the time to figure out how to sustain relationships with matter like water, lithium, and the sun.

See also: ELECTRICITY, GRIDS, INFRASTRUCTURE, OFF-GRID, PIPELINES, SERVERS.

Nigeria

Philip Aghoghovwia

To write about Nigeria in relation to ENERGY and culture is to write about the Niger Delta, the very theater of energy production. Indeed, one cannot write about energy culture in the Nigerian context without engaging the spectacle of violence it elicits, both in the public mind and in the sphere of creative imagination, precisely because the form of sociality that oil energy generates in cultural production is imagined and inscribed in idioms of violence. My concern here with this concept of violence is less with the destructiveness it inflicts (which is considerable) than with the intricacy of its operation in the Niger Delta, especially the AFFECT and spectacle that characterize its coexistence alongside oil.

Recently I attended an exhibition, *Delta Remix: Last Rites Niger Delta*, that showcased the post-Saro-Wiwa era of oil production in Nigeria. This slideshow exhibition was curated by Zen Marie and other colleagues at the Wits School of Arts at the University of the Witwatersrand, part of a series of events at the Johannesburg Workshop in Theory and Criticism (JWTC). It was a "representational intervention" that remixed images from a catalog commissioned and published by the Goethe-Institut, *Last Rites Niger Delta: The Drama of Oil Production in Contemporary Photographs* (Stelzig, Ursprung, and Eisenhofer 2012). The photographs in this text capture the violence that oil production perpetrates in the Delta, both on the landscape and in the life of the population at extraction sites.

The title of the book merits reflection. "Last Rites" evokes a sense of ritual and ceremony, with a loaded eschatological undertone surrounding this rite of transition, which precedes death. "Last Rites" projects a certain imperceptible presence and performance of immolation in the Delta, at once self-inflicted by acts of community protests and more devastatingly visited upon the ecosphere by the technologies and politics of oil production—the latter being deployed in objectionable and crude modes of petro-colonialism by the oil extraction complex.

The exhibition was an attempt to capture the quotidian realities of oil production and its manifest social and environmental repercussions in the Niger Delta. It provoked robust debates at the JWTC workshop around questions of environmental devastation, militancy by local communities as a means of protesting perceived injustice, and, finally, the vexing question of representation. What protocols of representation are available for negotiating and understanding the oil energy cultures in Nigeria?

In the Nigerian milieu as elsewhere, *representation* is a controversial term because it defies a clearly delineated epistemology. Perhaps this explains why Judith Butler understands *representation* as an "operative term within a political process that seeks to extend visibility and legitimacy" (1990, 1). Indeed, visibility and legitimacy are ineluctably entangled and brought into sharp focus in the Niger Delta context, yet the relationship between visibility and legitimacy is fraught, as is evident in *Last Rites*. In the Niger Delta, claims to legitimacy seem to be at odds with the modality through which it is made visible—violence.

Violence is a crucial yet complex trope in Nigeria's energy culture: violence in the Niger Delta takes multiple, intricately related forms and textures. It complicates the very notion of violence as a physical phenomenon. Its forms are both spectacular and subtle. In its subtle aspect, violence is, as Rob Nixon argues, slow and deficient in visibility. Slow violence permeates every aspect of oil production in Nigeria, which complicates the understanding of any particular instance of violence (2011). The situation of oil extraction in the Niger Delta obstructs and destabilizes the idiom in which violence is articulated. I want to examine two photographs from the exhibition in terms of these dynamics, with particular attention to how legitimacy (or legitimacy claims) coalesces with insurrectionary visibility and thereby complicates the forms of violence that operate in the cultural landscape of Nigeria.

Figure 10 depicts the signpost that marks the first oil well in the Niger Delta, yet the image lacks any sense of historical monumentality. The signboard is in a sorry state, the inscriptions on its rusty white board fading fast, increasingly less visible amid the encroaching bush. Yet this neglected signboard is in fact symptomatic of the invisibility of Oloibiri—and perhaps the entire Niger Delta—on the agenda of social development, despite the fact that this village bore the first fruits of commercial oil for the Nigerian state. A cursory look at the photograph reveals, along the lines of Butler's analysis of the operations of representation, how both legitimacy and visibility are entangled and at work in this site.

Figure 11 was released to press by the insurgent group Movement for the Emancipation of the Niger Delta (MEND) as part of their guerilla propaganda in February 2006. This photograph has since gone viral on the Internet, and it has been read as an iconic IMAGE of the militarized sociality of energy production and the spectacular violence that flourishes in the Niger Delta. In a broader context, this image can be read in terms of conventions of representation that depict Africa as a crisis-ridden outback of filth, poverty, sickness,

Figure 10. The first oil-wellhead also known as "Christmas Tree." On the timeworn, rusted sign is written "Oloibiri Well No. 1. Drilled June, 1956. Depth: 12,000 Feet." (Ed Kashi)

Figure 11. Macon Hawkins, an American oil worker held hostage by The Movement for the Emancipation of the Niger Delta MEND. (Michael Kamber, *Last Rites Niger Delta* 11)

violence, insurgents, war, and death. To be sure, the Niger Delta embodies all these! But why this obsession with such dramatic images of youth violence, which Michael Watts describes as "the masked militant armed with the ubiquitous Kalashnikov, the typewriter of the illiterate" (2011, 62)? What is left out of this quintessential image of the spectacle of violence, which the oil complex in the Niger Delta seems to elicit in the public mind?

More precisely, looking carefully at the photograph itself, one can observe that the white man does not appear to be frightened or agitated as would a hostage: he seems calm and collected. How are we to understand MEND's release of this image, in terms of the relationship between their claims to legitimacy and the hypervisibility of violence that has been produced by the photo's circulation and reception? In attempting to draw attention to their social and environmental plights through propaganda and guerrilla tactics, the "resource rebels" complicate their claims to legitimacy by falling into the trap of violence in its intricate operation—a violence that constitutes for itself and within itself "the news," an ideological operation that, as Edward Said notes, "determine[s] the political reality" of a phenomenon (1982, 24–25). The workings of such operations contaminate the youths' moral claims for justice as they are made visible through the instrumentality of representation.

The political reality of cultural expression in the Niger Delta, which seeks to make visible the felt experience of petro-injustice, stages violence as a form of performance. This performance of violence is the visible expression of the experience of violence that is felt but eludes visibility because of the nature of its operation in the Niger Delta. In any case, what is salient in Figure 2 is that the slow violence of environmental injustice is further elided by the spectacle that characterizes the protocols of representation by which it is protested and made visible. And this spectacle tends to become the violence in itself. Indeed, the spectacle of militancy has come to constitute itself in the public mind as the characteristic violence of the Niger Delta, especially given the dominance of militarized insurgency in Nigeria's petromodernity.

See also: CHANGE, IDENTITY, IMAGE, PETRO-VIOLENCE, PHOTOGRAPHY, RESOURCE CURSE.

Nuclear 1

Matthew Flisfeder

Nuclear names not only a prominent form of ENERGY but also myriad ways of being in relation to energy, society, and the world. *Nuclear* occupied a significant place in postwar politics and culture, as a source of great energy and great destruction. But recent concerns about the development of nuclear capabilities in Iran and North Korea, as well as the 2011 DISASTER at the Daiichi nuclear power plant in Fukushima, Japan, point to the salience of nuclear technology today. In addition to its political valences, nuclear themes recur throughout postwar and contemporary popular culture. Is "nuclear" still an adequate energy metaphor, or does our continuing enthrallment to the reality and metaphor of belonging to a nuclear world impede thinking about our collective futures?

Nuclear technology began with Marie Curie's discovery of radioactivity in uranium in 1898 ("Marie Curie" 2014). Not until 1933 did Jewish Hungarian-born physicist Leó Szilárd discover the possibility of obtaining large amounts of energy, and explosive capability, from nuclear reactions (L'Annunziata 2007, 240). Szilárd filed for a patent the following year, claiming that he wanted to prevent his discovery from being weaponized. With the rise of fascism in Europe, however, he altered his position and urged his friend, Albert Einstein, to warn President Roosevelt about the growing nuclear research program in Germany. Szilárd thus helped spur the American-led Manhattan Project to develop nuclear weapons. In 1943, a research laboratory was set up in Los Alamos, New Mexico, under the directorship of J. Robert Oppenheimer, where the first atomic bomb was developed. The only two atomic bombs ever to be used militarily were detonated by the United

States at the end of World War II, over Hiroshima on August 6, 1945, and Nagasaki on August 9, 1945. The first nuclear power plant went online in the small town of Obninsk, near Moscow, in June 1954, six months after President Eisenhower gave his "Atoms for Peace" speech at the United Nations, advocating the development of peaceful uses of nuclear technology (Eisenhower 1953) and precipitating amendments to the US Atomic Energy Act to allow for commercial development of nuclear power plants ("Atomic Energy Act" 2013).

This tension between energy and weaponry is inherent to the history of nuclear technology. In nuclear fission, the nucleus of a particle splits into smaller parts, producing free neutrons or photons and releasing large amounts of energy. In bombs, the fissile material must be capable of sustaining nuclear chain reactions. In nuclear reactors, however, the rate of the chain reaction is controlled by rods of material that absorb the neutrons and slow the fission. Commercial reactors contain only a small percentage of fissile material, bombs approximately 90 percent (Marder 2011).

Anxieties about the nuclear arms race and the commercial development of nuclear energy recur throughout global popular culture of the postwar period. In Ishirô Honda's 1954 film, *Godzilla* ("Gojira"), a gigantic monster/dinosaur comes to life as the result of FALLOUT from US atomic weapons–testing in the Pacific. Stanley Kubrick's *Dr. Strangelove or: How I Learned to Stop Worrying and Love the Bomb* (1964) depicts American nuclear anxiety after the 1962 Cuban Missile Crisis. The film's sardonic juxtaposition of the "cowboy" riding the nuclear missile with the iconic IMAGE of the mushroom cloud depicts contradictory nuclear impulses—a fear of destruction linked directly to the imperialist, "frontier-minded," machismo that might well bring destruction about.

A parallel development to the commercialization of nuclear technology in the postwar period was the emergence of the term "nuclear family," what Gilles Deleuze and Felix Guattari (1983) mockingly dub the "mommy-daddy-me" relationship. The conjugal family of early bourgeois culture spread to the middle strata of the working population in capitalist countries, as the "class compromise" of the postwar period made possible a comfortable, consumerist lifestyle, best encapsulated in the common imagery of the suburban middle-class family. We might consider whether the rise of the nuclear family in the postwar period was one way of managing anxieties about nuclear technologies, quite literally domesticating them.

The popular animated series *The Simpsons* (1989–present) offers a sharp commentary on postwar American society by satirizing both the nuclear family and the nuclear power plant. One episode, "Two Cars in Every Garage and Three Eyes on Every Fish," juxtaposes the suburban ideal with rising concern about the hazards of nuclear waste; no matter how loveable the bumbling patriarch Homer Simpson might be, he is a terrifyingly ineffectual nuclear safety inspector. Concerns over the safety of nuclear power are also evident in the eerily prescient film *The China Syndrome* (1979), which tells the story of a news anchor (Jane Fonda) and her camera operator (Michael Douglas) who witness and secretly film the technicians at a power plant responding to a nuclear meltdown. The film was released on March 16, 1979, twelve days before a malfunction in a secondary cooling pump at the Three Mile Island plant near Harrisburg, Pennsylvania, caused a partial meltdown and allowed some radioactive steam to escape. Box office sales for *The China Syndrome* increased following this event ("Timing Is Everything" 2013).

The Three Mile Island incident and the 1986 Chernobyl disaster in Ukraine were signal events in the growing antinuclear movements in Europe and the United States that helped bring GREEN issues to the forefront of progressive politics and found echoes in 1980s popular culture. In Robert Zemeckis's *Back to the Future* (1985) an early prototype of Doc Brown's DeLorean time machine was powered by a plutonium rod, causing Michael J. Fox's character, Marty McFly, to exclaim, "Are you telling me that this sucker is nuclear?!" Further tinkering on the time machine replaces the nuclear engine with one that runs on compost.

A contest between renewable and nuclear energy, and between nuclear energy and nuclear weaponry, is also at the heart of the often forgotten, and rather ill-conceived, *Superman IV: The Quest for Peace* (directed by Sidney J. Furie, 1987), in which Superman promises to rid the world of nuclear weapons. When Superman throws all of the nuclear missiles on Earth into the sun, his archenemy Lex Luthor secretly plants on one of them a device containing material that gives birth to the villain Nuclear Man. Superman himself is SOLAR powered, drawing his strength from Earth's yellow sun (not the debilitating red sun of his home planet, Krypton).

Nuclear and solar technologies are taken up as political metaphors in Langdon Winner's book, *The Whale and the Reactor* (1986). Winner draws upon Lewis Mumford's distinction between democratic and authoritarian technics. For Mumford, authoritarian technologies are powerful and system-centered but unstable. By contrast, democratic technologies are centered on the natural rhythms of humanity; they are relatively weak but more durable (Mumford 1964, 2). Winner uses this distinction to compare the politics attached to nuclear and solar technologies. Nuclear demands authoritarian, centralized, and systematized control; solar is more easily distributed and fosters more democratic and egalitarian forms of political organization. Nuclear energy is powerful but unstable. Solar energy is more accessible, comprehensible, and controllable than nuclear energy (Winner 1986, 32–33). Solar energy implies a commons—what Slavoj Žižek refers to as the "commons of external nature" (2009, 91).

As neoliberalism and austerity in postcrisis capitalism are centralizing power and concentrating wealth in the hands of the one percent, Winner's account of nuclear authoritarianism aptly describes current configurations of political power. Crises in capitalism disadvantage the working classes and the indigent more than the wealthy, and new forms of authoritarianism and plutocracy in liberal democratic countries include policies and policing that favor the rich. Given its association with authoritarianism, "nuclear" remains a useful metaphor for thinking critically about politics in the present. But in order to go (back) to the future, what we need is not merely new ENERGY REGIMES but also new energy metaphors for thinking beyond the limits of the present.

See also: AMERICA, FICTION, RENEWABLE, SUPERHERO COMICS.

Nuclear 2

Gabrielle Hecht

Fukushima: from the coasts of Tamil Nadu to the halls of the German Bundestag, the word now stands for danger and deception, contamination and vulnerability. Every day brings new distress. Cesium-137 clings tenaciously to the soil and buildings of northeastern Japan. Radioactive fish promenade across the Pacific. Over 40 percent of children examined by the Fukushima Health Management Survey have thyroid abnormalities; no one really knows yet what that means for their health. Contractors hired for cleanup operations rely on yakuza networks for a steady stream of disposable workers and toss contaminated debris into forest glens and mountain streams when they think no one is looking. This "cleanup" is projected to take forty years and cost 250 billion dollars. How could nuclear things ever be commonplace?

Rubbish, says the nuclear industry: we are shocked (*shocked!*) by such irrationality. What was exceptional at Fukushima were the circumstances, not the technology. An earthquake-plus-tsunami—who could have predicted it? The event was certainly unfortunate, but we will treat it as a learning experience. Nothing irreversible has happened! The area can be decontaminated. No one has died from radiation poisoning. The only thing people are suffering from is radiophobia, a well-documented form of hysteria also observed after Chernobyl. Why is everyone always picking on us? We are misunderstood. Unloved. Those reactors were old. The new ones are different, and we need hundreds of them to counter global warming. The survival of the species depends on nuclear power once again becoming commonplace.

Exceptional or banal? It is a perpetual question in the history of nuclear things.

In the beginning, there was The Bomb. It ended The War. Splitting the atom ruptured human history. Out with the age of empire, in with the nuclear age: so said the leaders of the "free world" and their weaponeers who hotly pursued planet-pulverizing potential. But fear not! Focus instead on imminent planetary UTOPIA. ELECTRICITY too cheap to meter, an end to hunger, cures for cancer. Working in the nuclear field was so much more fun than doing "conventional" science or engineering (and got so much more funding). Exceptional indeed, and in the best possible way.

Alas, there were spoilsports.

Breaking the building blocks of matter also created cracks in our chromosomes, dabbled with our DNA: so showed the geneticists who studied the generational effects of radiation exposure among the survivors of Hiroshima and Nagasaki (Lindee 1994). Photos of those cities, censored for two decades, trickled out to reveal horrifying burns, peeling skin, and ashy landscapes. As ever-larger thermonuclear weapons exploded in the Nevadan desert and the Kazakh plains, the giant ants and towering lizards stomping through the reels of B movies seemed harbingers of a radioactive FUTURE. Targeted cobalt-60 beams might treat tumors, but radiation exposure also caused cancer, leukemia, and other diseases. Citizens began to march: first against the weapons, then against the power plants. Exceptional, oh yes, but in the worst possible way.

Alarmed at the backlash, industry publicists did an about-face. Nuclear reactors were just another way to boil water, they insisted. They simply did so better and cheaper than other power plants: a mere twenty-five tons of uranium could produce the same amount of electricity as three million tons of COAL. Plus, it offered ENERGY independence. Europeans and Americans need no longer rely on fickle MIDDLE EAST oil suppliers. Measured in dollars (or francs, or pounds) per life saved, the industry spent more money on safety than any other (Hecht 2012, ch. 6). And frankly, the fuss about radiation was just silly. Radioactivity was a natural phenomenon. People were exposed to radiation from all kinds of sources: plane flights, granite countertops, kitty litter . . . not to mention X-rays, CT scans, and other medical diagnostic tools. Even food! Especially bananas, which contained traces of a radioactive potassium isotope. Oh yes, bananas. You would need to eat twenty million of them to get radiation poisoning. Some public relations genius even concocted the "banana equivalent dose" (BED) to offer a "friendly" way of explaining radiation doses. (Though let us pause to give radiation protection specialists some credit: they denounced the BED as totally misleading.) Still, you cannot get much more banal than bananas.

Exceptional, banal.

For a while, the nuclear power lobby felt certain of victory. Especially with global warming, the problem for which the nuclear solution had long been waiting. In January 2011, the French nuclear multinational Areva celebrated its tenth anniversary with an ad portraying nuclear power as the latest in humanity's long romance with energy. The last scene shows ecstatic young adults partying on a rooftop. The voice-over chirrups: "the history of energy is still being written. Let's keep writing it, but with less CO_2" ("pub

AREVA" 2011). Furious antinuclear activists filed a complaint with France's advertising ethics board, essentially arguing that the ad inappropriately banalized nuclear power.

Banal, exceptional.

Then came Fukushima. It brought the best bananization—oops, I mean banalization—effort yet: a four-minute animation (apparently created by private citizens) aimed at explaining the accident to Japanese children.

> Ever since the big earthquake, Nuclear Boy has had an upset stomach. [Squirt squirt.] "Unhh, unhh, my tummy hurts! I can't hold my poo any longer! Unhh!" Nuclear Boy is notorious for his stinky poo. It would surely ruin everyone's day if he pooped!
>
> We measured the stinky level around Nuclear Boy. Thankfully it wasn't that stinky, so we figured he had just passed some gas. [More squirting sounds.] ("Nuclear Boy" 2011)

The doctors come to visit. They work around the clock to ensure that Nuclear Boy does not poop again. He does fart some more. But the smell will dissipate after a few weeks. Besides, Nuclear Boy is not the first to have this problem. First there was Three Mile Island Boy. And then there was Chernobyl Boy! He "literally" pooped in the classroom, and it was diarrhea, and it went all over the place. Gross. Even if that happens in Japan—and let us hope it does not, ewww!—it will not be as bad as Chernobyl. Meanwhile, let us pray for the people in Fukushima. That is the least we can do for receiving Nuclear Boy's energy for so many years.

Turns out, there is one thing more banal than bananas.

Shit.

See also: COAL ASH, FALLOUT, RISK.

Off-grid

Michael Truscello

Should radical anticapitalists focus their efforts on sabotage and other forms of rupture designed to interrupt the flows of global capitalism, as many insurrectionary anarchists advocate? Or should they focus on sifting the debris from "the dead labours which crowd the earth's crust in a world no longer dominated by value," as Alberto Toscano argues (2011, 40)? Mike Davis describes the challenge for revolutionaries in a dying world dominated by capitalism: "Since most of history's giant trees have already been cut down, a new Ark will have to be constructed out of the materials that a desperate humanity finds at hand in insurgent communities, pirate technologies, bootlegged media, rebel science and forgotten utopias" (2010, 31). To sabotage or to scavenge? These are the options on offer in the radical Left literature on industrial and ecological collapse.

But beyond the dichotomy of sabotage or scavenge lies a third movement: living off the grid. The grid symbolizes modernity and materially perpetuates its project. While some radicals equate human survival with annihilation of the grid and the sociotechnical assemblages it enables, and others ask what can be salvaged from its components, a smaller group of radicals (as well as right-wing militias and religious fundamentalists) believe the only way to survive the collapse of industrial civilization is to leave its grids behind. I want to complicate these scenarios by suggesting that a planet now dominated by hyperobjects (T. Morton 2013) like global climate change, NUCLEAR contamination, and noxious industrial drift is always-already toxic. Our world is haunted by the remainders of modernity no matter what we do—sabotage the grid, recover and reassemble the debris, or simply bug out.

The popular concept of living off-grid brings together privileged fantasies of individual autonomy, racialized visions of wilderness as a place for whiteness, and reified notions of INFRASTRUCTURE as a self-contained set of materialities and practices. In its most pedestrian definition, *off-grid* refers to living detached from the electrical grid, which often also entails separation from industrial sewage treatment, food production, and other amenities of urban existence. As Hannah B. Higgins notes, "it is now possible to have electricity, phone, and water without the coordinating spatial system associated with wires overhead and pipes below, be they carriers of telephone calls, digital information, electricity, or water" (2009, 76). Off-grid is therefore not necessarily a form of anarcho-primitivism that identifies with hunter-gatherer societies; some forms of off-grid living include technological accoutrements such as SOLAR and wind ENERGY. In addition, actually existing GRIDS are rarely the enclosed material containers the term suggests: redundancies and tangential forces abound, to "guard against disruption" (ibid., 75). Modern grids often assemble public and private sector labor and materialities that traverse many forms of governance. The modern grid is a "mongrel assemblage" (Genosko 2013, 112), not a monument to modernist rationality.

While it is difficult to assess the number of people living off-grid by choice or circumstance, there may be as many as 2 billion people off-grid worldwide, and 180,000 or more in the United States (Higgins 2009, 76). TV shows such as *Doomsday Preppers*, *Apocalypse 101*, *Dude, You're Screwed*, *Naked and Afraid*, and *Off the Grid with Les Stroud* have fostered the popularity of off-grid living. There is even a glossy magazine, *Offgrid*.

Although off-grid living has been imagined as a panacea for a dying world dominated by capitalism, it is as enmeshed within the ruins of modernity as the worst kinds of urbanism. David R. Cole captures this problematic in his assessment of petrocultures: "We are petro-citizens. This statement is true even if we reject the everyday use of cars, if we only use public transport, or if we believe in a green solution to impending environmental catastrophe" (2013, 5). Similarly, the grid suffuses off-grid living by conditioning subjects who move between grid and off-grid, by striating off-grid spaces, and by toxifying off-grid environments through industrial toxic drift and other forms of contamination. One can no longer simply leave the grid when grid-enabled hyperobjects are ubiquitous. Today, whether one sabotages the grid or leaves for the protection of rural environs, everyone seems fated to keep living in some way on the grid or in its toxic wake.

Consider, for example, the First Nations communities living downwind from oil refineries in Northern Alberta. Measurements of air pollution inflicted on these communities indicated volatile organic compounds six thousand times higher than normal, and some chemicals registered in concentrations equal to those present in Mexico City during the 1990s, when it was the most polluted place on earth (Valentine 2013). Elsewhere, dioxins, polyvinyl biphenyls (PCBs), and other organochlorines have made the ARCTIC one of the most toxic places on Earth, and FALLOUT from the Chernobyl disaster spread over an area at least 62,000 miles. Such drifting horrors are beginning to amalgamate in ways detrimental to life: for example, scientists fear that channels of water opening in Arctic ice may increase the presence of mercury in Arctic food chains (Marshall 2014). Where will off-grid dreamers hide from the hyperobjects of the ANTHROPOCENE?

While some point to Thoreau's *Walden* as the urtext of off-grid living, most of human civilization has never lived on the grid; stories of life without the electrical grid are, therefore, plentiful. Perhaps the stories more indicative of contemporary off-grid living are those that examine what anthropologist Brian Larkin calls "the politics and poetics of infrastructure," which imagine infrastructure "as concrete semiotic and aesthetic vehicles oriented to addressees" (2013, 329). These texts figure survival in terms of infrastructure that is either present or absent. The PHOTOGRAPHY of Edward Burtynsky often captures the objects and landscapes involved in powering the grid. One could also examine "drowned town" stories in recent Canadian novels like Michael V. Smith's *Progress* (2011) and Riel Nason's *The Town That Drowned* (2011): such narratives explore the affective qualities of hydroelectric DAMS under construction. Off-grid aesthetics need not be understood exclusively as survivalist literature.

The discourse of off-grid living may be more accurately called a discourse of post-industrial exile—an attempt to escape the demonstrable physiological stresses of industrial civilization, the authoritarian creep of SURVEILLANCE, and the violence of the nation-state. While the grid form was never inherently authoritarian and existed prior to European modernity, it has participated in "the initiation and naturalization of property regimes" that enabled the proliferation and dominance of global capitalism (Rose-Redwood 2008, 51). Exile from the grid invites dramatic challenges, including the increasing proliferation of grids worldwide (especially through engineering megaprojects) and the relatively immobile artifacts that grids leave behind: "Poles, lines, and transformers are simply difficult to remove from the landscape, once installed. Moreover, linked to their staying capacity, dominating discourses continue to impose their interpretations on symbols after their manifestation as enduring objects" (Winther 2011, 218). This is one reason the nation-state appears in the popular imaginary as something indomitable: even if the desire for the state form could be curbed, its infrastructural relics, from large dams to highway systems, would haunt our landscapes for decades. This is also why modern colonial powers have always promoted the grid form—as in the neocolonial exploits of energy grid construction in contemporary Africa, which one scholar compares with "the 19th-century 'scramble for Africa'" (McDonald 2009, xvi).

Roy Scranton argues that the Anthropocene poses the philosophical challenge of recognizing that industrial civilization is "*already dead*" (Scranton 2013). "The sooner we confront this problem," he continues, "and the sooner we realize there's nothing we can do to save ourselves, the sooner we can get down to the hard work of adapting, with mortal humility, to our new reality." The challenge is "to learn how to die not as individuals, but as a civilization." Although he does not use the word, Scranton reframes the meaning of suicide in the context of attenuated environmental DISASTER. The petromodern state form may be something like Paul Virilio's "suicidal state" (1998), but its characteristic inanimate objects (oil PIPELINES, hyperboloid cooling towers) have even more agency than the officers of the Schutzstaffel or the Khmer Rouge. Ironically, Scranton's response to the suicidal trajectory of the Anthropocene remains a humanist one: this is his primary ignorance and also the short-sightedness of most off-grid living. Resistance to the suicidal state of neoliberalism requires more than a cabin in the woods, a casual understanding of permaculture, and a high-powered rifle. Resistance now requires that we learn how to die not only as a civilization but also as a species.

See also: CHANGE, DETRITUS, GUILT, INFRASTRUCTURE, NETWORKS.

Offshore Rig

Fiona Polack and Danine Farquharson

Huge Iron Island
37 stories high, two city blocks square
impervious
to the attacks
of an indignant ocean.
Mother Earth is victimized
by this man-made mechanical rapist,
and grudgingly surrenders her treasure
to the oil magnate with the
dollar sign eyes.
—GREG TILLER, "Rig"[1]

Despite being linked to the same global circuits of power as the land-based oil and gas industry, the offshore rig poses its own problems of conceptualization. These difficulties are largely attributable to the longstanding figurative history of the sea as "protean" (Raban 1992, 2). In its association with the ocean, the technologically sophisticated rig becomes embroiled in centuries-old discourses.

The sea "has long functioned as Western capitalism's primary myth element," according to Christopher Connery (2010, 686). Indeed, contemporary discussion of offshore oil and gas invokes the ideal of mastering the ocean for economic gain. In their use of the overdetermined language of the frontier, and their emphasis on human ingenuity and technological prowess, accounts of drilling in deep water and offshore ARCTIC locations resonate with Hegel's comment that "the sea invites man to conquest" (1956, 90). In the wake of catastrophes like the Deepwater Horizon SPILL, and with growing awareness that the world's oceans "hitherto thought virtually immune from human tampering, might be gravely endangered" (L. Buell 2009, 201), the notion of conquering the sea appears increasingly anachronistic (*Offshore* 2013). However, Western ideas of the ocean as a site for

1. "Three weeks after the first anniversary of the *Ocean Ranger* disaster, the *Evening Telegram* printed a poem by one of the men who died, Greg Tiller. Written on his last trip home and found by his family after the disaster, Tiller's poem takes on prophetic—and commemorative—meaning" (Dodd 2012, 110–11).

adventuring and economic gain persist in public and corporate discourse about offshore ENERGY.[2] Perhaps the tenacity of this figuration can be linked to John Gillis's point that "although fully half of the world's peoples now live within a hundred miles of an ocean, few today have a working knowledge of the sea" (2013). Frontier discourse also persists because, since the 1940s, the majority of offshore platforms have been built out of sight of land (Freudenburg and Gramling 1994, 20). This invisibility is a function of where oil and gas deposits are located but also of political, economic, and aesthetic considerations.[3]

In the absence of direct contact, the offshore rig lends itself to myriad reimaginings beyond the oceanic frontier.[4] In his quest for extreme and unfamiliar locations for *Breaking the Waves* (1996), filmmaker Lars von Trier found in the rig a space that is threatening (the men who WORK on the rig are alien to the people of the land), quotidian (the workers perform commonplace tasks including showering, eating, and dressing), but also, and most importantly, transcendent. Frequently shot with the god's-eye-view camera angle, the rig ultimately functions as a structure of faith, a weird sacred space wherein the miraculous or the ineffable are possible.

The rig's resemblance to an island magnifies its figurative potential. As Rod Edmond and Vanessa Smith note, "Islands are the most graspable and slippery of subjects. On the one hand they constitute a bounded and therefore manageable space . . . . On the other hand, they are fragments, threatening to vanish beneath rising tides or erupting out of the deep" (2003, 5). Both figurations are discernable in imaginative deployments of the rig. Lisa Moore's novel, *February* (2009), uses the sinking of the *Ocean Ranger* off Newfoundland in 1982 to consider the devastating emotional impact of "vanishing" rigs. The protagonist, Helen, endures years of anguished mental efforts to travel back through time and space "to be with [her husband] Cal when the rig goes down" (294). The narrator represents the difficulties of Helen's situation: "She is there. Helen is there with him. But she is not there, because nobody can be there. [The rig] is there and it is not there" (300–01). The rig vanishes as it sinks in the cold North Atlantic storm, but this narrative moment makes clear the offshore rig's continuing resonance.

In China Miéville's weird short story "Covehithe" (2011), drowned rigs, including the *Ocean Ranger*, erupt out of the sea and, in an inversion of the oceanic frontier narrative, threaten to conquer the land.[5] The *Rowan Gorilla*, the first such animated rig to come ashore, is a horrific sight as it

2. The pervasiveness of the frontier narrative is apparent in the title of Eric Skjoldbjærg's Norwegian oil-boom thriller, *Pioneer* (2013).

3. See the section on "Visual Aesthetics" (20.5) in *Devon Beaufort Sea Exploration Drilling Program* (Devon Canada Corporation 2004).

4. Photography on offshore platforms is strictly regulated. See Ryan Carlyle, "Taking Pictures on an Offshore Oil Rig is Serious Business" (2013).

5. The *Guardian*'s short fiction project "Oil Stories" commissioned writers from around the world "to drill down through layers of cliché and cant to explore the hidden reservoirs which fuel our dreams and power our nightmares" (Miéville 2011). The stories, including China Miéville's "Covehithe," were published online.

> walked through buildings, swatted trucks then tanks out of its way with ripped cables and pipes that flailed in inefficient deadly motion, like ill-trained snakes, like too-heavy feeding tentacles. It reached with corroded chains, wrenched obstacles from the earth. It dripped seawater, chemicals of industrial ruin and long-hoarded oil.

Miéville evokes both the horror story and the "petroleum sublime" (Hatherley 2011). The protagonist and his daughter respond to the "come-back" rig with exhilaration and, as the "petrospectral presence" comes close to the girl, she stops breathing. In a further terrifying twist, the rigs breed and spawn "unsteady six-foot riglets." For the characters in "Covehithe," the most compelling sights are underwater videos of "violent and revenant rigs" walking on the ocean floor. In Miéville's fictional world, oil platforms cannot be abandoned, eliminated, or ignored. The animated, offshore rigs in "Covehithe" end up as both tourist attractions and the subject of kids' clubs, coloring books, and games. The power of global capital to translate the horrific into the profitable is subjected to Miéville's satiric dystopian vision.

Miéville's literary fantasy about the afterlife of offshore oil rigs is countered in current utopian visions for platforms on exhausted oil deposits. The seasteading movement, which aims to create permanent ocean communities at sea, is currently contemplating using abandoned oilrigs as the basis for new colonies (seasteading.org). As Philip E. Steinberg, Elizabeth Nyman, and Mauro J. Caraccioli note with skepticism, "political actors have long drawn on utopian imaginaries of colonizing marine and island spaces as models for idealized libertarian commonwealths" (2012, 1532). Meanwhile, intense discussion continues among scientists, politicians, oil companies, and environmental groups about the potential benefits of converting defunct oil platforms into marine reefs (Hecht 2012), and abandoned rigs in the Pacific are being transformed into hotels for scuba divers (Sesser 2010). By contrast, some jurisdictions, including those governed by the Oslo-Paris Commission (OSPAR) in North Sea waters, insist that all "non-virgin materials" in decommissioned rigs must be removed from the ocean (Jørgensen 2012, 60). Conversation around the fate of rigs is unquestionably necessary. In the North Sea alone, 129 offshore installations had already been decommissioned by 2009; by 2025 over 220 UK production fields and their associated structures will have met the same fate (ibid., 57).

At the same time that ocean drilling for oil and gas is imagined as the work of adventurers on an oceanic frontier, existing sea-based rigs are rapidly becoming defunct. Abandoned rigs, in their resemblance to island spaces, generate what Greg Dening identifies as "near-magical inventiveness" (2003, 201). But the offshore rig also illustrates—perhaps is even the metonymic marker for—the transient nature of oil and gas exploration.

See also: FICTION, PETROREALISM, SPILL, SPILLS, UTOPIA.

Petro-violence

Michael Watts

There is something unsettling about the world of Big Oil, not least the overwhelming intellectual vertigo it produces. Secrecy, guardedness, defensiveness, and corporate ventriloquism are hallmarks of the industry. Despite its technical expertise and scientific sophistication—drilling in deep water is like putting someone on the moon, oil mavens like to say—there is a startling degree of inexactitude, empirical disagreement, and lack of (or lack of confidence in) basic data. Why are the simplest facts of the oil world so vague, opaque, and elastic? Epistemological murkiness greets seemingly mundane, banal questions of how much oil there actually is and how much enters the world market. Oil's finiteness notwithstanding—one recent estimate posits 1.2016 trillion barrels of proven crude reserves (OPEC 2016)—for every Malthusian like King Hubbert (of Peak Oil fame) there is a prominent doubter like MIT's Morris Adelman, for whom oil is inexhaustible (see T. Mitchell 2011, 188).

What is true for quantity is also true for price and the operations of the oil market—I use the term loosely. The dynamics of price determination remain utterly confusing, mindboggling in their irrationality and unpredictability. Oil, says Kent Moors in *The Vega Factor*, "does not respond as most other goods do to market factors . . . . To date there has yet to emerge a consistent theory . . . fully explaining how oil prices operate" (2011, 43). Relations between DEMAND and supply fail to operate in any predictable fashion and are unrelated in any determinate way to price. No single theoretical discipline in the social sciences, says Oystein Noreng in *Crude Power*, "has been successful in analyzing the energy markets" (2005, 8). At the center of neoliberal capitalism stands a commodity for which the terms *market rationality* or *market fundamentals* seem irrelevant to the operations of the oil and gas global value chain.[1]

Both industry and academic commentators argue that there is great deal of certainty about oil, a global commodity whose powers are often assumed to be Olympian. Oil has been vested with gargantuan, deterministic, and even magical powers—not least its capacity to generate conflict, wars, and violence. It has been called a "curse," the "devil's excrement," the source of the "Dutch Disease." Oil distorts the "natural" course of development. Michael Ross's (2012) dystopian account of how petroleum wealth shapes the development of oil-producing nations addresses the scale, instability, and secrecy surrounding petro-states—properties that collectively explain the so-called paradox of plenty: the state pathologies and human developmental failures of putatively oil-rich states. For Ross, oil produces all manner of pathologies: oil "dependency" is said to "hinder democracy" (as if copper might promote constitutionalism); oil revenues permit low taxes and encourage patronage, fund bloated militaries that enable despotic rule, and create a class of state-dependents employed in modern industrial and service sectors who are unlikely to push for democracy. *New York Times* columnist Thomas Friedman has even identified a First Law of Petropolitics: the higher the average global price of crude oil, the more free speech, free press, fair elections, an independent judiciary, the rule of law, and independent political parties are eroded. Hugo Chavez and Mahmoud Ahmadinejad have been the law's most devious exponents. Fundamental uncertainty about the basic empirical contours of the industry contrasts starkly with strange certainty about the purportedly deterministic effects of oil and gas resources.

In one influential line of thinking, the Promethean qualities of oil in oil-producing states usher in an economy of hyperconsumption and spectacular excess: bloated shopping malls in Dubai, corrupt Russian "oilygarchs," and waves of crime and social dysfunction. There is even a psychological appellation for this condition: Gillette syndrome. In the 1970s, psychologist ElDean Kohrs showed how a commodity BOOM in the COAL town of Gillette, Wyoming, brought a wave of crime, drugs, violence, and inflation. Gillette syndrome would afflict new gas fields of Wyoming, indigenous oil communities in Ecuador, and rough-and-tumble Russian oil fields of Siberia. Gillette syndrome also has a clear political manifestation. In *The Bottom Billion*, Oxford economist Paul Collier argues that greater oil dependency produces an increased likelihood of civil war and violence: oil revenues captured by rapacious political elites foster autocratic rule, "the survival of the fattest" (2007, 46). Those revenues are predated or looted by rebels for whom oil finances not an emancipatory politics (social justice, self-determination) but instead organized crime

1. Over one billion barrels of oil can be traded in a day on the New York Mercantile Exchange and the Intercontinental Exchange, much of this being "paper oil" (never delivered physically as oil), which is to say part of the booming commodities futures (and derivatives) market. See Ahn 2011.

conducted as violent rebellion and war. Here *oil* means oil *money* and *oil politics* means *rents* captured by state agencies and the political class; the agency of oil corporations, oil service industries, or financial institutions is almost nonexistent.

Michael Klare's *The Race for What's Left* (2011b) exemplifies another approach to petro-politics and petro-violence: a Malthusian account of corporate oil and global geopolitics (what Klare calls the US global oil acquisition strategy). Resources like oil are finite, while industrial capitalism's ENERGY appetite continues to grow across the globe; more than half of the oil consumed between 1860 and the present was used since 1980 (T. Mitchell 2011, 233). The specter of Peak Oil amplifies the geopolitical pressures and violent struggles precipitated by tight oil markets, slower rates of discovery, and difficult operating environments (the "end of cheap and easy oil," as the oil industry puts it); states and corporations embark upon a desperate scramble for oil. Precisely because of oil's strategic aspects, exploration and development have a praetorian cast: a frontier of violent ACCUMULATION working hand-in-hand with militarism and empire. In this account, we are about to enter a new "thirty-years war" (Klare 2011a, 1) for resources, characterized by market volatility, ruthless resource grabs, and a sort of military neoliberalism. Petro-politics here is less a violent struggle over oil rents than the post–Cold War power politics of a Big Oil-Big Military-Big Imperial State triumvirate that understands oil as a matter of national security: the 2003 invasion of Iraq is, in this account, paradigmatic.

Each of these accounts posits a different relationship between oil and violence. In the modern period—and particularly since 2011—petro-violence has been associated with two powerful figures: the mercenary and the Muslim terrorist. An example of the former is the so-called Wonga coup in Equatorial Guinea led by Simon Mann, in a striking replay of Frederick Forsyth's novel *The Dogs of War* (Roberts 2006). In 2004, mercenaries (organized by mainly British financiers) attempted to replace President Teodoro Obiang Nguema Mbasogo with exiled opposition politician Severo Moto, who was to be installed as the new president in return for granting preferential oil rights to corporations affiliated to the conspirators. The result was a fiasco—a "swashbuckling fuck up" in Mann's account (2011, 46); he and other mercenaries were arrested in Harare and imprisoned in Zimbabwe and Equatorial Guinea. In its lethal combination of tyranny, corruption, psychosis, mercenary interests, and nasty geopolitics, the Wonga coup is petro-violence in its purest and most distilled popular form.

Petro-violence has another recent avatar: a lethal, combustible mixture of oil, radical Islam, American empire, and the bulge in populations of youth, as exemplified by Osama bin Laden (son of an oil contractor) and al Qaeda. As Timothy Mitchell shows, oil was never able to create a political order for its own purposes. Among various forces across the MIDDLE EAST, organized Islam had its own purposes and dynamics that, in the context of oil profits, arms, and corrupt comprador classes, provided a nonsecular alternative—political Islamism—capable of undercutting the "political control of Arabia" that Big Oil (and imperial states) required (2011, 213). The rise of violent global Salafism was one expression of this conjuncture.

The world of oil and gas is and has been saturated with violence: symbolic, cultural, political, ecological, and economic. Oil bears the hallmark of what Hannah Arendt, echoing Marx in volume 1 of *Capital* (1867), called "the original sin of primitive accumulation," dripping with blood and dirt (1957, 36). The annals of oil are an uninterrupted chronicle of naked aggression and the violent law of the corporate frontier. This violence is palpable and existential. To enter the shale gas lands of North Dakota, or the oil fields of Ecuador, Siberia, or the Niger Delta, is to experience petro-violence directly; these spaces are securitized and surveilled, rough and tumble, confrontational and uncomfortable. They bear the marks of violent enclaves everywhere. A distinct genre of PHOTOGRAPHY—what one might call "petro-imagery"—powerfully captures this latent or actual violence around the world: Richard Misrach's *Petrochemical America*, Edward Burtynsky's *Alberta Oil Sands #6*, Ernst Logar's *Invisible Oil*, Rena Effendi's *Pipe Dreams*, and George Osodi's *Delta Nigeria: The Rape of Paradise*.

It is one thing, however, to register the association between oil and violence but quite another to chart the complex traffic between them. The challenge is to acknowledge the violence that attends the upstream and downstream sectors of the oil industry yet to not presume that violence inheres in the commodity itself (a sort of commodity determinism). The politics of oil cannot be deterministically reduced to the survival of the fattest, patrimonial politics, and oil wars. To locate the root of conflict and particular sorts of politics in a global commodity—even a resource as indispensable as oil—tells us very little about how patrimonial regimes can deliver very different political and economic orders, how to think about the uneven institutional and governance capabilities across petro-states, and why violence occurs in different forms in different places—and, in some parts of the world, not at all.

See also: ADDICTION, CORPORATION, ENERGOPOLITICS, IMAGE, NIGERIA, RESOURCE CURSE, STATISTICS, SURVEILLANCE.

Petrorealism

Brent Ryan Bellamy

In ecological thought, thinking big is back in a big way. And why not? The twin problems of global warming and global pollution are intensified by an energy-reliant system of ACCUMULATION and dispossession that operates at a global scale. Thinking big seems to match the scale of solution-seeking to the size of the problem. Announcing the emergence of an interdisciplinary field they call "energy humanities," Dominic Boyer and Imre Szeman frame this problem as an ecology-energy impasse: "It is not an exaggeration to ask whether human civilization has a future. Neither technology nor policy can offer a silver-bullet solution to the environmental effects created by an energy-hungry, rapidly modernizing and expanding global population" (Boyer and Szeman 2014). The discursive mode arguably most interested in coming to terms with the scope of our ecology-energy impasse is that of theory, with examples ranging from Eugene Stoermer's and Paul Crutzen's theorizations of the ANTHROPOCENE (Crutzen 2002; Crutzen and Stoermer 2000) to Timothy Morton's attempt in *Hyperobjects* to furnish a language suitable to new materialism and to what he calls the "ecological emergency" (T. Morton 2013). But how do we begin to think *between* the proliferating big ideas of geology, climate science, new materialism, and the energy humanities?

I argue that the most salient risk in contemporary ecological theorizing is not in trying to think too big; rather, it is a problem of taking too easy a path toward that bigger picture. Our attempts to imagine social and ecological relations as totalities can be expressed only in complex and indirect ways, lest we fall back into what Hegel called "picture-thinking"

(*Vorstellung*; Hegel 1977). In *Realism in Our Time*, Georg Lukács warns against "a tendency to overemphasize *abstract* aspects of new subject-matter" to the detriment of "concrete realities" (1971, 115–16). To avoid the pitfall of mistaking the abstract whole for the sum of its concrete parts, I posit *petrorealism*—literary, cinematic, and gaming narrative forms, for example—as a way to mediate the scalar problem between thinking big and the specific situations and contexts of petromodernity. *Petro-* is meant to posit that all texts produced within petroculture are functionally marked by the ontology of oil, even as they anticipate a world after oil. *Realism* emphasizes the forms of mediation of the various scales simultaneously implicated in specific instances within a larger whole. *Petrorealism* aims to offer a better grasp on the ecology-energy impasse.

Petrorealism (or its absence) is what is really at stake in Amitav Ghosh's seminal essay "Petrofiction," where he observes that the "oil encounter" has not produced an equivalently rich corpus of novels as the spice encounter (Ghosh 1992a, 138). Extending Ghosh's desire for big thinking, Peter Hitchcock suggests that sugar and coffee are commodities analogous to oil (Hitchcock 2010, 81). But if we understand Ghosh to be not merely marking a paucity of oil fiction but also expressing a desire for petrorealism, then these commodities are not so easily substituted for one another. Attention to the formal strategies for representing the oil encounter would reveal that thinking big is itself among the subjects of this fiction. Realism, in its varied forms and modes, has a penchant for narrating structure and marking scale without losing site of specificity. Indeed, Abdelrahman Munif's *Cities of Salt* quintet (1984–1989) and Upton Sinclair's *Oil!* (1927) fuel Ghosh's and Hitchcock's desires for a realistic petrofiction. For Ghosh, the slow and careful details of Munif's story make it stand out: for instance, few of the oil developers from the United States are named and instead are simply referred to as "the Americans," one exception being Sinclair, who leaps out from the page like oil gushing from a well. Hitchcock emphasizes Upton Sinclair's realistic portrayal in *Oil!* of the beginnings of US oil production and dependence, a seminal moment in oil's centrality to the twentieth-century American political and cultural imaginary. Yet as Hitchcock observes, oil's centrality manifests primarily in its invisibility: "it is oil's saturation of the infrastructure of modernity that paradoxically has placed a significant bar on its cultural representation" (2010, 81). Petrorealism could elaborate the near omnipresence of oil in everyday life in an attempt to defamiliarize or to make strange our petrosubjectivity.

Examples abound of novels, films, documentaries, and other kinds of texts that instance what petrorealism could be and do. Situated within distinct formal mechanics, the following examples think big without falling into the trap of picture thinking. They fall into five provisional categories:

1. *Maps of energy presents that do not foreground energy*: Consider Noël Burch and Allan Sekula's exploration of container ships and the global circulation of commodities in *The Forgotten Space* (2010), Max Brooks's depiction of social totality through circulation and exchange figured as contagion in *World War Z* (2006), or Steven Soderbergh's chart of global flows and borders, figured

through the drug trade or the spread of disease and the development of vaccines, in *Traffic* (2000) and *Contagion* (2011).

2. *Postcolonial film and writing*: In the recent short film *Pumzi* (2009), water sovereignty and labor as a clean energy source clash with the protagonist's discovery of uncontaminated soil. Jennifer Wenzel's description of *petro-magic-realism* in Ben Okri's story, "What the Tapster Saw," combines "the transmogrifying creatures and liminal space of the forest in Yoruba narrative tradition" and "the monstrous-but-mundane violence of oil exploration and extraction, the state violence that supports it, and the environmental degradation that it causes" (Wenzel 2006, 456).
3. *Science fiction energy futures*: When Kim Stanley Robinson discusses terraforming in the *Mars Trilogy* (1993, 1994, 1996), he shows that petrorealism can not only be about oil but also hold together the complex of various forms of energy, their scales, and temporalities.
4. *Actual accounts of the petro-present*: James Marriot and Mika Minio-Paluello's travelogue *The Oil Road: Journeys from the Caspian Sea to the City of London* (2012) maps the oil present spatially, economically, and ecologically. Their figure of the "oil road," reviewer Adam Carlson notes, "gives us a powerful tool for representing the totality, for seeing through the haze, to make sense of both the physical Oil Road, and the Carbon Web—the political, social and economic, the superstructure of the infrastructure" (Carlson 2013).
5. *Interactive documentary and documentary/videogame hybrids*: *Offshore* (2013) and *Fort McMoney* (2013) offer an immersive petrorealism. The former depicts an oil rig modeled on the *Deepwater Horizon*, which viewers self-navigate through an eerie maze of stations and compartments. In the latter, viewers travel to Fort McMurray, Alberta, and explore the town—they can follow bottle collectors, visit the Oil Patch, and vote on important town issues.

Following these examples, petrorealism does not operate in terms of longing for a return to a time before oil. Instead, it follows Stephanie LeMenager's (2012) insistence on the irreversibility of petro-capitalism and looks to futures that take the infrastructures and imaginaries of petromodernity into account with ingenuity and rigor.

Petrorealism attempts come to terms with petromodernity from within. There is no external vantage from which to write about its flows and LIMITS. Lukács makes a useful distinction between the view critical realism had from outside socialism versus the view socialist realism had from within it. Despite enabling the critical realist to better grasp his or her own age, this perspective "will not enable him [sic] to conceive the future *from the inside*" (Lukács 1971, 95). But this is precisely the task before us. To quote another mid-century Marxist, "Petroleum resists the five-act form," and so we must embrace the new styles and forms that resist petroleum (Brecht 1977, 29). My hope is that by learning from petrorealism, we might draw as near to the root of the energy-ecology impasse as possible, tracing spatial connections between capital's energy demands and effects and the temporal possibilities of reaching beyond our energy-dependant, growth-based system of social relations to a FUTURE in which ENERGY is no longer mere metaphor or cause for speculation but the actual driving force of our creative endeavors to overcome such crises. By maintaining a moment of narration within the elaboration of a larger totality, petrorealism sharpens our focus on the task at hand: we must accept the logic of the impasse without overemphasizing its abstract qualities. It is here that the work of petrorealism stands revealed, not merely as an existing archive but as a critical task for the future.

See also: ENERGY, FICTION, INFRASTRUCTURE.

Photography

Georgiana Banita

Photographs are the results of a diminution of solar energy, and the camera is an entropic machine for recording gradual loss of light.

—ROBERT SMITHSON, "Art Through the Camera's Eye"

Photography and thermodynamics are linked by their simultaneous rise to cultural prominence in the 1820s and by the reliance of the photographic apparatus on SOLAR energy. In 1824, Joseph Nicéphore Niépce placed lithographic stones coated with bitumen at the back of a camera obscura and produced a fixed image of a landscape. Heliography, the name for this method, emphasizes long exposure to natural light, although the process also consumes fossil fuels. In his 1844 *The Pencil of Nature*, William Henry Fox Talbot describes his calotype prints as "impressed by the agency of light alone," as "sun-pictures themselves" (1968, 3). With the advent of digital photography, few traces of this primitive solar apparatus remain. Yet the origin of photography in the inscription of light and its continuing reliance on energy-intensive technologies invite a material approach to the photographic IMAGE that highlights how ENERGY gives shape to visible reality and rescues it from the passage of time.

As earthwork artist Robert Smithson notes, a photograph marks the presence of light for as long as it takes to impress an image on the photographic surface (1971, 373). Photographic entropy inheres in the medium, since photography both memorializes reality and exposes the temporality (and temporariness) of exposure. It does so never more convincingly than in energy photography, a thematic subset in the history of photographic art. Energy photography aligns solar photo technology (and photographic film as a petroleum byproduct) with explicit visual renditions of fuel landscapes and economies along an axis of temporal anxiety and melancholia. Such images correlate the role of photography in capturing residual traces of reality with various forms of energy EXHAUSTION in the fossil-fuel age, during which the camera became one of the modern era's most beloved PLASTICS.

Theorizations of photography often employ geological metaphors that pivot on the material aspects of the photograph. In W. J. T. Mitchell's words, "a photograph is fossilized light" (1994, 23)—which adds another necrotic metaphor to photo theory's obsession with the photographic imprint as a death mask. From Susan Sontag and Roland Barthes to Jean Baudrillard, photo theorists focus on how we resurrect these "fossils" in the act of seeing. But such fossilization also implies an entropic loss that applies both to the regress of photographed reality irretrievably into the past and to the momentary use of light in the photographic process. What such photographic metaphors miss is the literal, material reliance on fossil fuels throughout the history of photography. By positing the category of energy photography, I want to explore how photographers interested in energy depletion and peak oil uncover the energetic unconscious of photography itself. Their meditations on photography's entropic genealogy invite us to reimagine energy in terms of futurity and impending DISASTER as well as a dialectical oscillation between power and idleness, inscription and erasure, exposure and darkness. Stasis and exhaustion do not signal the inevitable impasse of late petroculture but have, in fact, long permeated the imagination of energy: they have played a key role in the materiality, temporality, and aesthetics of visual technologies.

An entropic survey of photography would begin with *Power*, the series of precisionist (photographically precise) paintings by Charles Sheeler published in *Fortune* in 1940. To counter post-Depression vulnerability, Sheeler seemed to capture in the water wheel, railroad, electrification, and aviation the power and majesty of American technology. Yet this impression is misleading. In his equally photographic *American Landscape* (1930), Sheeler had envisioned a triumphant future where NATURE has been overpowered by technology and economic enterprise. Despite its evocation of tremendous power, his landscape also evokes a sense of powerlessness. Leo Marx interprets this immobility in an overly optimistic light: "Sheeler has eliminated all evidence of the frenzied movement and clamor we associate with the industrial scene . . . . This 'American Landscape' is the industrial landscape pastoralized" (Marx 1964, 356). Rather than fueling powerful machines, energy is depicted as pooling in these frozen shapes: power at rest, not in motion. The painting is not so much a pastoral as an entropic dystopia that the *Power* series, despite its superficial techno-buoyancy, only confirms. By reducing energy contraptions to their basic forms, Sheeler traces the inevitable disintegration of these assemblies into their component parts, into the primitive forms from which they evolved, and finally into the wilderness that preceded them and is never entirely absent from the pictures.

The minimalism of Sheeler and other precisionists stirred controversy in their day. Since then, however, the widely acclaimed work of Mitch Epstein and Richard Misrach has entrenched the deeper minimalism—dwindling fuels, economic bust, and material obsolescence—underneath the maximalism of a resource-intensive technological society. If Sheeler expressed a tentative skepticism about technological power at the height of the Machine Age, more recent photographic work that responds to resource scarcity and energy insecurity implies that art and politics, photography and power collide, in Leo Marx's words, not only when the (fossil-fuelled) machine enters the garden, but also when it exits.

Like Sheeler, Mitch Epstein pairs colossal energy infrastructures with compositions that evoke stillness and fatigue. In *American Power* (2009), Epstein's photograph of the Amos Power Plant in Raymond, West Virginia, updates Sheeler's *Classic Landscape* (1931). Sheeler's diagonal line traces the evolution from America's agricultural past, represented by grain silos, to its industrial future. Ambivalence appears in the line's obliqueness, how it meets the detumescent slope of the sky, and the tomblike grain piles. Less ambivalently, Epstein's perpendicular axis sucks the viewer in, only to dead-end between ashen refuse piles and skeletal machinery. To emphasize this erosion, Epstein mobilizes a photographic nostalgia that disentangles the structures from their contemporary contexts to align them with similar images from the past. An abandoned gas station in Snyder, TEXAS, recalls Ed Ruscha's cheerily pumping gasoline stations, circa 1962. Now overgrown with moss and weeds, the fuel pump resembles a derelict tombstone in bluish twilight. The monumental materiality of energy emblems is destabilized through increasingly abstract, hollow, and haunted representations.

Misrach's *Petrochemical America* (2012) similarly revolves around petro-ruins and debris. Oil tankers and refineries appear incongruous with the surrounding landscape, even as they excrete malodorous chemicals into land and water. A camper van and a cemetery evoke the tension between availability and exhaustion, amplitude and atrophy, gain and loss, characteristic of energy photography. The photos present toxic swamps and eroded coastlines in solid, palpable hues as if, like Niépce, Misrach fixed the image in a film of bitumen exposed to sunlight. The underlying question is how photography might use the material textures of petro-contamination as a photosensitive surface in order to document (or otherwise express) the ravages of fossil fuels.

Robert Smithson's ruins, deserts, and iconographies of exhumation paved the way for 1970s post-oil embargo photo aesthetics by yoking the material energies of photographic practice with Sheeler's critique of modernity's excessive enthusiasm for mass and speed. Where Epstein and Misrach draw on human presence to delineate a descent into disorder, in Smithson's *Partially Buried Woodshed* (1970) and *Asphalt Rundown* (1969), chaos has already occurred and humanity has been, as Michel Foucault concludes in *The Order of Things*, "erased, like a face drawn in sand at the edge of the sea" (1973, 387). In keeping with this nonanthropomorphic epistemology, Smithson's essay, "Entropy and the New Monuments," testifies to his "almost alchemic fascination with inert properties" (Smithson 1966, 12). His formulation aptly describes both his own work and the images discussed above: "one perceives the 'facts' of the outer edge, the flat surface, the banal, the empty, the cool, blank after blank; in other words, that infinitesimal condition known as entropy" (ibid., 13). The partially buried woodshed collapses under the fossilizing mass of a landslide that wipes off human endeavor, smothers the potential of WOOD as an energetic material, and presses upon the viewer its cultural obsolescence as a fuel. *Asphalt Rundown* does not stage collapse as much as deceleration, retardation, and entropy in synecdochic space and time: a truck is slowly unloading its asphalt load down the side of a sand dune. The regression is twofold: a derivate of crude oil returns to its site of extraction, the desert sand, while the asphalt's viscous trickle paves over the buoyant springs of human mobility and consumption. Smithson's entropic narrative is driven by a violent inversion of energy currents that run their propulsive power into the ground to carve out a landscape of inward, submerged ruins: slopes, craters, sinkholes.

Rather than looming on the horizon as a prospective threat, exhaustion has always posed an inherent limitation and counterforce to all ENERGY SYSTEMS. In line with this dialectic, the photographic imagination of energy culture finds its constitutive moment in the scattered American consciousness of the machine in the garden: seduced by triumphant mechanism and recoiling from it, celebrating the breach of the frontier pastoral and secretly longing to be reconquered by unruly wilderness. Politically, this aesthetic signals the rise of a chronopolitics working to supplant geopolitics as a key paradigm for political judgment and response to environmental CRISIS and peak oil. The archive sampled here conveys thematically and embodies formally the constitutive function of resource diminishment for photographic production in order to foreground the erosion of available time for political decision-making. It questions the current political servitude to incrementality (gradual, pending dissipation, warming, and decay) for understanding both climate change and energy depletion. And it intimates that we already are inhabitants of a partially buried shed that only a swift intervention will save from a creeping but catastrophic asphalt rundown.

See also: CHANGE, DETRITUS, ENERGOPOLITICS, EXHAUSTION, IMAGE, PETRO-VIOLENCE.

Pipelines

Darin Barney

When thinking about pipelines, the temptation to revert to one's native Heideggerianism is almost too great to resist. In a 1955 address, when Heidegger tried to concretize his view of the essence of modern technology as *Gestell*, or "enframing," he turned to petrochemicals rather than his stock example of hydroelectric DAMS. Under the regime of technology as Gestell, Heidegger says, "Nature becomes a gigantic *gasoline* station, an energy source for modern technology and industry" (1969, 50). ENERGY was central to Heidegger's conception of technology as enframing, in which the world is set upon as standing-reserve, a mode of being "which puts to nature the unreasonable demand that it supply energy that can be extracted and stored as such" (1977b, 14). If pipelines are anything at all, they are surely instruments of this "unreasonable demand."

But not just instruments. This, too, we learn from Heidegger, long before we hear it from the actor-network theorists, object-oriented philosophers, media archaeologists, and new materialists: more than just instruments for transporting oil and gas, pipelines are *things*, of the sort that condition the possibility of *Dasein*, or being in the world. Dasein, Heidegger writes, is originally mediated: "it never finds itself otherwise than in the things themselves, and in fact in those things that daily surround it. It finds itself primarily and constantly in things because, tending them, distressed by them, it always in some way or other rests in things" (1982, 159). Pipelines are things that daily surround us, distress us, to which we must attend. They are MEDIA in, with, and through which we come to be in the world as the sort of beings we are.

To describe oil and gas pipelines as media is to confirm, rather than strain, the substantive meaning of this category. As media historian Lisa Gitelman writes, "media [are] socially realized structures of communication" (2006, 7). In this sense, transportation infrastructures are the oldest, most enduring media, especially if we accept that communication is not limited to the circulation of meaning via symbolic representation but also includes forms of "organized movement and action," by which "social reality . . . is built and organized" (Sterne 2006, 118). Transportation is communication, and its infrastructures are media by virtue of the materialities of circulation, distribution, and interaction they make present, not because of the semantic meaning of their contents. Consider the difficulty of expressing in words the magnitude of the enterprise currently underway in Alberta's oil sands. Extraction alone consumes "enough natural gas every day to heat four million homes," "enough water to supply two cities the size of Calgary . . . the same amount of water going over Niagara Falls in an eight hour period" (Nikiforuk 2010, 4, 62). Such equivalencies are so fantastic they almost fail to signify, but, standing near an array of transmission pipelines at Hardisty, Alberta, you experience communication of a different sort. You get the message. Following Hans Ulrich Gumbrecht, we might say that transportation infrastructures are media because they "produce presence" and, in so doing, communicate "what meaning cannot convey" (2004, 16).

Nevertheless, before they are even built, pipelines prompt a profusion of meaning-making in the conventional sense. State approval and regulatory processes for pipeline developments are media for the production and circulation of contested scientific, technical, economic, and political knowledges about what pipelines are, what they do, and what they mean. Studies are made, presented, contested, and archived. There are blooms of data and information. Discourses are mobilized, claims are made, and languages are translated. State and corporate public relations machines are swung into high gear. There are demonstrations, occupations, and protests. Moving and still images, graphics, text, voice, and sound proliferate via a similarly diverse array of media that together make up a network of which the pipeline-to-come forms the trunk. Almost none of this activity would be possible without petroleum and the pipelines that communicate between its source and many destinations.

Pipelines accomplish a range of mediatic functions: they contain, store, convey, conduct, transmit, connect, distribute, span. In these respects, they are like rivers, canals, railroads, and highways, and also like telegraphy, telephony, portable print media, and wired and wireless digital NETWORKS. In none of these functions are pipelines simply neutral. It matters how and what they contain and convey, between where and whom they conduct it, for what purpose, to whose benefit and detriment, under whose control, with what foreseen and unforeseen consequences. As with all media, these contingencies make pipelines "political machines," which is to say they are technologies of and for the condensation, exertion, and contestation of power (Barry 2001). If digital networks are the INFRASTRUCTURE of the information society, then, as Timothy Mitchell (2011) shows, pipelines are the infrastructure of actually existing "carbon democracy." Pipelines mediate a "form of life" in which we are invited to become beings whose lives are always-already materially and politically organized around the production and consumption of petrochemicals (Winner 1986, 12).

Like all media, pipelines aspire to the dream of invisibility and the fantasy of immediacy. Just as it is best when digital networks deliver us images, sound, and text wherever and whenever we want them without bothering us to register the infrastructure at all, it is best (at least from the perspective of energy capital, energy states, and energy consumers) when pipelines deliver energy without anybody noticing them (Barry 2009). That pipelines are often remote to major population centers and are usually mostly buried makes them largely unseen but—again, like all media—they leap to visibility and demand attention at crucial moments: when they are being planned and built; when they are disrupted; and when they fail. These moments, in which the oil and gas cannot simply be contained, are when pipelines exceed their ready-to-hand instrumentality and become present-at-hand as things with agency, acting in the world as part of complex assemblages with other human and nonhuman things (Bennett 2010).

The networks of which pipelines make up one part are spatially expansive, intermedial, and temporally disorienting. There is no inhabited continent free of existing oil and gas pipelines, and major pipeline expansion projects are underway everywhere, with major concentrations in North America, Eastern Europe, and the Asia-Pacific region (Petroleum Economist 2012). Pipelines traverse national borders, linking typically remote, often impoverished RURAL and quasi-colonial peripheries to concentrations of industrial production and urban consumption, continuing a tradition whereby the movement of staple commodities materializes an imperial network of uneven communication that undergirds and outstrips the globalizing influence attributed to information technologies (Innis 1970). Pipelines do not accomplish this task on their own: their functionality relies on other media of transportation, including storage tanks, ships, ports, railways, and tanker trucks, without whose interoperable multimodality pipelines would be useless. The same goes for the various media of transformation that make up the "network of bodies" forming the "carbon web" that makes petrocultures and petroeconomies possible—energy, manufacturing and retail corporations, financial institutions, state regulatory agencies, the military, and the scientific complex (Marriott and Minio-Paluello 2012, xii). Far from being simple lines on a map, pipelines materialize a complex spatiality that both ramifies, in the sense of passing on effects from one site to another, and is itself a ramification, in the sense of being one subdivision of a more complex process of mediation whose boundaries are difficult to discern.

The same quality attends to the temporality of pipelines. As Carola Hein (2009, 35) points out, the finiteness of their nonrenewable contents means that oil and gas pipelines are essentially temporary infrastructures, knowingly designed to eventually join the pile of modernity's technological refuse as "residual media" (Acland 2006). But pipelines mediate a span of time whose sheer extent is almost impossible to conceive: extracting resources from deposits formed hundreds of millions of years ago to produce effects that reach infinitely forward into unimaginable futures. In the short-term calculus of industry, pipelines are a sure thing: an investment that promises healthy returns. For the rest of us, they are a wager in an uncertain game. Media structure our temporal attention, and pipelines are no exception. They are, however, exceptional in that the time to which pipelines ask us to attend is forever. Now.

See also: ARCTIC, DISASTER, LIMITS, NETWORKS, UNOBTAINIUM.

Plastics

Gay Hawkins

It is difficult to consider plastic as fuel when you confront its ubiquity as urban litter or ocean waste. It seems so passive and inert, the dead stuff of disposability denied even the biological momentum of decay. The eternal persistence of plastic seems to fuel only apocalyptic visions of ecological DISASTER: petrochemical cultures buried in their own DETRITUS.

From a different angle, however, plastic represents not the end of NATURE but rather movement or process. This understanding of plastic recognizes what Manuel DeLanda calls "the expressivity of materials," or their morphogenetic potential (1995). For DeLanda, materials lack inherent fixed qualities, and their multiple forms are not the outcome of externally imposed structures. Rather, the capacities of materials emerge as they participate in new relations; they are both shaped by and shape those relations in distinct ways. They inevitably "have their say" (DeLanda 1995).

How does plastic have its say in contemporary culture? Through what processes has this material become a force in the world, fueling new economic, ecological, and political realities? And in what ways are the forces of plastic, its variable capacities to intervene in the world, realized? Thinking of plastic as process is more than just a material turn; it recognizes the active role of the more-than-human in assembling the social. Material and nonhuman elements do not simply express culture or mechanistically enable human being; through ontological alliances they participate in making realities.

The particular reality I want to investigate is disposability: the cultural and ecological implications of more and more things produced for a single use. Disposability entails a fundamental interdependency in petrochemical cultures that drives the ACCUMULATION of plastics in environments and bodies, human and nonhuman. Most understandings of disposability trace a narrative from resource extraction of finite oil reserves to industrial production, to global supply chains, to fleeting use, to excessive waste. Disposability's cascading effects seem to unfold according to a linear and destructive logic. But the most striking thing about disposability is that it forces one to consider the interrelationships among markets, resources, materiality, consumers, and environmental degradation *all at once*: not as a sequence of effects but as an "agentic swarm." This is Jane Bennett's term for the agency of assemblages, where multiple actants interact in uneven, unpredictable ways (2010, 32). Understanding disposability as assemblage, rather than narrative, allows one to perceive processes of change and emergent causation as multidirectional rather than teleological. One can investigate how materials act, how they shape relations in expected and unexpected ways, and thus ask new questions about how plastic, once considered so durable, emerged as the definitive material of disposability.

In *American Plastic*, Jeffrey Meikle (1995) argues that after World War II, plastic shifted from being a synthetic replacement for natural materials into something made neither to last nor as a substitute but instead *made to be wasted*. This massive shift had several aspects: economically, the fundamental integration of the plastics and petroleum industries and the expansion of mass consumption; culturally, the material density of everyday life and the equation of plastic artifacts with modernity and consumer choice; environmentally, in escalating amounts of urban waste. Plastic began to express itself as the promise of a gleaming, ever-new FUTURE. It shifted from being ersatz synthetic substitute to standing on its own feet and "having its say." It acquired an identity Meikle calls "plastic as plastic."

The rapid growth of disposability was central to this transformation. Single-use objects—plastic bags, spoons, lids, straws, food containers, the list goes on and on—facilitated the emergence of new cultural formations. As Meikle says, what was remarkable about disposable plastic things was that they appeared stylistically as rubbish from the very beginning (1995, 186). In postwar consumer culture, plastic's capacity to express itself as a throwaway changed how "waste" and wastefulness were understood—materially and morally—and how people experienced movement. The anonymity and ubiquity of ever more plastic generated cultural consciousness of an increasing flow of plastic in everyday life. Disposable plastic things seemed to come from an inexhaustible source; they arrived from a "continuous infinite" (Boetzkes and Pendakis 2013). Relations with plastic were fleeting; the unbearable lightness of disposable being denied any sense of permanence. The rise of plastics and disposability reverberated in culture and economy by changing perceptions of social time and space. Movement no longer meant simply passing through particular times and spaces; it was also an ordering of continuity that worked in many directions. If things could appear as waste before they were used, then production and consumption did not create waste; rather, waste emerged as intrinsic to plastic ontologies, present from the beginning.

Consider, for example, the plastic lid on a takeaway coffee. It is a market device facilitating consumption on the move, a plastic object with unique design and physical properties, and waste—pretty much simultaneously. Although you might think that the lid is packaging that performs an essential function, and only becomes "useless" garbage after you remove and trash it, this linear narrative misses how these qualities and calculations are folded into each other. They do not emerge in a series of transitions and shifting valuations; rather, they animate each other. The lid's future as waste is anticipated even before it is used. This quality does not appear later as a result of consumption but is inscribed in its very form and function, its plastic materiality.

In this vignette of disposability, the disposable object shapes how things move, not just how they are apprehended. The lid passes quickly from barista to consumer within a distinct spatiality and temporality. This emergent time-space is not an accelerated product life cycle—from production to consumption to disposal—but rather a horizontal network of relations in which waste is immanent. Waste is not displaced to another time or space after consumption, nor is it an effect; it is a material presence animating and ordering interactions in particular ways.

This distinct form of movement that disposability engenders does not mean simply "rapid turnover" or the intensification of commodity circulation. It indicates the dynamic emergence of new social forms out of assemblages of differential relations. Becoming disposable revealed novel capacities or forms of expressivity for plastic. For DeLanda, capacities are virtual unless they are being used. Material capacities are not fixed or hidden within the material as a limited number of possibilities awaiting human discovery. They are emergent and invented in the ongoing, contingent processes of materials interacting in the world.

While there is no doubt that considerable money and effort went into disciplining the biophysical properties of plastic to be amenable to numerous economic and industrial applications, plastic made its own suggestions. It revealed surprising capacities. Some of these capacities were molecular, others cultural (Hawkins 2013). The capacity crucial to plastic's emergence as "disposable" was its willingness to perform as both market device and mass material. As a market device, plastic reconfigured numerous industries. It offered a pragmatic solution to the need for containerization and the logistics of distribution and mobile consumption. In economic terms, it helped reconfigure and expand markets in ways that shaped new forms of production. And culturally, plastic reconfigured how consumers apprehended products (especially food and pharmaceuticals), how they shopped, and how they became comfortable with an overpackaged world and a series of fleeting encounters with plastic in more and more spaces of daily life.

In this way, plastic expressed itself as the archetypal mass material, quantitatively and topologically. The quantitative sense refers to the "plastics explosion" of the 1950s and 1960s when plastic emerged as a massive industry that transformed other industrial processes. The topological refers to the reverberation of this event across multiple registers. With the proliferation of plastic things, plastic emerged as the ultimate marker of consumer democracy, a culture of bewildering abundance and choice. And it was the material experience of seriality as disposability. The constant flow of plastic, the sense that plastic was always "ready to hand," added to the effect of disposability as a cultural and bodily experience. The senses were captured by plastic's evanescence and transience, by its abundance and endless replaceability. At the same time moral imaginations were learning to be untroubled by its presence as always-already waste.

See also: CHINA, DETRITUS, ENERGY SYSTEMS, METABOLISM, NETWORKS, PLASTIGLOMERATE, TEXTILES.

Plastiglomerate

Kelly Jazvac (artist) and Patricia Corcoran (scientist)

> Environmental problems are socially constructed via public campaigns that legitimate claims and build support for reform and change. Rationality is crucial to this process, insofar as science is enlisted to provide evidence for this or that harm. However, it is often the emotions that go with environmental issues that can win the day for specific campaigns. Thus, affective elements . . . are essential in how issues are socially constructed.
>
> —ROB WHITE, *Crimes Against Nature*

What if the material evidence from a scientific finding were also presented as art? What could be gained from such re-presentation?

For one, it puts the evidence firmly in the category of *made by human* as well as *observed by human*. When you are making something, as an artist or fabricator does, you are as close to inside it as you can possibly be. A sculptor making an armature for a sculpture figures out the inside support so the exterior can stand on its own. Conversely, studying a phenomenon, instead of making a phenomenon, implies being outside, carefully and objectively looking in.

Art is a subjective activity, from the maker's point of view and the viewer's. Something curious occurs when a readymade object is appropriated by an artist and declared to be art under her authorship: what had been an external phenomenon now becomes incorporated into her oeuvre. The object, now also an *art* object, has new power: it tells viewers it is okay to look at this thing subjectively because it is art. Emotion and speculation are encouraged. Duchamp himself did not think this expressive strategy would work well if used willy-nilly; therefore, he limited his readymade output to a careful selection over the course of his career. Had he always put urinals in the gallery, they would have quickly lost their affective power (Duchamp 1973, 141–42). Like the urinal that troubled the categories of both art

Speculations by an artist, fact checked by a scientist: Interdisciplinary collaboration in the Anthropocene.

and plumbing, could the category of "scientific evidence as readymade" be better socially and affectively employed than either category on its own?

In collaboration with geologist Patricia Corcoran, I have appropriated scientific evidence as an art object. Our collaboration began in line for coffee, where I saw a poster for a talk about PLASTICS pollution by Captain Charles Moore, who discovered the Great Pacific Garbage Patch. I was both fascinated and horrified by the talk, and I exchanged emails with Corcoran, who had organized it. She invited me to accompany her to the beach in Hawaii touted as the "dirtiest beach on Earth" to investigate a curious geological phenomenon that Moore had observed. He hypothesized that the interaction of molten lava with plastic garbage was creating a sensational new substance (Corcoran, Moore, and Jazvac 2014).

Our fieldwork in Hawaii led us to discover of a new type of stone.[1] We called it "plastiglomerate." Plastiglomerate is a composite of melted plastic debris, sand, basalt rock, coral, organic debris, and various other fragments typically found on a Hawaiian beach. It turns out that its formation process is more banal than sensational: it is not made by lava but instead completely fashioned by humans. Plastiglomerate is an anthropogenic substance three times over: first, it is composed of human-made plastics; second, because inadequate disposal brought the material through the ocean currents of the world to Hawaii; and lastly, because human-made fires on the beach melted the plastic debris, thereby fusing it with natural materials to create plastiglomerate. It is a hybrid that crosses categories and challenges measurement as either nature or as pollution.

As an artist, I saw in these aesthetically compelling objects a heartbreaking story of humans prioritizing their role as consumers over their role as citizens of the planet. I also saw objects that I could never make myself, given all of the forces, distances, and politics at play in their creation. Yet I had inadvertently and passively contributed to plastiglomerate's making. Some (or perhaps most) things that are anthropogenic cannot be forged by a single human working individually. And considering the aesthetic appeal of plastiglomerate, I wondered about the ability of artists and other humans to make beautiful things through environmentally destructive means. Is this what art should be? And if not, what other models or systems are available to artists who strive for sustainable practices?

Scientifically speaking, plastiglomerate offers material evidence of the irrevocable impact of humans on the environment. Plastiglomerate is not merely a discrete form of hybrid pollution; it also has the strong potential to become embedded into rock (Corcoran, Moore, and Jazvac 2014). This anthropogenic substance can be fused into the earth and become preserved within its FUTURE rock record.[2] It reminds me of Spike Lee's movie *Inside Man* (2006), in which bank robbers make their hostages dress like them so that the

1. Technically, a rock is a naturally occurring material. Since plastiglomerate is not naturally occurring, we refer to it as a stone.

2. The rock record is any geological formation available for study. It provides information concerning Earth's internal, surficial, and atmospheric processes, including the EVOLUTION of life and its effects on the natural environment.

police cannot tell the good guys from the bad guys. What's a well-meaning detective to do?

Like the fusion of Earth/pollution or hostage/bank robber, our plastiglomerate samples have led a hybrid existence since their extraction from Kamilo Beach: they were key evidence for a scientific manuscript; they have served as readymade art objects presented in a gallery among other plastic artworks; they have been employed as pedagogical tools and subjects of educational outreach programs in a Canadian museum; and they have become a commodity whose sale supports research and volunteer activism.[3] In short, plastiglomerate has played both subjective and objective roles across different disciplines and environmental efforts, too. This hybrid resists being contained by any single category.

"Ambiguities of definition," according to Rob White, can work against finding solutions for environmental problems (White 2008, 37). If the stakes of an environmental issue are entangled with other serious issues across broad geographic distances (as with, for example, the way that PIPELINE projects entangle issues of job creation with issues of environmental harm to the atmosphere or ecosystems along the pipeline's path), it becomes difficult to pinpoint who should address it, how, and where. Environmental CRISIS is itself a hybrid state. Like plastiglomerate, Earth and our impact upon it are irrevocably intertwined. Our planet, now host to superstorms, radioactive caribou, and "managed" tailings ponds, has become a mixture of it and us. This predicament demands a more hybrid, collaborative, cross-disciplinary model of environmental and political thinking: instead of pegging blame and responsibility, an expanded field in which acting, making, thinking, feeling, and dissecting can occur simultaneously. If we were to approach hybrid phenomena through as many forms of knowledge as possible, more people would have a more complex understanding of the environment.

To mobilize caring about a hybridized scenario, we will need cross-disciplinary systems of navigation and inquiry that operate objectively and subjectively, while still being held accountable to the criteria of the the relevant disciplines. Rather than ignoring or erasing evidence of a mess that has been made (as with the Canadian federal government closing 157 environmentally related research facilities and programs; *The Fifth Estate* 2014), the objective and the subjective, working in tandem, can inspire action and the development of new systems that operate both rationally and emotionally. For Henri Bergson, "the intellect is characterized by the unlimited power of decomposing according to any law and of recomposing into any system" (1992, 141). The conventional systems, laws, and categories available to us do not seem to foster emotional clarity with respect to environmental crises. But if not, let us work with that: let us find each other in line for coffee and collaboratively recompose how we address the messy realities of an environment composed as much of plastiglomerate as of grass and trees.

See also: AFFECT, ANTHROPOCENE, CHINA, COAL ASH, DETRITUS, METABOLISM, PLASTICS.

3. *GSA Today*; Oakville Galleries (Oakville, Ontario); Louis B. James Gallery (New York); Visualizing the Invisible research project at Western University; Hawaii Wildlife Fund.

Renewable

Werner Hofer

Two news items in 2013 should have alerted every CEO or senior manager of an energy utility company, every politician with responsibility for the economy of a country, and every leader of a large manufacturer to recent changes in how the public imagines ENERGY, specifically the place of renewable energy in the mix of fuels we burn through on a daily basis.

The first reported the result of a popular vote in Hamburg, where citizens decided, against the wish of their local government, to buy back the energy grids of their city, mostly from large European utility companies ("Initiative" 2013). The second was an article in the German magazine *Der Spiegel*, which portrayed the slow demise of the once-mighty energy utility RWE. After years of failed investments, a dwindling market share in private households due to community energy NETWORKS and self-generation, and the phasing out of once immensely profitable NUCLEAR generation, RWE now barely survives and must service a staggering 35 billion Euros in accumulated debt.

Germany is undoubtedly the most successful economy in Europe and one of the most successful globally. Its unique blend of dual education and high-value manufacturing in cars, planes, machinery, and chemicals, together with its extremely successful *Mittelstand* (small to medium size companies that are world leaders in vital niche markets), make it highly resilient in the face of CRISIS.

To understand where Germany is today, it is instructive to sketch its path over the last thirty years. Three decades ago, political discourse in Central Europe was determined by Cold War paradigms, national economies were focused on heavy industries and vanity projects in war technology, and information was restricted to print MEDIA and government-owned radio and television.

Two crucial events defined this period in Germany. The first was the deployment of US Pershing II missiles on German territory in 1983. These missiles could reach RUSSIA within minutes and were seen as proof that the United States at least contemplated waging a war of aggression against the Soviet Union. The deployment, which Germans fiercely resisted to the very end, created a powerful alliance of left-wing and church-based critics of the official government line on international politics.

The second event was growing evidence of the destruction of the German forest by acid rain from heavy industry and power stations. In 1983, it was estimated that 30 to 40 percent of the (then West-) German forest had been damaged by acid rain (Park 2013, 63). This fact, and the realization that trees were vital for CO_2 regeneration, quickly attracted national media attention.

These two events created a powerful force that spanned the political spectrum, capable of overturning the country's consensus on almost anything from international relations to educational policy to economic relations. The German Green Party emerged out of this moment in the early 1980s. While it took more than fifteen years (until 1998) for the Greens to become a member of a national government, their thinking influenced a generation of Germans who are now in their forties and fifties and who are largely responsible for the dramatic change in both German energy INFRASTRUCTURE and the strongly felt desire in Germany to increase the proportion of renewables in Germany's energy mix.

Even so, the translation of this desire into a full-on shift to renewables has been blocked to some degree by the positions that various social groups take toward energy. Today, energy is perceived by most people as a commodity, provided by large multinationals and readily available on DEMAND. The limiting factor in its use is cost. Energy has tended to be considered value neutral: using more or less of it does not correspond to either an improvement or deterioration of quality of life in general. This means that people adjust their consumption by balancing benefits and costs individually. Any interference in this balance will be seen differently by different parts of the population.

Those who are not overly affected by energy costs—the affluent part of society—will react negatively to attempts to reduce their energy consumption: for them, the SUSTAINABILITY movement or the GREEN agenda consequently tend to be seen as enemies. However, those whose personal comfort is severely affected by rising energy prices—people who suffer from fuel poverty—will see energy companies as the enemy and aim to increase their energy consumption via reduced costs.

Both groups neglect two fundamental facts about energy. First, limited supplies will always increase prices; therefore, attempting to reduce costs within existing supply chains in an environment of increasing demand is futile. Second, increased spending on energy nearly always has a negative effect on an economy locally, as it leads to capital exports and reduces available funds for investment. Additionally, using existing hydrocarbon reserves to the full extent will probably mean the extinction of life on Earth on a massive scale, which, in turn, makes human survival questionable.

In order to confront these facts and puncture this class-based position-taking, rational debate needs to be framed in terms of economic and ecological value generation. Such debate would lead to the conclusion that large international utilities, with their outdated business models, are doomed to fail because economies of scale do not work for renewable energy. Instead, FUTURE energy networks will likely depend on community-based and fully integrated networks.

Model towns and villages such as Kempten and Wildpoldsried in Southern Germany demonstrate how this approach works in practice. Fifteen years ago Wildpoldsried started to transform its energy production by hiring energy consultants and assembling a team for bulk sourcing SOLAR power generation on individual dwellings. The uptake to date involves about 20 percent of all buildings and generates close to four megawatts peak ELECTRICITY. In addition, investment pools were created to finance the construction of five wind turbines, capable of generating fifteen megawatts. A further development involved biogas generation from farm waste material, which feeds combined heat and power generation. The total energy generation at present is, over one annual cycle, more than 300 percent of electricity demand ("Welcome in Wildpoldsried" 2014).

Small-scale production on individual properties, which removes demand from the market, is bad for utilities because it reduces their consumer base. It cannot be incorporated into a utility business model because the individual ownership of property prevents production from being scaled up. There is no incentive for a utility to invest in small-scale renewables.

But the small scale is precisely where renewable energy is produced and used most efficiently. And here is the main obstacle for the energy revolution: consumers of renewable technology need to organize themselves to bargain for better prices in renewable energy technology. Similarly, in order to reduce losses in the generation and distribution of heat, local heat networks need to go beyond a minimum size, which is only possible if individuals form larger interest groups.

Strange as it may sound, a revolutionary turn to renewable energy depends on a political decision by individuals to form energy COMMUNITIES. Only in this way can the promise of the changes in Germany be fully realized. If the struggles of RWE and the decisions by the citizens of Hamburg with respect to their energy GRIDS are any indication, this may be one revolution we will not have to wait long for; it has already begun (Provost and Kennard 2014).

See also: CHANGE, CORPORATION, ENERGY REGIMES, GREEN, MIDDLE EAST, RISK, SUSTAINABILITY.

Resilience

Susie O'Brien

"In our current (and long overdue) efforts to drastically cut carbon emissions, we must also give equal importance to the building . . . of resilience" (Hopkins 2010). This is the premise of Rob Hopkins's *The Transition Handbook: From Oil Dependency to Local Resilience* (2010), which outlines the process by which communities can work toward local autonomy, reducing their dependence on fossil fuels.[1] Environmentalists hailed the publication, with one review proclaiming: "At last, here is a book about our common future that we don't have to be afraid to read" (L. Wallace 2010). In a 2011 speech titled "Resilience and the Energy Sector," Shell CEO Peter Voser invoked a different view of the FUTURE, centered on the challenge of "help[ing] Shell to remain resilient in the face of future challenges," thereby "enabl[ing] us to make a positive contribution to the resilience of society" by "[responsibly] increas[ing] energy supplies for a growing world population." (Voser 2011). It is worth analyzing how the concept of resilience figures in these divergent ENERGY futures—not to embrace one and condemn the other but rather to consider what work resilience does in debates about energy. To what end is it deployed? What possibilities does it allow us to imagine, and what does it foreclose?

Defined as the capacity of a system to undergo disturbance and retain its basic structure and function, resilience offers a model of CHANGE based on complex adaptation. Resilience

1. Beginning with Totnes, England, in 2005, the Transition initiative encompassed 475 communities by 2013 (http://www.transitionnetwork.org).

theory, developed by forest ecologist C. S. Holling in the 1970s, challenges traditional models of SUSTAINABILITY based on an optimal balance of resource use and conservation, by highlighting the crucial role of change in ecological-social systems. Efforts to rationalize system function through regimes of efficient management are inimical to resilience, which depends on the interaction of diverse, multiscalar processes, moving at different temporal cycles through stages of growth, conservation, release, and reorganization. Melinda Cooper and Jeremy Walker (2011) demonstrate a fundamental congruence between resilience ecology and neoliberal economic theory, each of which challenges the capacity of top-down management approaches to forestall CRISIS in complex systems. Generally optimistic about the capitalist market's capacity for self-regulation, neoliberal economic theorists, led by Friedrich Hayek, denounced the Club of Rome's *The Limits to Growth* (Meadows et al. 1972), which spearheaded efforts to formulate international environmental regulations (Walker and Cooper 2011, 150).

Despite the anti-ecological bent of neoliberalism, its metaphorical resonances with resilience theory concretized over time, as resilience ecologists sought to expand its parameters to a general model of "social-ecological resilience." The ecological model supports the normative thrust of Hayek's economic theory, according to which "perturbations are not only inevitable; they are also necessary to the creativity of organized complexity" (Cooper and Walker 2011, 150). As resilience thinking proliferates in areas ranging from defense to education, it tends to stress the adaptive aspect of complex adaptive systems and possibilities for leveraging turbulence to enhance growth. In the alignment of ecological resilience theory with neoliberal philosophy, one of Hollings's early insights is often forgotten: that natural systems are not infinitely self-correcting, that "resilience can be and has been eroded, and that the self-repairing capacity of eco-systems should no longer be taken for granted" (Folke et al. 2004, 558).

In the oil industry, the discourse of resilience ecology is increasingly invoked to foreclose the meaning of environmental politics. As Voser's speech suggests, Shell is an enthusiastic proponent of resilience thinking, which informs its *Shell Energy Scenarios to 2050* (2008). These scenarios are notable for their frank acknowledgment of the realities of climate change and the growing public awareness "that energy use can both nourish and threaten what [people] value most—their health, their community and their environment, the future of their children, and the planet itself" (10). In contrast with Shell's 2001 scenarios, which bore the Thatcherian subtitle "TINA" (There Is No Alternative), the 2008 document, subtitled "TANIA" ("There Are No Ideal Answers"), integrates free-market philosophy with the idea of ecosystem interdependence, as defined by the Stockholm Resilience Centre (quoted in *Signals and Signposts* 2011, 46).

Other changes in Shell's scenarios are worth examining. The 2001 scenarios were framed around tensions between three forces: market incentives, coercion and regulation, and community, the last category including sovereignty and environmental movements in sites like the Niger Delta and the Alberta oil sands. The preferred scenario, not surprisingly, sees market incentives as the dominant force; the worst case scenario, "Flags," represents the dominance of community, with resource sovereignty movements asserting a violent stranglehold over the forces of law and the market (*Shell Energy Scenarios to 2025* 2005, 15). But in the 2008 version of *Scenarios to 2050*, "sovereigntist demands . . . are replaced by ecological concerns" (Zalik 2010, 557). What had been identified as community is reframed more positively as grassroots, encompassing a broad constituency as described in the favored scenario:

> While international bodies argue over what environmental policies should be . . . and many national governments worry about energy security, new coalitions emerge to take action. Some bring together companies from different industries with a common energy interest . . . . Individuals effectively begin to delegate responsibility for the complexities of the energy system to a broader range of institutions besides national governments. (*Shell Energy Scenarios to 2050*, 2008, 26)

Thus, as Anna Zalik notes, "*the grassroots* . . . includes companies, private capital itself, thus leaving outright opponents to industrial expansion undifferentiated from those seeking market-led (de)regulation" (Zalik 2010, 557). COMMUNITY, along with the associated values of sovereignty and environmentalism, has vanished from the scene.

Shell's idea of resilience differs in key respects from the Transition initiative, which envisions enhanced local autonomy and reduced dependence on fossil fuels as means of nurturing the capacity of communities to "respond creatively to change and shock" (Hopkins 2010). While Shell folds the local into the global, dropping "community" and "sovereignty" from its vocabulary, Transition emphasizes diversity, modularity (self-organized micro-systems), and tight feedbacks—all of which are maximized at the local level.[2] Where Transition and Shell's views of resilience align is in their assumptions of turbulence to come, and in their beliefs in the incapacity of nation-states. They both see grassroots innovation as the best guide for navigating that turbulence.

Resilience provides the conceptual scaffolding that links market deregulation with ecological responsibility, against a backdrop of volatility—another dominant theme in Shell's 2050 scenarios. Several critics are troubled by resilience theory's acceptance—even embrace—of turbulence as a predicate of growth (Evans and Reid 2014; O'Malley 2010; Zebrowski 2009). In this model, disasters are "not necessarily completely undesirable—indeed they are opportunities to enhance resilience and test the morphogenetic properties of society" (Zebrowski 2009). The implications for fossil capitalism are significant. Naomi Klein argues, "The oil and gas industry is so intimately entwined with the economy of DISASTER—both as root cause behind many disasters and as a beneficiary from them—that it deserves to be treated as an honorary adjunct of the disaster capitalism complex" (2007, 512). As Zalik notes, the uncertainty that attends disasters, both real and imagined,

2. Critics note that the Transition movement's spatialization of oil politics is based on a "narrow view of how the impacts of peak oil will be distributed between nations and sectors of society" (Bailey, Hopkins, and Wilson 2010, 602).

not only invigorates the oil futures market (which influences contemporary pricing), it also intensifies a DEMAND for innovation, including the exploitation of "unconventional" methods of energy extraction such as FRACKING and tar sands (2010, 26, 57).[3]

Although it has been around for decades, a hallmark of resilience thinking is the claim that it represents a new path, a clear advance over the previous model of sustainability as an approach to resource management. The idea of resilience can be bracing, even energizing, in a way that sustainability (in its obsession with conservation and LIMITS) never was. Hopkins (2010) enthuses, "we may find it hard to get to sleep due to our heads buzzing with possibilities, ideas and the sheer exhilaration of being part of a culture able to rethink and reinvent itself in an unprecedented way."

In our insomniac hours, we might want to ponder the rush to resilience as the new beacon of our energy futures. For all the limitations of sustainability ("an elusive goal," as Roger Keil notes, "as long as fundamental processes of uneven development are not brought under control" [2007, 43]), its proponents, including *The Limits to Growth* (Meadows et al. 1972) and the 1987 Brundtland report, *Our Common Future*, "were potentially revolutionary, as they proposed an immanent critique of capitalism" (Keil 2007, 43), albeit one that was quashed by GREEN development, in paradoxical combination with neoliberal backlash. Resilience theory harbors no such critique, drawing resistance and the logic of dialectics smoothly into the flows of the complex adaptive system. Focusing on the flows, we miss the friction between and within scales; focusing on the system, we miss the arena of the political, that "decisive material and symbolic space . . . from which different socioenvironmental futures can be imagined, fought over, and constructed" (Swyngedouw 2007, 38). Resilience must be thought within this space if it is to serve the imagination of just, as well as viable, energy futures.

See also: AFFECT, CORPORATION, CRISIS, FUTURE, NIGERIA.

3. This relationship goes unacknowledged in Shell's scenarios, an omission that is doubly significant given their role in shaping the global futures markets (Zalik 2011, 554).

Resource Curse

Janet Stewart

We might expect that countries where significant oil reserves are discovered would experience unfettered socioeconomic progress. Yet in many parts of the world, oil wealth is no guarantor of prosperity. Instead, as is documented, for example, in Michael Watts and Ed Kashi's photo-essay book *Curse of the Black Gold* (2008), oil wealth often seems to herald political instability and economic CRISIS. The title of Watts and Kashi's book on oil extraction in the Niger Delta alludes to a phenomenon described by some economists as the "resource curse," a key analytic tool in numerous studies of countries rich in natural resources. Yet this title belies the nuanced account of the resource curse idea in Watts's introduction, which criticizes many of the assumptions underlying the term. These contradictions epitomize the complexity of the debate surrounding the idea of resources being a curse.

The term "resource curse thesis" first appeared in 1993 as the subtitle of *Sustaining Development in Mineral Economies*, in which Richard Auty traces the history of developing mineral-exporting economies from the 1970s to the 1990s. Observing that "resource-rich countries . . . may actually perform worse than less well-endowed countries," Auty focuses his attention on countries rich in hard mineral resources (1993, 1); other analysts have noted similar trends in oil-rich states. In *The Oil Curse*, Michael Ross explores the economics of the "curse" in order to offer advice on how to avoid the negative effects of resource wealth. The strength of his approach is that, unlike economists who are seduced by a simplistic version of the resource curse thesis, he analyzes the full social and ecological costs of oil production. He emphasizes the international political efforts made by states and NGOs to counter the "curse" by promoting the "use of oil, gas and mineral resources for the public good" (2012, 246).

In developing the idea of the resource curse in this way, Ross takes up a line of argument familiar to political scientists, although his critique is markedly less radical. In *The Paradox of Plenty*, for example, Terry Lynn Karl investigates the impact of oil on a number of important oil-producer states (or "petro-states" as she terms them), in order to understand why these states chose similar modes of development that consistently produced "generally perverse outcomes" (1997, xv). Karl's work is important not least because she draws connections across a diverse range of states; she eschews a simple narrative that would make essentialist distinctions between Global South and Global North. In her analysis, even Norway, so often held to be the model of a successful petro-state, does not entirely escape the potentially problematic effects of oil wealth on the political landscape, even if it is better placed than developing countries to weather the eponymous "paradox of plenty" (214–16). Conceived as an alternative to the resource curse thesis, this concept allows Karl to criticize the assumptions underpinning much of the literature following Auty, which, she argues, failed to understand economic effects as the outcome of particular political and institutional structures and specific forms of decision-making (ibid., 5–12). Related objections appear in Timothy Mitchell's *Carbon Democracy*. Offering an important and nuanced set of reflections on the political relations engendered by the global reliance on oil, he notes that studies that assume a resource curse are generally flawed in that they distinguish between oil-producing states and non-oil states. This approach, Mitchell argues, conceals the nature of the cycle of oil's production and consumption: the "limits of carbon democracy" affect producer and nonproducer states alike (2011, 6).

Today, Mitchell and Karl are hardly lone voices raised against the resource curse thesis; the growing critical literature falls into two camps. On the one hand, there are works like *Beyond the Resource Curse* (2012) that, although seeking to overcome the limits of the concept, do not fundamentally reject it; on the other, works like *Oil Is Not a Curse* (Luong and Weinthal 2010) reject entirely the resource curse thesis and the thinking that underpins it. In it, Pauline Jones Luong and Erika Weinthal conclude their analysis of the fate of Soviet successor states with a chapter on the "myth of the resource curse," in which they debunk the assumptions underlying this way of thinking that, they claim, assume an overly truncated time frame (2010, 322–36).

This practical exercise in debunking fails, however, to draw out fully the implications of identifying the resource curse as myth: one might raise a further set of questions about the oil curse that attend to language itself. In this case, the crucial question to ask is not whether the assumptions that undergird the resource curse thesis are empirically justifiable but rather what does it mean to imagine oil as a curse or as a resource? To meditate upon these questions, as, for example, Rob Nixon (2011, 69–76) so eloquently does, is to reflect on our relation to the object—on our disposition toward oil itself. Understanding how dispositions are constructed and how they are put to work is, as Jane Bennett (2010) and others argue, a crucial aspect of the ethical-aesthetic turn that investigates mechanisms through which knowledge leads to action. To reflect upon the ways in which we imagine our relation to oil is, then, in itself already an ethical act.

Considering what it means to imagine oil as a curse is the approach Watts takes as he meditates upon evocative descriptions of the fetishistic nature of oil, in which narratives about oil, whether invoked in scholarly literature or by the popular imagination, appear in the guise of the cautionary fairytale (2004b, 51). This approach also informs Murad Ibragimbekov's award-winning short film *The Oil* (*Neft*) (2003), which deftly recycles archival footage of Azerbaijan's oil history to offer critical reflection on oil's status as both blessing and curse. The performative language of a "curse" imbues oil with agency: a curse, after all, is an utterance that does something. It acts. And by attributing agency to oil, the language of "curse" appears to shift responsibility for the consequences of oil extraction to the substance itself. In other words, assigning agency to the object in this way brings with it, as Fernando Coronil (1997) suggests, the RISK of positing that the significant inequalities characteristic of many petro-states as natural, magical, and, in any case, inevitable.

While Coronil warns of the dangers of an overemphasis on the power of things, the language of the resource curse is also employed to rob oil of its materiality. If oil is imagined as a mere resource to be exploited, it becomes an economic abstraction (Mitchell 2011, 1–3; Weszhalyns 2013). Understanding oil and other fossil fuels primarily as resources is part of the general disposition toward NATURE at the heart of global capitalism. As Marx famously maintained, the ACCUMULATION of value—the very logic of capitalism—degrades the "natural and social characteristics" of human and nonhuman nature (Marx 1959, 77; Moore 2003, 325). A price can be placed upon oil once it is imagined as a resource. This abstraction, which masks the human labor that transforms natural resources into commodities, presupposes the exchangeability of oil, which allows oil to be conceived as an isolated entity that can be extracted from the ecosystem without consequence (Harvey 1993, 6).

To describe oil as a resource is to dematerialize and objectify it, while naming oil a curse mobilizes its imagined agency to maintain and justify the political and economic status quo. In the juxtaposition of "resource" with "curse," in other words, lies a tension between "rational" control over oil and the "irrational" power of oil. This tension can perhaps best be described in terms of Max Horkheimer and Theodor Adorno's "dialectic of Enlightenment": the complicit relationship between Enlightenment and myth, between reason and its irrational other. The resource curse formulation reproduces this dialectic by combining the abstractions of commodity capitalism with awe for the magical power of the material object. Recognizing how this linguistic structure functions offers a productive way of grasping and, potentially, countering the self-destructive yet seductive course of contemporary culture in socio-environmental terms.

Realizing this potential, however, entails nothing less than a fundamental rethinking of dispositions toward oil and an attendant reimagining of conceptions of nature, following the critical approach of Bruno Latour (2004), Jane Bennett (2010), and others. Their emphasis on the artificiality of a divide between nature and culture offers a basis from which to develop alternative dispositions in which oil and other fossil fuels are imagined as actants in a global ecosystem: things that produce effects and demand responses in their own right. This form of thinking requires a fundamental shift from an economic model of ENERGY production and consumption based on ownership of land and resources to an ecological model that places coexistence at its center.

See also: BOOM, ELECTRICITY, ENERGOPOLITICS, FRACKING, GRIDS, NIGERIA, PETROVIOLENCE, UNOBTAINIUM, UTOPIA.

Risk

Karen Pinkus

Risk, as I have written elsewhere, is a risky term.[1] We do not know exactly where or when the term originated—in early maritime law, perhaps. More significantly, we do not know quite how to use it properly. *Risk* in its most general sense (*Webster's*: "the chance of injury, damage or loss") contrasts significantly with its meaning in the context of investment: price volatility. In common speech, *risk* has a primarily negative connotation, as something to be avoided. But in the market, *risk* has a positive, aspirational sense so long as prices move upward, even if in fits and starts (Warren Buffett: "I'd rather have a lumpy 15% return on capital than a smooth 12%" [Loomis 2001]). If risk veers toward the negative in the market context (i.e., lower prices, away from profit), it risks losing its definitional force and means simply "a bad investment."

In either sense of the term, *risk* hovers around many of the keywords in this volume; it could be a meta-entry. Risk is calculated in the complex forms of monetization and indemnity that ENERGY companies (formerly known as oil companies) develop around practices of exploration (including the risk that a given exploration will not yield commercially valuable fields), drilling, extraction, transportation, refinement, delivery to end-users, and consumption itself. As Timothy Mitchell shows in *Carbon Democracy* (2011), oil and price volatility (that is, risk in the market sense) are so intimately tied together that we can no

1. In "The Risks of Sustainability" (Pinkus 2011), I find an early link—not an origin—in language around seafaring and veering (the origin of *environment*).

longer speak about supply and DEMAND. In the energy industry (or rather, the fuel industry, a distinction I discuss below), successful participants immunize themselves against risk in the general sense while they may also cultivate risk in the market sense. And it is not merely fossil fuels that are associated with risk; a similar list might well accompany future fuels or RENEWABLE fuels. Think, for instance, of the risk to birds by the operation of wind turbines.

Perhaps neither the common sense nor the market definition of *risk* is adequate to our current situation. At first glance, there would seem to be only slight risk (in the common sense) in individual efforts to conserve energy or to achieve energy autonomy (going off the GRID, for instance). What is the harm, after all, if I can generate my own power, in the Rifkinian fantasy of a distributed hydrogen commons? What is the broader social harm in the conservation or non-use of energy? Such questions become more complex, however, when considered at a broader scale—say, for instance, attempts to prevent drilling at the Yasuní oil field in Ecuador or to block proposed infrastructural projects like the Keystone XL pipeline. Individual stakeholders may choose not to use (up) fuels, but the risk of such activism (or better, passivity) is that my choice removes me from a global network/tragic commons that is not in the least diminished by my absence from it. Regardless of individual decisions of non-use, industry and lobbying groups like the Potential Gas Committee, for example, are busy searching for and assessing the risks of new drilling sites (http://www.potentialgas.org).

By its very nature, *risk* in the market sense is incompatible with non-use of fuels because growth is the only mechanism that can drive risk (i.e., price volatility leading to profit). And the risks associated with fossil fuel development are often dismissed as mere bad investments and thus not risky. To take just one example of how the two senses of *risk* are at odds in energy markets: the Dallas-based Breitling Energy Corporation offers individual investors with a minimum of $100,000 the opportunity to own mineral rights to, and receive royalties on, fractions of oil and gas properties. An ad for Breitling in *Fortune* notes reassuringly: "Investors are not responsible for costs or risks associated with monthly upkeep, exploration, or development of their property. These are handled by the energy companies." Breitling offers the prospect (if not quite the promise) of receiving monthly royalty checks without having to touch the fuel itself, avoiding all risk in the common sense. The royalty company acknowledges, however, that "these [lease fractions] are not liquid assets," perhaps aware of the double entendre. The notion that such an investment could be risk-free in the market sense seems highly disingenuous at best. As one expert claims, "Money is the easiest thing to get in the energy industry right now" (Beard 2014); by NECESSITY, there is risk involved.

These distinctions may seem wholly unphilosophical, perhaps even aggressively so. From Aristotle (2004), we could muster a complex set of ideas about *dynamis* and *energeia*, which Giorgio Agamben (1999) translates as *potentialità* (potentiality) and *attualità* (actuality or energy), respectively. We might say that fuels (as distinct from energy) are pure potentiality: prior to or separate from the many inventions, constructions, machines, and technologies that dazzle us with their complexity. The potentiality of fuel, its *dynamis*, implies power yet also describes a state prior to the use of that power. (By contrast, *energeia*, for Aristotle, implies action or enactment.) These distinctions shed light upon the risks involved in using or not using fuels. Fossil fuels still in the ground would be, in Agamben's reading of Aristotle, *dynamis* or potentiality. But as groups like the Potential Gas Committee seek out and assess the potential risks of new drilling sites, such potential is stripped of the aura of potentiality—in Aristotle's sense of the capacity to do but also to not do something. Instead it comes to mean simply "that which is waiting," that which stands in wait for exploitation—in Heidegger's terms, as *Bestand*, standing reserve (1977, 17). This distinction between fuel and energy can open up thought around questions of potentiality and hope in the face of climate change.

We can think about fuel, but regardless of our thought—indeed, with absolute indifference to thought and language—fuels will be used (up). When they are used (up), our thought about them is reduced to history, or death. In the case of fossil fuels, we have nothing to do but mourn their loss and the greenhouse gases they emitted into the atmosphere. On the other hand, to engage in a forensics that elevates some fuels over others (as more efficient, cleaner, transitional, greener) risks not grasping their essence. In this case, thinking fuel could be subjected to the same sort of critique that Antonio Negri makes about Agamben's concept of inoperativeness (*inoperosità*): the risk that thinking amounts to passivity (see Attell 2009). It is not clear, however, what sort of activity is proper to counter such risk. In risk, between the common sense and the market sense, we confront an aporia, emblematic of the times, and for that all the more vexing.

See also: DAMS, ENERGY, GREEN, GUILT, NETWORKS, OFF-GRID, RESILIENCE.

Roads

Deena Rymhs

Traffic is not only a technique; it is a form of consciousness and a form of social relations.
—RAYMOND WILLIAMS, *The Country and the City*

On May 31, 2012, Musqueam[1] residents and supporters blocked traffic to Vancouver's Arthur Laing Bridge to protest further development in the vicinity of an ancestral village and burial site. Earlier that year, human remains from the four-thousand-year-old village, Cəsnaʔəm, had been unearthed during construction of a condo development near the bridge's on-ramp. While construction was ordered to halt, debates ensued over title, compensation, and jurisdictional authority. For motorists traveling that morning, the protest offered a salient reminder that the bridge connecting Vancouver to its international airport runs through one of the most significant ancient sites in the Pacific Northwest. Just as poignantly, this moment revealed the durability of cultural memory and the intertwining of place with people, history, and knowledge, even as such places are paved over. Highways and motor routes are such totalizing spaces that they often elide the social and physical worlds that prefigured them. The modes of vision conditioned by roadways rely on a particular "geometry of perspective," to borrow Lawrence Barth's words (1996, 482), that is not just a way of seeing but also a form of ignorance.

Roads are spaces of thick signification whose ties to colonial violence and expropriated territory undermine their mythologization as places of freedom. For many Indigenous communities, the road has not often been so open. Rather, roads frequently led to the cur-

1. The Musqueam trace their ancestral occupation of present-day Vancouver back several thousand years.

tailment of Indigenous peoples' territory and their containment into segregated spaces like reserves and inner cities. While Indigenous reserves may be on the outer limits of public consciousness in CANADA, the lands on which they are situated are often crucial nodes in the economic flows and everyday circulation of the country. Industrial and extractive demands are frequently prioritized in the construction of roads through or near reserves in Canada. The Indian Act, Canada's statute of laws regulating nearly every aspect of Indigenous peoples' lives, establishes the federal government's mandate over the planning of roads on reserves. In some instances, industry access roads have been constructed on lands still under territorial negotiation, as in the Taku River Tlingit First Nation v. British Columbia ruling (2004), in which the Supreme Court of Canada allowed a mining road on Indigenous traditional territory before title had been established.

Yet roads also throw notions of ownership into question: roadblocks, a longstanding mode of political demonstration for Indigenous groups in Canada, expose the porousness of boundaries and the imbrication of territory. At their most effective, blockades task people to see roads differently by performing the road as a border—by making legible the regulation, discipline, and control of road spaces that are less recognizable when such practices are exercised by the state. As fluid entry points into multiple racialized spaces, roads can be sites of anxiety, spatial interfaces whose meeting and leaking of boundaries, bodies, and geographies contain the potential for violence.

As sites of social memory, roads evoke histories of political contestation along with personal and collective trauma. The disappearances of several Indigenous women on British Columbia's Highway 16, also called "The Highway of Tears," and numerous other highways in Canada points to a topography of violence where roads come to signify, among other things, immobility. One might link these ongoing instances of violence to the racialization of space in settler-colonial history, where practices like the pass system intensified geographies of segregation between reserves and surrounding areas by restricting the mobility of Indigenous residents who were required to obtain the Indian Agent's permission to leave the reserve. While the enduring effects of these genealogies of segregation are evident today, Highway 16 is also part of a neoliberal landscape where capitalism's logic of uneven ACCUMULATION comes into sharp relief. Without access to vehicles of their own, many of the Indigenous women who have gone missing on Highway 16 resorted to hitchhiking along this stretch of road that serves as a crucial transport route for the forestry, gas, and oil industries. The contrast between the restricted AUTOMOBILITY of these women and the turbo-mobility of the trucks passing daily through this corridor makes it necessary to view such instances of gendered, racialized violence as moments when abstract economic and social processes materialize in local landscapes.

This tension between mobility and confinement emerges in the work of Métis playwright Marie Clements, whose plays, *The Unnatural and Accidental Women* (2005) and *Burning Vision* (2003), join emergent theoretical formulations of space, biopolitics, and states of exception. Clements's writing answers the representational challenges that Rob Nixon sees "slow violence" posing for artists, critics, and theorists (2011). *The Unnatural and Accidental Women* connects racial and sexual violence to the industrial history of Vancouver. In this work, which is loosely based on the real-life murders of several Indigenous women in Vancouver, Clements links the female characters' experiences of displacement to the logging industry's violent alteration of landscapes. Fixing its gaze on the Vancouver that grows out from the first "skid road"—a term that originates in Pacific Northwest logging towns[2]—*The Unnatural and Accidental Women* depicts a city that bears traces of its frontier past. Its grid of streets represents a different relationship to the land than the trails and traplines intimately remembered by Aunt Shadie, the ghost of one of the murdered women whose pathways are reinscribed by logging roads.

Burning Vision scales its reflection on roads and mobility to a larger geopolitical sphere. Tracing the passage of uranium from its extraction in the Northwest Territories to its deadly arrival in Nagasaki and Hiroshima, this play questions the outcome of new NETWORKS, movements, and flows. Multiple mobilities—water routes, barges, railways, log tracks, capital—figure in *Burning Vision*'s map of uranium's movement along what is evocatively called the "road to the atom." *Burning Vision* explores recent histories of ecological and human destruction and the role of technologies of mobility in this violence. These networks of destruction are not contained in the past: depleted uranium was used in the Persian Gulf War against retreating Iraqi military and civilians on what was called the "Highway of Death."

Technologies of mobility and the geographies they create are crucial considerations in light of recent, insistent claims to Indigenous lands by projects like the proposed Keystone XL pipeline, the Northern Gateway Project, and the Mackenzie Valley Pipeline. The Keystone XL project, a 2,000-mile pipeline designed to transport oil from the Alberta tar sands to the Gulf of Mexico, would cut through vast tracts of Indian country in Canada and the United States, contaminating water sources, threatening local ecosystems, and irrevocably affecting the social and cultural lives of residents living alongside these developments. Several Indigenous groups in Canada and the United States have organized transnationally to oppose a project that is itself without borders. As routes of transport, PIPELINES might be likened to roads in that they connect disparate communities in often-imperceptible ways. Their connection goes further: roads and automobiles create a dependence on fossil fuels that necessitates pipeline projects.

Kanien'kehaka scholar Taiaiake Alfred observes, "ongoing indigenous struggles against colonialism consist mainly of efforts to redress the fundamental injustice of being forcibly removed from the land or being denied access to the land to continue traditional cultural activities." Alfred notes, however, that there is another, more immersive, dimension to colonialism: "near total psychological, physical and financial dependency on the state" (2009, 42). The administrative, industrial, and economic circuitry in which roads

2. Loggers built "skid roads" of old railroad ties or wooden planks to move the felled trees down to the mill. These skid roads eventually became associated with the places frequented by loggers, and by the 1930s, skid road (or "skid row") had become a term for a rough part of town. This etymology contains an interesting set of substitutions and reversals: while skid row is first associated with roads and movement, it later comes to describe a despairing terminus of violence, poverty, and immobility. See Turner (1986).

function—networks of transport, resource extraction, state power—and their attendant socioeconomic inequalities, entrenchment of petroleum-based economies, regulation of land, and deterritorializing of place are part of the material and ontological dependency on the state that Alfred describes. Roads operate within larger networks of industrial and global economies, making it necessary to recognize forms of violence that mask themselves as neutral space.

See also: AUTOMOBILE, GENDER, IDENTITY, INFRASTRUCTURE, NETWORKS, SPIRITUAL.

Rubber

Andrew Loman

The petroleum industry emerged alongside the late-Victorian imperial romance, a genre of FICTION in which white men have violent adventures at the Empire's periphery before returning, usually enriched, to its core. Because petroleum exploration was global, it would not be surprising to see cross-pollination between the romance and discourses on petroleum, in the form, for instance, of adventure novels about oil explorers.[1] Yet no such subgenre of Victorian petro-romance exists.[2] Still, even if the major imperial romances do not make oil an explicit subject, certain well-known works are demonstrably interested in petromodernity. Edgar Rice Burroughs's *Tarzan of the Apes* and Arthur Conan Doyle's *The Lost World*, both published in 1912, express this interest through references to late Victorian human catastrophes attending the extraction of wild rubber.

1. Timothy Mitchell summarizes the standard history of oil in the MIDDLE EAST as a "tale of heroic explorers discovering unimagined wealth in a desolate territory" (2011, 43), which might equally summarize the imperial romance plot of Robert Louis Stevenson's *Treasure Island*.

2. This circumstance may be attributable to the physical properties of oil itself: as Franco Moretti notes, the treasure characteristic of imperial romance is "a fairy-tale entity . . . where the bloody profits of the colonial adventure are sublimated into an aesthetic, almost self-referential object: glittering, clean stones: diamonds, if possible" (1998, 63). Petroleum may well be the opposite of Moretti's clean aesthetic object. And perhaps oil exploration was too modern a subject—too closely linked to contemporary bloody profits—for the taste of the authors of imperial romance.

Rubber has an extravagantly bloody history. As John Tully (2011) argues, rubber is essential to modern industrial society, and DEMAND for it grew alongside industrialization. But demand exploded with the invention of the car in the 1880s, since the AUTOMOBILE rests (figuratively and literally) on the rubber in pneumatic tires. At first, rubber only grew wild, presenting opportunities for enormous profit to those who controlled the regions where wild rubber grew. In Africa, the BOOM's chief beneficiary was Leopold II of Belgium, sovereign of the Congo Free State. In South America, so-called rubber barons included both local businessmen and foreign concessionaires (among them the Vanderbilts). Their wealth depended upon brutal exploitation. As John Loadman notes, "The . . . South American and Congo rubber production industries [have] one common thread—the deaths of millions of natives" (2005, 145). Millions: the anti-Leopold activist Edmund Dene Morel estimated that ten million Congolese died in the rubber regime (Morel 1920, 109). Roger Casement placed the death toll in the Putumayo River region of South America at thirty thousand, which "the British Parliament concluded . . . was only one . . . instance of what was probably happening over much of the rubber-producing area of South America!" (Casement in Loadman 2005, 163). The rubber death toll that midwifed the petroleum age is a shattering vindication of Walter Benjamin's melancholy apothegm that a document of civilization is always a document of barbarism (1977, 256).

Tarzan of the Apes

The two-part structure of *Tarzan of the Apes* is sutured by two figures: Tarzan and rubber. In the first part of the novel, Tarzan grows from infancy to adulthood, gathering strength until he is king of the apes. Alongside his conventional rise to power are multiple references to rubber atrocities. Tarzan was born in Africa because his parents traveled there when the British Colonial Office asked his father, John Clayton, to investigate

> conditions in a British West Coast African Colony from whose . . . native inhabitants another European power was known to be recruiting soldiers for its native army, which it used . . . for the forcible collection of rubber and ivory from the savage tribes along the Congo and the Aruwimi. (Burroughs 2010, 5)

Tarzan later encounters refugees "fleeing from the white man's soldiers who had so harassed them for rubber and ivory that they had turned upon their conquerors one day and massacred a white officer" (ibid., 65). Rubber also haunts the second part of the novel, when Tarzan travels to AMERICA in search of Jane Porter. His possession of a car emphasizes his fantastical transformation. One character marvels, "The last time I saw you you were a veritable wild man, . . . and now you are driving me along a Wisconsin road in a French automobile" (ibid., 236). But the automobile symbolizing Tarzan's modernity is also linked to Africa through tires whose rubber may have been collected there.

The parallel between Tarzan and rubber informs the novel's perverse appropriations of reformist discourse. Burroughs borrows moral authority from the anti-Leopold campaign by making Tarzan's father a proxy, of sorts, for Roger Casement, whose famous report indicted Leopold II's management of the Congo Free State. Orphaned because of his father's expedition, Tarzan suffers collateral damage of rubber atrocity. One could envision Tarzan allying himself with the Congolese. But the novel immediately erects a cordon sanitaire: Tarzan kills the first Congolese he meets. Worse, at the climax of the African chapters, French soldiers attack the refugees' village and butcher all the men. The soldiers, one might say, do Leopold's work for him—and that of Tarzan, absent from this massacre only because he anticipates it, having arrived sooner, lynched a man, and rescued an imperiled French officer. The novel is equally perverse in describing Mbonga, the refugees' king, who sports "a necklace of dried human hands" (Burroughs 2010, 77). This necklace appropriates a key symbol in the anti-Leopold campaign: the severed hand, a mutilation common in Leopold's Congo as a ghastly means of accounting for bullets expended (for each bullet, a hand). Photographs of survivors mutilated by soldiers seeking hands enough to account for bullets shot became important evidence of atrocities. Hence the necklace conflates Mbonga with the perpetrators his community has fled. In these instances, the novel marks its inability to reconcile its own contradictions. The transformation of Tarzan from his rubber self—raw African resource—into his automobile self—American modern—is the novel's object; hence the refugees, having refused to serve modernity as its slaves, must die. Yet the novel also acknowledges modernity's brutality, which draws the Claytons to Africa and pushes Congolese refugees to Tarzan. The village massacre, Mbonga's necklace—these are sites of struggle between Western self-regard and Western self-reproach.

The Lost World

The rubber terror likewise shadows the original return-of-the-dinosaurs novel, *The Lost World*. The novel's narrator, Edward Dunn Malone, is modeled on the anti-Leopold campaigner Edmund Dene Morel, and Lord John Roxton, the novel's great white hunter, is another Roger Casement proxy. Casement's campaign against rubber barons in South America consisted of a report to the British Parliament, the so-called Blue Book, documenting atrocities perpetrated in the Putumayo River region of Peru. In the violent "heroic" style typical of imperial romance, Lord John wages war against these "slave-drivers"; as Malone reports,

> Lord John had found himself some years before in [the] district [where] the wild rubber tree flourishes . . . . A handful of villainous half-breeds [*sic*] dominated the country, armed [some] Indians . . . and turned the rest into slaves, [forcing] them to gather the india-rubber . . . . Roxton . . . formally declared war against . . . the leader of the slave-drivers, . . . which ended by his killing with his own hands the notorious half-breed [*sic*] and breaking down the system which he represented. (Doyle 2008, 58)

The Lost World is like *Tarzan of the Apes* in that the rubber terror figures anamorphically—the reader never sees it directly—but it is nonetheless crucial, and not only for provoking key incidents in the plot. Like *Tarzan of the Apes*, *The Lost World* identifies modernity with the automobile—with the taxicabs taking Malone to his appointments and the electric broughams carrying public speakers to their lectures. And like *Tarzan of the Apes*, it links the modern, metropolitan center with an exaggeratedly primitive milieu at the imperial periphery, in this case, the dinosaur plateau. The automobile is the crucial link: the plateau's first discoverer, Maple White, is an artist from Detroit, Michigan—which was already, in 1912, the center of the American automobile industry.

Fictions of empire routinely deploy the trope of anachronistic space that casts the imperial core as the present and the periphery as the past, making the protagonists of imperial fiction de facto time travelers. *The Lost World* extends this trope to its limits, allegorizing petromodernity's dependence on the ENERGY stored in the remains of creatures long dead. The novel registers the uncanny relation between the genre's temporal logic and fossil fuels, which, as Timothy Mitchell puts it, "are forms of energy in which great quantities of space and time, as it were, have been compressed into a concentrated form" (2011, 15). In the novel's final chapters, the protagonists return to London with a live pterodactyl, which promptly escapes. The animating spirit of modernity, the pterodactyl disappears in the heart of London.

References to rubber in *The Lost World* and *Tarzan of the Apes* painfully dramatize the ambivalence engendered in the beneficiaries of petroculture when faced with the colonial neo-slavery on which it depended. Both novels deprecate the rubber slavers. But both also identify their heroes with massacres. Tarzan's French allies butcher Congolese fugitives. Lord John Roxton, scourge of neo-slavers in the anterior action of *The Lost World*, exterminates a tribe of "ape-men" that menace him on the dinosaur plateau at the novel's climax. The ambivalence dramatizes a double bind: repugnance at a grossly exploitative system, on the one hand, and complicity with it, on the other. The novels denounce. They also confess.

See also: AFFECT, EXHAUST, GUILT, PETROREALISM, PETRO-VIOLENCE, RESOURCE CURSE, ROADS.

Rural

Erin Morton

The postwar modernization project rendered "the rural" as a space of technological backwardness oriented toward subsistence (Mardsen 2008; Samson 1994). Even the linguistic root of its concomitant category, "the country" (in French as *contrée* and in Latin as *contrara*), points to the idea of rurality existing in opposition to something, which, more often than not, means modern progress (Muecke 2005). Connected to this historical understanding of rural places is the more recent idea that opposes rurality with the global city, the former typically understood as a terrain of perpetual struggle and the latter as a site of what Saskia Sassen calls "the geographic dispersal of economic activities that mark globalization" (2001, xix). Yet if the global city is typically understood as a space fueled by rural migrants who WORK there for the promise of a better life defined by wage labor, modern industry, and cultural development, how can we think along the same lines about the global countryside?

One way of approaching this question is to unpack the shifting ideology of rurality itself, especially in a globalized age in which the country fuels the city through migrant workers, food products, and ENERGY sources drawn from COAL, lumber, and natural gas, at the same time that the city fuels the country through the creation of such rural-urban interspaces as suburbs. At Canada's West Edmonton Mall, for example, Alexander Wilson notes that part of the attraction of the suburb's artificial landscape is "control over nature [by] bringing everything into its climatized, commodified space, especially objects and species from the natural world" (1992, 198). Even through such interspaces, rural communities continue to provide a crucial link between people and the natural environment as urban sprawl reaches out to meet the countryside—and to incorporate many of its ideals into everyday suburban life. Moreover, rural resources fuel the high-consumption lifestyles of many (sub)urbanites whose dependency on such life-sustaining amenities as ELECTRICITY and water remain contingent on the SUSTAINABILITY and survival of the spaces of global countryside (Sumner 2005).

Periodizing rurality necessitates probing the longer history of its critical dependency on and contribution to urbanity. For many urbanites, as Paul Cloke summarizes, "the rural stands both as a significant imaginative space, connected with all kinds of cultural meanings ranging from the idyllic to the oppressive, and as a material object of lifestyle desire for some people—a place to move to, farm in, visit for a vacation, encounter different forms of nature, and generally practice alternatives to the city" (2006, 18). Whether in the nineteenth century or today, urbanites often see rural communities as confronting a choice between adjusting to the inevitable reach of the global neoliberal economy or facing inevitable annihilation (Sumner 2005). This attitude, resonant with Margaret Thatcher's oft-cited neoliberal epitaph, "There Is No Alternative" (TINA), belies a pervasive ideology that casts the rural as backward, parochial, and irrational. Yet histories of rural communities tell us that they have been enmeshed within modernization for over a century and have fueled the capitalist expansion that such processes require. The idea that liberal capitalist societies emerged as purely industrial phenomena disregards the fact that rural industrialism was essential to the ideal of transnational modernization that began in the late nineteenth century and continues to shape the conditions of neoliberal energy consumption well into the twenty-first century.

Rural energy projects of the first industrial revolution were deeply integrated with urban modernity. In the rural Canadian province of New Brunswick, for example, historian Bill Parenteau argues that new transnational industries such as pulp and paper were resource and capital intensive from the early twentieth century onward and spurred increased urban development elsewhere (1992, 8). As the province's first expansive modern industry, pulp and paper development required both vast forestry resources and the creation of mega-projects, such as hydroelectric DAMS funded with external capital, in order to sustain rural and urban industrial expansion. As a result, local state control of forest and waterways increased but also led to a shift from public ownership of provincial hydroelectric development toward corporate control by pulp and paper industries (Parenteau 1992). As Parenteau makes clear, "Under the management of an interventionist provincial state prepared to meet the requirements of multinational capital, New Brunswick appeared to be well placed to participate in the second industrial revolution [of the postwar period], which was reshaping the North American economy" (1992, 43).

The New Brunswick government subsequently ushered in a "power for industry" model of rural development, which was meant to capitalize upon the North American trend of continental resource exploitation that began in the second half of the twentieth century (Kenny and Secord 2001, 84). The goal was to tap into the US need for hydroelectricity in order to expand the market beyond the Canadian border, which in turn promised to bring both economic prosperity to a rural province plagued by its association with underdevelopment and the modernization ideal of electric power to rural residents. Initially controlled by transnational pulp and paper companies, by the late 1950s hydroelectricity transformed into a public utility that created New Brunswick's first modern provincial power grid (Kenny and Secord 2001). Yet if this moment of capitalist growth in North America demanded rural industrial development through such energy channels as hydroelectric expansion, it also demonstrated the extent to which rural communities adapted and even embraced postwar modernization.

Rural resistance to the modernization ideal of electricity in New Brunswick was also widespread, however. The industrial strategy of hydroelectric damming throughout the 1950s and 1960s was advanced as a social betterment strategy to improve rural living. But it also reinforced settler-colonial claims to Wolastoqiyik lands along the Saint John River valley. Lauded as a symbol of modernity and progress, the damming of the Saint John River resulted in the forced relocation of Indigenous Wolastoqiyik and rural settler communities alike. According to historians James Kenny and Andrew Secord, "Whereas in the past the river had been viewed largely in 'local' terms—as a source of fish or for personal and commercial transport—now the river was presented as an engine to power regional economic growth more generally" (2010, 6). The result was rapid organization against the mega-project by local Indigenous and settler populations who protested engineers' and policymakers' plans to flood churches and graveyards (including Loyalist settlements and Meductic, an ancient Wolastoqiyik burial ground and national historic site) as well as fertile riversides in a province where 85 percent of the land was either forest or infertile soil (Kenny and Secord 2010). Despite the concerted efforts of rural protesters, the Mactaquac Dam opened on the Saint John River in 1968. The project's long-term effect in New Brunswick was to reveal the ways in which hydroelectric expansion in the late twentieth century involved contested ideas about ecology, heritage, and political economy there, as it did elsewhere (Kenny and Secord 2010).

Into the late twentieth and early twenty-first centuries, rural places such as New Brunswick remain sites of neoliberal energy industrialism, if only because of their uneven reliance on natural resource and agricultural sectors for employment. While pseudoscientific rationales for such projects as hydroelectric dams abound in rural communities, so too does the social and economic cost of such expansion for global city and global countryside alike.

See also: CANADA, CHINA, ELECTRICITY, LIMITS, NETWORKS, SPIRITUAL, URBAN ECOLOGY, UTOPIA.

Russia

Alexei Penzin

"I Love Oil" was an ambitious and provocative song released in 2013 by a well-known Russian comic team. An accompanying music video released in 2014 made the song so popular that it could be heard everywhere in Russia.[1] As it spread across the country, "I Love Oil" transformed a blunt comic parody into something sharper and more profound—an index of the place of oil in Russian social and political life today.

The video for "I Love Oil" begins with a shot of a petroleum worker, dressed in a blue uniform and wearing a yellow helmet. He presents a metallic barrel to an upper-middle-class woman and her small daughter in their shiny luxury apartment. The kid and her mother both look happy and applaud the worker and his precious gift. Then a series of surrealist and humorous images point to the array of possibilities engendered by the mysterious substance in the barrel. The woman pays for her purchases at a shopping mall by nonchalantly pouring a black liquid into the cash register. Then follows a wild dance party with revelers swimming in and drinking the same oily liquid and using it as a substitute for lipstick. Another scene portrays the dangers of the liquid. When a partygoer pulls out a cigarette lighter, everyone looks shocked and horrified. But a brave, quick-fingered DJ throws an album that cuts off the flame and makes it safe to resume dancing in the black goo. This hilarious sequence of images shifts to an idyllic farm, where Russian peasants

1. The song can be viewed at http://www.youtube.com/watch?v=EucLgHzuZaw. Googling the Russian segment of the Internet gives more than 500,000 references to the video.

dressed in folkloric costumes drink from a samovar—not tea but the same oily stuff. We then see bank robbers who cram their loot bags with oil, which leaks out when police shoot bullets that puncture the bags. Finally, a little girl looks sympathetically at a homeless person, who collects the black liquid rather than coins in his overturned hat.

The song's lyrics are just as remarkable. Its syllogistic refrain (accompanied in the video by images of shopping in Milan) can be translated as "While there is oil in Russia, I am in Milan. And if I am in Milan, there is still oil in Russia." In one sequence of the video, the female singer converses with her boyfriend, dressed as the perfect corporate "petro-male." "Do you love me?" asks the petro-male. "I love oil," responds the female singer.[2] He replies, "I *am* oil!" "I love you!" she exclaims. This impeccable logic is then disturbed by a potential inconsistency when the male singer continues with inimitable self-confidence: "I also have gas!" The female singer enthuses, "I love gas!" "But you love oil!" the petro-male responds. The female subject looks embarrassed: the unity of the substance reveals an internal split. She cries out desperately "Oil!" "Gas!" "Oil!" "Gas!" To stave off this fragmentation, she declares, "I love Russia!"

At the video's climax, we return to the idyllic farm where Russian peasants dance ecstatically. But something has changed. A modern oil derrick now appears within this premodern setting. The female singer triumphantly concludes, "I live in Russia! I live in Russia! I live in Russia! Russia is the land of opportunity!"

The video depicts the "real abstraction" of money collapsed into a very real, smelly, and oily substance; the commodity form of oil translates its monstrous contents into phantasms that invade everyday life, as reminders of the far-from-idyllic places that make it all possible. The troubling division of the substance into oil and gas is papered over by the ideological IMAGE of a united Russia as an alternative object of love. The song implies an "oil, *ergo sum*" formula—Russia has oil, therefore I am (a hedonistic ecstatic consumer)—which all too many Russians are hoping to make their reality. Even in this over-the-top pop video, the phantasmagoria of one very special commodity and its human subjects reveals the realities of contemporary petro-capitalism.

"I Love Oil" provides an entry point into the complexity of post-Soviet Russia and the international crises provoked by its rising geopolitical ambitions. Booming oil and gas prices in the past decade have transformed the country politically, socially, and culturally. After an era of relative social equality, income inequality in the new Russia is among the highest in the world (Credit Suisse Global Wealth Report 2012). Russian oil means extraordinary wealth for a tiny minority and social distress for the majority. Financial capitals like Moscow and Saint Petersburg may appear to be wealthy modern cities, but devastation reigns on their margins, as in most other Russian cities and towns, which resemble siblings of Chernobyl—with the devastating culprit being oil rather than NUCLEAR power.

2. There is a pun in this dialogue. Oil in Russian is *neft*, which is homonymous with the word *nyet*, which means "no" or "not." The woman's reply could also imply negation through hesitation: "I love . . . no (*neft*). To resolve this ambivalence, the petro-male shouts, "I *am* oil!"

This new socioeconomic reality is reflected in three discourses that criticize the problematic dependence of Russia on natural resources. The nationalist and conservative argument laments that resource dependency reduces the national economy to oil money and financial speculation. They mourn the more diversified industrial production of the Soviet Union as a lost arcadia. The second discourse, simultaneously liberal and neoliberal, also laments oil's overdetermination of all spheres of society, but it locates the ideal alternative to this regrettable state of affairs not in the (totalitarian) USSR but rather in the contemporary West, a society of knowledge and highly developed "human capital." Industry in contemporary Russia, by contrast, demands relatively few workers for the oil fields and deems the rest surplus population, useful only for staging mass support for the actions of elites. These variants of (neo)liberal and conservative thinking are widespread throughout mainstream and oppositional MEDIA and the academy. They produce strange combinations that are surprisingly resistant to critical challenges to their basic presumptions. Thirdly, Left intellectuals discuss Russian peripheral or semi-peripheral capitalism and its historical place as a natural resource supplier to the economic world-system—a minority opinion on oil in Russia with little influence on public opinion or government policy.

While I am sympathetic to standard Leftist positions about peripheral capitalism, it is worth challenging the stereotypical view that an economy based on gas and petroleum is necessarily a backward mode of production that excludes an intellectual labor force (i.e., human capital) from its key industries. Oil and gas require a hybrid economy and specific forms of human capital; its vast administrative apparatus is saturated by intellectual and communicative labor. The key struggles in this economy occur not only in oil and gas production but also in transportation, information services, and logistical work on new PIPELINES and other oil INFRASTRUCTURE.[3] These activities require political articulation, all the way up the ranks of Vladimir Putin's government; the oil economy is nothing if not a form of contemporary capitalism based on communication and the subjective qualities of workers—greed, cynicism, opportunism, etc. Shadow economic gurus navigate sophisticated and inventive schemes of bribes and kickbacks production. The proliferation of public relations spin doctors produce images of stable, prosperous life under corporate top manager Putin. This corporative managerial model has spread throughout Russian society, culture, and politics.

At the core of post-Soviet life is a post-shock subjectivity produced through brutal processes of primitive ACCUMULATION in the 1990s. One key moment of this foundational period was the privatization of oil and gas resources, intertwined with massive violence and social processes that were aimed at reshaping subjectivity according to the demands of profitability and the market. All previous professional roles, social attitudes, affects, and beliefs have been displaced; intellectual labor has migrated from the abandoned cultural and academic field into media, pop culture, public relations, and the cognitive architecture

3. For example, recent tensions between Russia and Turkey in 2015–2016 implied the question of building a new pipeline to transport the resources from Russia to EU consumers via Turkish territory; this project was finally abandoned.

of the oil and gas industries. Just to survive in the new Russian oil culture, many former academics, writers, and artists have begun composing inventive, obscure, and delirious public relations scenarios generously sponsored by their new masters, investing their talents and training in the emergent opaque texture of post-Soviet reality. "I Love Oil" is just one product of the revengeful and cynical general intellect that sends out secret SOS messages on the conditions of Russia's contemporary oil and gas economy and society.

See also: CANADA, ENERGOPOLITICS, RESOURCE CURSE, RURAL.

Servers

Mél Hogan

In 2011, Greenpeace released the first report on the energy choices made by Big Tech companies, calling attention to "dirty data" and the ENERGY that powers cloud computing. They found that if the Internet were a country, it would be the fifth largest consumer of energy and is projected to use an estimated 1.9 billion kilowatt hours of ELECTRICITY by 2020, an amount that exceeds the current consumption of CANADA, France, Germany, and Brazil combined (Ladurantaye 2011). In 2012, Greenpeace reported once again that Big Tech was rapidly expanding without adequate regard to source of electricity and relied "heavily on dirty energy to power their clouds" (Greenpeace 2012). In 2014, they reported on Big Tech's overall commitment to powering data centers with renewable energy, living exclusively off solar, wind, hydro, and geothermal power (Green Energy Investing 2015). In 2016, claims are being made about having achieved this goal, using 100 percent renewable sources to power their server farms (i.e., data storage centers composed of clusters of computer servers). But is this commitment to renewable energy by Big Tech really a green turn?

Server farms are being built in both urban and RURAL areas, often repurposing and adapting old city buildings or built anew on the perceived endless expanses of the rural landscape. Server farms consume so much electricity because the machines require not only a power supply but also reliable and constant cooling to mitigate the heat that

Thank you to Sabine LeBel, M. E. Luka, Tamara Shepherd, and Andrea Zeffiro.

they generate, particularly as servers are arranged to fit together in clusters and stacks. Although the companies often cite cool climates that favor energy efficiency as the logic that determines the location of server farms, the more important criteria tend to be tax breaks and a proximity to pre-existing fiber optic cable routes (Dilger 2013; Fehrenbacher 2013). However, local inhabitants tend to receive little return on their investment of tax incentives in terms of generating intellectual communities through job creation or social services.

The location of server farms is neither neutral nor independent of structures imposed on land and water. Countries with more favorable conditions for server farms than the United States include Sweden, Finland, Switzerland, and Canada. "The Node Pole," located close to the Arctic Circle in northern Sweden, for example, boasts natural cooling that saves companies millions of dollars annually in energy bills (thenodepole.com). Server farms built in such opportune locations, which foster the "greenest, most cutting-edge facilities" (Facebook Luleå 2014), demonstrate how wasteful it is to locate server farms in places that do not offer natural cooling or a renewable water supply. But the environmental issue is much larger than this.

These much-vaunted favorable conditions, however, obscure attention to how the server farms themselves, regardless of their location, contribute to environmental waste and affect people and planet on a global scale through unsustainable energy DEMAND. The growing reliance on server farms exacerbates cycles of pollution and waste that are attendant upon our increasing dependence on networked communication and the piles of e-waste that are also its byproduct.

As physical locales largely inaccessible to the general public, the server farms of large corporations enforce a guarded distance between users and their data: this is the promise of storage in the cloud. Users are disconnected not only from physical data storage but also from a proper understanding of networked culture and from the environmental repercussions of mass digital circulation and consumption. This distance serves companies' interests insofar as it maintains, on the one hand, an illusion of fetching data on demand, in and from no apparent space at all (the seeming immateriality of the virtual or digital, epitomized by the ethereal metaphor of the cloud), and, on the other hand, a solid material base that conjures notions of a secure and efficient system to which we can entrust our digital lives.

Google is among the companies that offer virtual tours of their data centers and glossy renditions of their inner workings through carefully curated brand interfaces (Holt and Vonderau 2015). Facebook uses its own social networking platform to publicize its environmental efforts at various data center locations, thus creating a quasi-official timeline of its environmental stewardship (Facebook 2014). Apple hosts a renewable energy website that documents their progress toward "100% renewable energy" (Apple 2012). These sites reach toward some mythical place beyond materiality, where the tangible aspects of server operations are consistently remediated as flat and virtual. Deflecting attention to massive energy consumption and pollution, these company interfaces tell a story of progress within a larger discourse of efficiency and economies of scale, where server farms justify their existence, even their dire importance. This powerfully visual discourse reinforces the disconnect between the seemingly GREEN, weightless, and immaterial flows of data and the massive material INFRASTRUCTURE that manages it at a distance (Cubitt 2013).

NATURE—or more specifically, so-called natural disasters—plays an important role in revealing the easily forgotten materiality and precarious positioning of these storage sites. Disruptions caused by storms, hurricanes, high winds, tornadoes, heat waves, droughts, and floods have begun to create a shift in public understandings of Internet storage as material rather than merely virtual. In 2012, Hurricane Sandy caused massive flooding in New York and New Jersey that left "wounded servers in its wake" (Thibodeau 2012). In June 2012, a heat wave caused by violent storms in Virginia took out power for various social media outlets (Hazlett 2012). Beyond these major weather events, more chronic atmospheric conditions create challenges for data storage. In India and CHINA, high levels of air pollution shorten the lives of server components; sulfur corrodes copper circuitry, and filtering contaminants from the air to cool servers in hot and humid locations is proving impossible (Rogoway 2013). Rather than tackling air pollution as the underlying cause, however, engineers are crafting more durable components to survive those difficult environments (Miller 2013).

Stormy weather and other environmental disruptions push the metaphor of cloud computing to its limits by reminding us of the stubborn material realities and vulnerabilities of data storage. *Slate*'s 2012 article, "Americans Think Cloud Computing Is Disrupted by Bad Weather," intended to point out how little people know about remote hosting and the operations of the Internet more generally, but it failed to recognize the brilliance of the metaphor. The "cloud" is an extremely suggestive concept for understanding the where, how, and when of our data in an increasingly wirelined and wireless society, where materiality is carefully concealed by metaphor and yet revealed by infrastructural failures and atmospheric disruptions (Mattern 2013; Mackenzie 2010). Cloud storage reinforces the illusion that our technology—wireless and unplugged—is at once physically self-contained, as devices and servers, yet also virtually networked and connectible. Memory is, for the most part, no longer localized (saved on the device itself) but rather stored remotely (in the cloud) as a means of providing multiple access points, communication among all of our wired devices, and synchronicity among users. As a crafty metaphor, storage in the cloud serves a psychological function as well. It distances us from our data and from awareness of the environmental implications of MEDIA consumption practices (Maxwell and Miller 2012; Notley and Reading 2013).

Digital media storage is containment-in-motion, activating the "tubes" of the Internet as subterranean and subaquatic infrastructure, interconnected carriers amassing cables in and through metropolitan regions, often retrofitted to meet current technological and communication needs, and thus doubling as traces of empire (Starosielski 2015; Blum 2012; Mendelsohn and Chohlas-Woos 2011). No longer mere containers of data, servers in this framework and at this scale become, as new media scholar Jussi Parikka (2011) argues, part of our consumptive practices: each gesture toward the database requires energy. Each search and click triggers server after server to connect, circulate, and deliver data.

Yet the experience of the Internet is seamless, simultaneous, instantaneous—a feeling that host companies bank upon to reinforce a culture of digital dependency and online urgency. Because servers do so much—store so much—for a perceived so little, the Internet is understood to be "efficient"—a term that energy expert Daniel Yergin (1991) theorized as "too precious to waste." Yergin's conservative sense of efficiency is at odds with current (mis)conceptions of servers as capable of delivering a perpetual virtual feed, which lack a holistic understanding of media ecology. As with other natural resources previously underestimated and mismanaged—food, oceans, and forests, for example—digital information is a new kind of resource prone to a different kind of excess, which nonetheless confronts LIMITS to growth. Under current rates of expansion, data might eventually need to be rationed, in part because the reigning model is inherently set to fail because it denies its own limitations (Bobbie Johnson 2009). The idea that we can continually match the growth of data to new physical storage centers is emblematic of an increasingly global culture of waste that rewards planned obsolescence and short lifespans for the devices through which we access the Internet (Sterne 2007). The irony is that Internet efficiency helps to generate environmental consequences and natural disasters that disrupt its illusion of seamless immateriality, while pushing it beyond ethical, sustainable, and RENEWABLE energy uses. Cloud computing may be the epitome of blue sky thinking.

See also: DETRITUS, GRIDS, NETWORKS, RENEWABLE, SUSTAINABILITY.

Shame

Jennifer Jacquet

In the late nineteenth century, John D. Rockefeller established Standard Oil, the first multinational CORPORATION to get rich off fossil fuels—so rich that his grandsons started a large charitable trust with some of the money. In the early twenty-first century that same charity—the Rockefeller Brothers Fund—announced that it would divest in the very origins of its endowment: fossil fuels.

What happened in the intervening century? Fossil fuels powered the Industrial Revolution and globalization, among other things, but most of society (with some important exceptions) also realized their use was leading to dangerous global climate change. Yet the United States, the world's largest emitter at the time, announced it would not ratify the Kyoto Protocol, the 1997 agreement to commit countries to binding international emissions reductions. At the close of the twentieth century, in the wake of failure to come to an international agreement, the need to do something about climate change started to be addressed not just as a political cause but also as a moral one.

Into this new moral cause entered shame—both an emotion and a tool. Shame exposes transgressors to an audience that expresses opprobrium and helps establish and enforce new norms of behavior. Shame is like the militia for punishment—allowing citizens to express reproach, often when the formal political system is not.

Many uses of shame for climate change in the post-Kyoto world were aimed at individuals for their patterns of consumption. "Tell your parents not to ruin the world you will live in," urged the 2006 film about climate change *An Inconvenient Truth*. In January 2008, *New*

York Times reporter John Tierney called for "mood rings, bracelets, lapel pins or anything else that could change color" depending on how much ENERGY an individual used, so that those mood-ring-wearing individuals could be shamed for inordinate amounts of energy use. An acquaintance went around leaving note cards that said "asshole" on parked Hummers. Both shame and shaming were at work to instill a new norm to make people more motivated to reduce emissions. However, the focus of shaming in these last years of the Bush administration remained on individual patterns of consumption rather than other individual behaviors, or institutions, or systems of production.

Then people got smarter about how to use shame. The target shifted from individuals in general toward specific individuals and corporations that had disproportionate impacts in terms of climate change and climate policy. Researchers and activists exposed a network of climate denialists—pundits as well as academics—who, alongside a complicit media, had helped undermine climate policies to reduce emissions.

As some key examples, Greenpeace (2001) published a report titled "A Decade of Dirty Tricks" and would continue publishing similar work, including the web project ExxonSecrets, which showed the company's funded network of outspoken climate denialists. Greenpeace (2010) also exposed the Koch Brothers, owners of Koch Industries (with a lot of its operations in fossil fuels), for giving almost $80 million to groups denying climate change. Reporter Jane Mayer (2010) wrote about the billionaire brothers in the *New Yorker* and put an end to David Koch's assertion that Koch Industries was "the largest company that you've never heard of" (Weiss 2008). Researchers and activists exposed a network of climate denialists—pundits as well as academics—who, alongside a complicit MEDIA, had helped undermine climate policies to reduce emissions. In 2015, the nonprofit group Organizing for Action made it easy to call out (privately over e-mail or publicly over Twitter) members of the US Congress who deny climate change (Goldenberg 2013).

The shaming has not stopped there. New research continues to expand the way we see who is to blame and, therefore, who is to shame. A paper published in 2013 showed that ninety corporations are responsible for roughly two-thirds of historic carbon dioxide and methane emission (Heede 2013), and follow-up work has discussed the role of investor-owned corporations in funding and rejection of scientific evidence on climate change (Frumhoff et al. 2015). E-mails from one of ExxonMobil's own scientists showed that ExxonMobil knew about climate change in the early 1980s, years before it became a public issue, but spent the next three decades funding and promoting climate denial (Mulvey and Shulman 2015). Chevron, the single greatest contributor to emissions, gave $2.1 million to US political groups in 2013–2014 (see opensecrets.org). It is clear that fossil fuel corporations have much more influence over the US government position than individual consumers do (who, given the choice, might actually express a different set of preferences, like traveling by SOLAR car or public transportation).

Groups may not feel shame, but evidence suggests they can CHANGE their behavior in response to being shamed, although certain groups are more sensitive than others. Beginning in 2010, a group of nonprofit organizations began singling out banks that funded mountaintop removal to extract COAL. As a result of the repeated shaming campaign, several banks have cut ties with mountaintop removal (Sorkin 2015). Banks rather than coal companies were the chosen targets because they have greater concern for their reputation.

The spotlight of shame has now turned toward banks that finance coal projects. Recently, a Dutch NGO (supported in part by the Rockefeller Brothers Fund) compiled a list of commercial banks' financing of the coal industry worldwide from 2005 to 2014 (see coalbanks.org). JPMorgan Chase tops the list, although there is a growing role of Chinese banks, and bank financing for coal is increasing despite contrary claims by the same banks that they are socially responsible and "transitioning to a low-carbon economy." Shaming campaigns target these banks to get them to change their ways. The World Bank and the European Investment Bank have recently introduced policies to phase out financing for coal power.

While shaming has gone after the reputations, other efforts have attacked finances, through divestment in fossil fuels. Urged on by many nonprofits, including 350.org, hundreds of institutions have pledged to divest in fossil fuels. Alan Rusbridger, editor in chief of the *Guardian* newspaper, started a campaign to address climate change, which included covering the homepage and the print version of the paper with what looked like an oil slick and highlighting the logos of companies deemed most responsible. The *Guardian* also started petitions urging institutions, like the Gates Foundation, to divest (Howard 2015).

Shaming, divestment, boycotts, and demonstrations all serve the same end: stigmatization. The stigmatization of climate change denial is underway, as is the stigmatization of fossil fuels, particularly fossil fuel investments. In the absence of clear and binding international policy, with few coercive, formal means of enforcing laws, informal sanctions such as these take over.

Fossil fuel companies have a lot to lose, and will not take attacks on their reputations without a fight. In September 2015, Australia's coal industry attempted a "coal is amazing" campaign (which backfired; Milman 2015). The website divestmentfacts.com ("A project of the Independent Petroleum Association of America") commissions and compiles research on why institutions should not divest. The fossil fuel–loving US politicians that stood in the way of a binding international emissions agreement very well may continue to do so. As long as there is opprobrium for limited international action and coordination, as well as multinational corporations aiming to explore and exploit untapped fossil fuels, there will be a role for shame.

In addition to having smart targets, shame is a needy tool that requires the attention of a concerned and rowdy audience. A survey in 2015 of ten thousand citizens across seventy-nine countries found that that two-thirds of people believe that UN negotiators should do "whatever it takes" to limit global warming to two degrees, and 80 percent of those surveyed believe that their country should reduce emissions even if other countries do not ("Worldwide Views" 2015). One key ingredient to how well shaming will work in putting downward pressure on emissions is how much we each pay attention to climate change and how well we continue to DEMAND action.

See also: AFFECT, CORPORATION, GUILT.

Solar

Amanda Boetzkes

> Gazing intently at the gigantic sun, we at last deciphered the riddle of its unfamiliar aspect. It was not a single flaming star, but millions upon millions of them, all clustering thickly together like bees in a swarm.
>
> —JOHN TAINE, *The Time Stream*

The history of solar power invites us to consider the difference between a form of ENERGY that shapes cultural exchange and a resource that merely fuels production. In the past century, solar power has been touted as a clean alternative to oil and COAL. It has also inspired visions of new social, ecological, and economic systems it might generate. Solar energy is imagined as fundamentally heterogeneous, characterized by how it precipitates complex transactions and conversions that nonetheless preserve homeostatic LIMITS. Indeed, this aspect of solar energy is often touted by critics of the industrial capitalist model that seeks to accumulate a uniform type of energy with the goal of indefinite economic expansion.

The Prodigal Sun

Among the most influential accounts of solar energy is Georges Bataille's *Consumption*, volume 1 of *The Accursed Share*, written after the Second World War. Bataille connects solar energy's abundance to economies of exchange in which luxury and squander are integral and inevitable. Solar energy is fundamental to the exuberance of all living things: organisms amass energy until they reach a maximum limit of growth and then undergo processes of intensified energy consumption, of which Bataille privileges predatory eating, death,

Thanks to Tanner Jackson, who spent a term with me researching solarity.

and sexual reproduction. Thus, by virtue of its ceaseless prodigality, endlessly dispensing energy, illuminating and heating the earth without any need of return, the sun generates in the biosphere rhythms of energy ACCUMULATION, growth, and squander. The most basic form of expenditure is the propagation of vegetation: plants collect energy from the sun to regenerate and extend their spread over the earth's surface. Animals harbor more surplus energy than they expend in the growth of the individual organism. When the calf reaches its maximum growth, its surplus energy is devoted to "the turbulence of the bull" or to "pregnancy and the production of milk" (Bataille 1991, 28). Wild beasts demonstrate an even greater prodigality than agricultural animals, expending energy to hunt other animals. The carnivorous tiger possesses a tremendous power to consume life and so exists at a point of "extreme incandescence" (34). Its inborn solarity calls William Blake to ask, "In what distant deeps or skies burned the fire of thine eyes?" (74). To which Bataille responds, "This incandescence did in fact burn first in the remote depths of the sky, in the sun's consumption" (34). Solar energy is for Bataille the premise of an ecological system whose limits are maintained through violent expenditure.

Bataille's transhistorical account of solarity produces a theory of "general economies" in which forms of wealth, social formations, and cultural practices emulate the sun's glorious expenditure. Aztec rituals of sacrifice, the potlatch of Northwest Coast Native peoples, and Lamaism are forms of social regulation defined through patterns of energy accumulation and radical burn, or "*la dépense*." Bataille posits world wars and atomic detonations as expenditures of energy at a global scale, linked to what he sees as the restricted economy of bourgeois capitalism. Capitalism's failure to acknowledge our innate solarity, and its fundamental prohibition of expenditure, results in the extreme pressure to accumulate energy without waste (in the form of profit) and a collective drive toward planetary destruction.

If Bataille's solar economy offers critiques of Marxism for its focus on utility (it seeks a good *use* of economy) and of capitalism for its incapacity to expend energy, it has nevertheless naturalized the principle of sacrifice, according to Jean Baudrillard (1998, 191–95). While it appears that the sun gives without receiving, in Aztec cosmology the sun gods demanded sacrificial blood: the sun's gift comes at a price and with an expectation of return. The governing principle of Bataille's general economy is not sacrifice, after all, but an incessant process of challenge—"generous" giving provokes the obligation of a counter-gift, driving social exchange toward maximal excess, namely death. Solarity, the demand to expend, persists alongside the sun's expectation; it waits expectantly, burning bright like Blake's tiger. The question remains, in what form will we carry out this return of the prodigal sun?

Solar Community

In the late twentieth century, social theorist Murray Bookchin imagined that solar energy could enable communities to break free of mainstream energy management and its technologies that demand toilsome labor in the service of centralized production and corporate control. Bookchin advocated technologies that would satisfy material needs and produce ecological forms of association, which entailed restoring equilibrium between humans and the natural world and making the natural world a living, visible part of cultural life (Bookchin 2004, 94). In these decentralized communities, agriculture would be a primary site of intellectual, scientific, and artistic activity. "Ecotechnologies" would counteract the alienation caused by mechanization and dependence upon fossils fuels as inefficient energy sources that require labor-intensive techniques of extraction.

Bookchin privileges solar power because it is an inexhaustible source of energy, freely and equally available. Moreover, solar energy could power liberatory technologies that would eliminate dependency on an exploitative, polluting, and centralized energy complex. In 1971, when Bookchin's *Post-scarcity Anarchism* was published, many of the technologies he cites were in early phases of development, among them solar panels, solar-powered stoves, furnaces, and water heaters, and batteries that convert solar energy into ELECTRICITY. Today, solar technologies rank among the most viable alternatives to fossil fuels. However, solar energy seems to have been uncoupled from the decentralized postscarcity society that Bookchin imagined; it remains nested within dominant systems of energy production and distribution and must confront the formidable challenge of grid parity.

Renewable Energy and Grid Parity

The study of solarity led Bataille to conclude that while we are driven to acquire energy, it nevertheless will be and must be expended. A not unrelated dynamic helps to explain why current solar technologies struggle to compete with more profitable but environmentally detrimental sources of power like oil, coal, and natural gas. Recent technological advances in solar power were spurred by late-twentieth-century concerns about peak oil. One of the top oil producers, the United Arab Emirates, inaugurated the world's largest solar power plant, the Shams 1, in Abu Dhabi in 2013. But the fate of solar energy is still intertwined with that of other energy sources and determined by its profitability relative to other new technologies like FRACKING and oil sands extraction. The economic viability of solar energy is constrained by the problem of grid parity: the cost of a solar photovoltaic (PV) system to produce electricity as measured against the retail cost of grid power.

Obstacles to reaching parity include the fluctuating price of other energy sources, government subsidies, and the market value of PV system parts. While the International Energy Agency predicts that solar energy could supply a third of the world's power by 2060 (2015), a glut of solar panels and component parts has led to a rapid market downturn. Investors and manufacturers have taken extreme losses since 2008, as the bankruptcy of Chinese manufacturer Suntech Power in March 2013 shows (Bradsher 2013). Solar ENERGY SYSTEMS have gone from being an expensive but necessary alternative fuel to a risky investment because their cost is too low to recoup a return of profit. A "carbon price" that factored in the costs of emissions and other environmental impacts for all fuel sources remains little more than an idea in the rudimentary stages of implementation. But it appears that solar energy requires such a general economy to gauge its objective value. As Bataille's account of solarity implies, a global INFRASTRUCTURE that drew from a freely available source is inimical to capitalism's restricted energy economy. One can still only imagine a world in which seven billion people had equal access to free power and could thereby take hold of their inborn solarity.

The Future Burning Bright

For three nights in May 2009, the campus of the University of Central Lancashire was lit up by a solar-powered screen projecting animated images of the sun taken by NASA's Solar Dynamics Observatory. This public artwork by Chris Meigh-Andrews, entitled *Sunbeam*, identifies the crux of solar energy (Gere 2011). The digital screen was installed on a solar tracker array purchased by the university as part of an energy-efficiency initiative. The projection therefore showcased the virtuosity and profitability of solar technology. But the NASA photographs disclosed a deeper awareness of solarity, as visions of a multicolored ball of fire punctuated the night sky. The prodigious double of the earth roiled and burned against the grid of solar panels. This sun is more than just a fuel. It is a persistent DEMAND—a social contract to take what is given and to return through expenditure. The FUTURE of solar energy burns bright, but how it unfolds cannot yet be gauged.

See also: CHANGE, ENERGY REGIMES, EXHAUSTION, GRIDS, MIDDLE EAST, OFF-GRID, RENEWABLE, SUSTAINABILITY.

Spill

Antonia Juhasz

On November 8, 2013, I received an e-mail inviting me to contribute to this book, with a list of potential topics. *Spill* was the obvious choice, given the years I have spent trekking across the United States and much of the world following a trail of dirty, smelly, deadly crude oil unleashed through the mass negligence, greed, and hubris of men (the vast majority of the industry is male).

That same day, I received another e-mail, this one with a photograph of a massive black cloud billowing above red flames shooting out of nearly one hundred blackened train cars lying on their sides in Aliceville, Alabama. A train carrying crude oil fracked from fields in North Dakota had derailed, causing a series of explosions, releasing some 750,000 gallons of crude into a wetland (Sturgis 2014) and bringing about the largest rail oil DISASTER in US history (French 2016). It was just one of a series of such incidents that began the year before; more crude oil was spilled from trains in the United States in 2013 than in all the years since 1971 *combined* (Tate 2014).

As I began writing, "spill reports" just kept coming. Keeping track of the oil and natural gas disasters occurring daily was virtually impossible. The numbers are increasing everywhere, from all sources, and for good reason. The age of "easy oil," if it ever existed, is long over. The oil that is left is located in areas previously deemed too sensitive, secluded, important, risky, expensive, or technologically complex to pursue. The risks associated with exploiting it are growing—as confirmed by the UN Intergovernmental Panel on Climate

Change, which reported in September 2013 that in order to avoid the worst impacts of climate change, the vast majority of the world's oil must remain in the ground (Morales 2013). But also growing are the rewards for the few poised to cash in.

As oil companies delve ever deeper into oceans, frack their way through earth and stone, pry oil from tar, move more (and more volatile) oil over greater distances, and make greater use of militaries to fight wars on behalf of their black gold, disaster follows increasingly in their wake.

Spill is the word used to describe these events. To better understand how the public responds to this term, I posted a question on my e-mail listserv and Facebook and Twitter pages: "In the context of oil, what does the word *spill* mean to you?"

One resounding theme emerged from the dozens of responses: *spill* is far too innocuous to capture the depth, breadth, and frequency of the destruction taking place. Moreover, it creates a passivity toward these events, as it brings to mind, my responders repeatedly told me, *spilled milk*: a child knocking over a small glass, an innocent, simple, and careless mistake, easily rectified with the wipe of a cloth and just as easily forgotten.

The word also creates a powerful evasion of responsibility, many argued, by implying a one-time accident rather than a series of calculated decisions resulting in easily anticipated disastrous results. Rather than innocents, the individuals, corporations, and even governments at fault are more often criminals, all of whom must be held liable for their actions. Rather than being passed over, these events should lead to fundamental changes in public policy to ensure they are never repeated.

Everyone, it seemed, then had a personal story to share, or events that felt personal to them that reached a level of destruction wholly inconsistent with a "spill" and that, they insisted, required action. Yet, they shared an enormous frustration at the inability of governments to rein in the industry and at the MEDIA for adopting and repeating the *spill* terminology.

It is a word with which I have personally struggled for some time, but particularly after BP and Transocean's *Deepwater Horizon* drilling rig collapsed in April 2010, releasing 210 million gallons of oil into the Gulf of Mexico. In 2011, I released my book, *Black Tide: The Devastating Impact of the Gulf Oil Spill*. I had wanted *disaster* in place of *spill*, but my publisher argued that oil companies, the government, and the media had repeated the phrase "Gulf oil spill" so frequently that it had become iconic. In so branding this event, the phrase had the desired effect; the public understood this *spill* as a one-time fluke, easily washed away, with limited long-term consequence, though the opposite was true. Within approximately two years, for example, CHINA, NIGERIA, CANADA, and Michigan all experienced oil SPILLS of over one million gallons, each breaking national (and state) records (Wong 2015).

The *spill* nomenclature, moreover, makes it difficult, if not impossible, for harmed communities to attract an adequate level of attention from the public and policy makers to address the scope of the disasters they face. A particularly gruesome competition is emerging between affected communities that struggle to distinguish their spill as worthy of notice, by virtue of being the largest, involving the highest number of fatalities of people and/or other living things, or exacting the economic toll. It is a battle over scarce resources in which few come out victorious.

My respondents tried to come up with alternative terminology, but this is only part of the solution. The oil industry is constantly working to do the same, with the considerable advantage of money, time, resources, and access to ensure that its definitions and word choices dominate any discourse.

We want to do more than better define oil-related disasters; we want to stop them. Working for better regulation and oversight of the industry to achieve this end is critical yet has its own limitations. For though fossil fuel extraction can certainly be made "cleaner," "safer," and more "humane," it can never be clean, safe, or just.

It is therefore time to change our frame of reference: if fossil fuels are not pursued, they can never be spilled. Communities around the world are taking part in a new movement declaring their right to get out of the fossil fuel game altogether. They are passing laws to keep their "oil in the soil" and refusing to allow extraction where they live.

In Ecuador, citizens gathered over 680,000 signatures to put a referendum on a national ballot to require the government to leave in the ground indefinitely the nation's largest source of oil, located in the Ishpingo, Tambococha, and Tiputini (ITT) area of the Yasuní National Park (Valencia 2013). In so doing, they aimed to hold their government to the Yasuní-ITT Initiative, an effort begun in 2007, whereby the government pledged that if the global community put up half the value of this oil, some 3.6 billion dollars, then the oil would remain untouched. Although this effort has been set back by the government's rejection of the referendum and the lack of international financial support, it nonetheless demonstrates the potential of collective effort that is finding success elsewhere.

France and Bulgaria have both passed national laws banning the use of hydraulic FRACKING for oil and natural gas. And in the same month that I began writing this chapter, three cities in Colorado voted to ban hydraulic fracking, following dozens of other cities and towns across the United States and around the world (Coffman 2013). In April 2013, Mora County, New Mexico, voted to permanently ban all oil and natural gas development, becoming the first US county to do so (Associated Press 2013). The potential threat that such efforts pose to the industry are evident in attempts by states in the United States (Texas foremost among them) to bar local municipalities from prohibiting extraction in their communities.

The boundaries that set some regions and methods as off-limits for fossil fuel extraction are shrinking, while the environmental harms of extraction are growing. Increasingly, we all live on the front lines of the most extreme consequences of fossil fuels. Rather than simply opting out of "catastrophe discourse," why not opt out of fossil fuels altogether?

See also: BOOM, CATASTROPHE, COAL ASH, DISASTER, ENERGOPOLITICS, FRACKING, OFFSHORE RIG, RISK.

Spills

Stephanie LeMenager

As a noun, *spill* denotes a downpour of liquid, with early usages tending toward blood and rain, while the term *oil spill* dates from the era of industrial modernity, after the gushers of the 1890s and the manufacture of the Ford Model T. The Lumière Brothers' short film *Oil Wells of Baku: Close View* (1896) documents an Azerbaijani oil field fire that could be euphemistically described as a spill. Gushers have been described as spills insofar as they involve uncontained spillage of oil that blankets miles of surrounding land, destroying plants, crops, and soils. The Cerro Azul No. 4 gusher in Veracruz, Mexico (1916), became the subject of films sponsored by the Pan American Petroleum and Transport Company. The company recognized its industrial accident as a spectacular display of ENERGY, not pollution (Archer 1922, 102). Similarly, the Lakeview gusher in California (1910), resulting in a massive on-land spill of 378 million gallons, became a tourist attraction; special trains delivered sightseers to the small town of Taft (S. Harvey 2010). Whether a gusher ought to count as a spill remains an open question, while oil fires—common enough in mid-twentieth-century oil fields to be referred to as "oil weather"—dubiously qualify as spills, too, especially where no explicit human cause can be named. One could debate whether naturally occurring oil seeps, such as those on the seabed of the Gulf of Mexico, are spills. Natural seeps account for the majority of spilled oil normally present in California's Santa Barbara Channel ("Natural Oil 'Spills'" 2009). Synthetic Aperture Radar imaging has turned satellites into tools for detecting unreported spills or spills whose extent might best be grasped from space (Topouzelis 2008, 6643). An unreported or unseen SPILL is still a spill. The word *spill* connotes no particular agency, a blurring of cause and effect, the possibility of both human error and atypical natural occurrence.

Conventionally the oil spill has been conceived as taking place upon water, where some of the most historically significant spills have occurred. From the 1930s, oil spills have been recognized as a common source of pollution in shipping. As early as 1907, the wreck of the steel-hulled schooner the *Thomas W. Lawson* was categorized as a marine oil spill, because of the 58,000 barrels of light paraffin it leaked near the Scilly Isles. In the 1960s, when the *Torrey Canyon* foundered in the English Channel off of Cornwall (an event culminating in the Royal Navy bombing the tanker), the oil spill became an IMAGE of widespread ecological damage and the pyrotechnic impotence of state and corporate response (L. Clarke 1990, 67). (The bombing failed to burn off the tanker's oil cargo, which led to the application of several million gallons of detergents at sea and on coastal land.) Oil spills on the ocean may spread hundreds of thousands of miles, enter the water column and poison sea life, destroy branching corals, and embed in mangrove forests that provide invaluable coastal protection. When the Liberian-owned tanker *Prestige* foundered, its oil spillage affected 81 percent of Galicia's coastal waters, undermining the region's family-based fishing economy and preventing local people from celebrating the Christmas holidays with traditional foods like goose barnacles (García Pérez 2003, 213). Writer and photographer Allan Sekula's series *Black Tide/Marea Negra* documents responses to the *Prestige* spill. Its title invokes a common phrase in romance languages ("*marea negra*," "*la marée noire*") that indexes oil spills as black water—a more visually evocative word-image than *spill*, if equally removed from questions of blame or cause.

With its spilled-milk innocence veiling the fact that oil spills are toxic events, the noun *spill* derives from the verb *to spill*, which is fraught with agential knowledge. Several obsolete meanings of *to spill*, a transitive verb, relate to willful destruction, for example, "to destroy by depriving of life; to put (or bring) to death; to slay or kill" (*OED*). While these meanings are not prominent within contemporary usages of *oil spill*, they suggest a historical and possible world where the downpour of liquid might not be naturally occurring or accidental in the sense of blameless. Transitivity implies acting upon another; to spill is to do something to someone, a person or thing. The noun *spill* can signify spilled blood. The largest oil spill in history describes an act of war. The Kuwaiti oil fires of January-November 1991 resulted from actions of Iraqi troops who ignited over seven hundred wells as they retreated from Kuwait. Upward of 42 million gallons of crude oil were lost—burned or pooled into roughly three hundred "oil lakes" in the Kuwaiti desert. Subsequent air pollution in Kuwait caused respiratory problems there and in neighboring countries like Bahrain. The constellation of health problems affecting US and British troops, known as Gulf War Syndrome, has been connected to the oil fires and the burning of other chemicals used in oil production (Walker 1994). Ultimately, Iraqi troops contaminated over 40 million tons of sand and earth in one of the most intensive scorched earth strategies in modern history and, as it is remembered, the world's most massive "spill."

Does *spill* attach to acts of war because it points to a lost commodity—spilled milk as dark metaphor in a culture more concerned by the global oil market than warfare and often tangled up in warfare? Might *oil spill* be reconfigured within contexts of war, recognized as part of war's increasingly commoditized repertoire? The Hollywood war film *Jarhead* (2005) makes the environmental devastation of Kuwait a thematic device in its treatment of Operation Desert Storm, while Werner Herzog's documentary-style *Lessons of Darkness* (1992) decontextualizes the Kuwaiti oil fires. Much to the dismay of some early viewers in Berlin, Herzog follows private contractors charged with extinguishing the Kuwaiti fires with a voyeurism that uncomfortably equates filmmaking with pyromania.

If *spill* is as an ethically irresponsible word for the Kuwaiti oil fires, what of smaller-scale incidences of negligence, if not overt misconduct, that might be categorized as spills? Consider the Lac-Mégantic explosion and oil fire that killed forty-seven Canadians and may have resulted from the mislabeling of Bakken crude from North Dakota as a class three "dangerous good" rather than a more volatile class two product (Robertson 2013). If Bakken oil is more volatile than other types of crude, then different tanker cars must be used to transport it safely. Oil trains, the US Department of Homeland Security warns, are potential bombs—although their detonation might result from industrial accident or reckless cost-saving, not terrorism. The inadequate or negligent containerization of oil figures strongly in spill events, speaking to the broad connotations of *spill* as an inability to contain. After the *Exxon Valdez* spill of 1989, industry claimed it was too expensive to implement the same regulations on oil tankers as on liquid natural gas (LNG) tankers, including double bottoms, double side shells, and double decks; such claims came under scrutiny (L. Clarke 1990). While an LNG tanker might contain more energy than was released by the bomb at Hiroshima, an oil tanker holds potentially decades of contamination—a bomb exploding slowly, over time. Problems of containment plague the relatively new oil product known as dilbit or diluted bitumen, originating in CANADA's Athabasca oil sands. In the context of the proposed Keystone XL pipeline from the Great Plains to the Gulf of Mexico, the US Environmental Protection Agency recommended that bituminous sands pipelines no longer be treated like PIPELINES carrying lighter crude. When it spills in waterways, dilbit acts distinctively, briefly floating and then sinking, which makes cleanup difficult without dredging that might be as injurious as oil pollution (Frosch 2013).

The material problem of how to move oil, in what sort of container, complements the communications problem of containing oil spills within transmissible stories when the spill's effects persist over lengthy time frames, as they always do. The difficulty of broadcasting the middle-term effects of dilbit spills has become acutely evident to residents of Mayflower, Arkansas, where approximately five thousand barrels of tar sands crude burst from an ExxonMobil pipeline in 2013, and of southwestern Michigan, where in 2010 dilbit spewed from an Enbridge pipeline into a tributary of the Kalamazoo River, in one of the largest on-land spills in petro-history (Caplan-Bricker 2013; McGowan and Song 2012). Mayflower residents created the Mayflower Facebook Page to document nausea, headaches, and exhaustion caused, most probably, from the off-gassing of toxic chemicals into the air, typically after a rain event flushes oil from lake-bottom sediments (Kistner 2013). Michigan residents also employ social media to keep Enbridge "honest," they hope, and to describe the persistent decline of their property values. In the Niger Delta, over forty years of oil mining has decimated mangrove wetlands and severely polluted air, rivers, and creeks—a long-term, underrepresented "event" streamlined into imagistic narrative by novelist Helon Habila and poets Tanure Ojaide and Ogaga Ifowodo. The BP blowout on the US Gulf Coast produced poetry, song ("The BP Blues"), and long narrative forms such as Bryan D. Hopkins' film *Dirty Energy* (2011) and David Gessner's book *The Tarball Chronicles*. The mediation of a spill in print, film, or digital MEDIA returns to the problem of the spill as an a-causative "downpour." More explicitly, the spill figures in narrative media as a happening resistant to coherent plot. Spills are often multiply sited, difficult to grasp as precise causes and effects. The spill categorically defies endings, persisting in space and time through its effects on ecosystems and bodies.

See also: DISASTER, OFFSHORE RIG, RESOURCE CURSE.

Spiritual

Lisa Sideris

In 2013, when the Williams and Boardwalk Pipeline Partners announced plans to snake a pipeline of natural gas liquids through a large swath of Kentucky, they set themselves against two bulwarks of Southern culture: religion and bourbon.

The Bluegrass Pipeline would connect FRACKING operations in the northeast United States with chemical processing plants in the Gulf of Mexico. Natural gas liquids (NGLs) are hydrocarbons like ethane, propane, and butane that are removed from natural gas and used for a variety of purposes, often as petrochemical feedstock. NGLs may end up as plastic bags or antifreeze, tire RUBBER or lighter fluid. The underground pipeline would route these hazardous and highly flammable byproducts through the heart of the Kentucky Holy Land, home to a cluster of religious orders and the former residence of the Cistercian monk Thomas Merton, whose writings express an abiding spiritual connection to the forest monastery he inhabited for thirty years. "In this wilderness I have learned how to sleep again," Merton wrote. "I am not alien. The trees I know, the night I know, the rain I know" (2003, 43).

There is indeed something special about the land Merton knew so well. The unique geological features—porous limestone, known as karst, and abundant subterranean caves and springs—add beauty and interest to the region but also make it an ideal spot for local distilleries that depend on the iron-free, limestone-rich groundwater for the production of

Thanks to Robert Crouch, Constance Furey, Winnifred Sullivan, and especially to James Capshew for a helpful discussion of *genius loci*.

bourbon whiskey. Unlike many resources that humans laboriously extract from the earth, limestone is benign. "Swallow some," writes Scott Russell Sanders, "and your body will know what to do with it" (1985, 74). That is precisely what Kentucky thoroughbreds have done for ages. Calcium carbonate infuses the grassland and strengthens their bones, making Kentucky racehorses the fastest in the world.

Native Spirit

Throngs of spectators who consume the customary mint juleps on Derby Day also imbibe the limestone. Distilling is a key industry in Kentucky, alongside horse farming and COAL mining. Bourbon whiskey—corn liquor—is AMERICA'S only native spirit, Kentucky its proud birthplace. Legend has it that the distinctive process of aging bourbon in charred oak casks was perfected in the late eighteenth century by the Reverend Elijah Craig, a Baptist preacher who fled religious persecution in Virginia. Craig became a successful distiller and entrepreneur, eventually helping to establish Georgetown College, a Christian liberal arts institution near Lexington. Today, "Elijah Craig" is the name given to the premium oak-aged bourbon produced by Heaven Hill Distillery in Nelson County, one of many Kentucky counties that would be traversed by the pipeline ("Birth of Bourbon" n.d.). The tale of Reverend Craig's heavenly bourbon is likely apocryphal, but it captures the sacred quality of the land and what it produces.

The production of bourbon requires large quantities of water at virtually every stage. Cool spring waters are important in the distilling process (and were indispensable before the advent of refrigeration), while fermentation is aided by water with a somewhat high pH, which karst produces. The geological features that attract distilleries also make this region highly prone to sinkholes, which have been known to rupture PIPELINES. In February 2014, in the midst of heated debate about the pipeline project, a spontaneous sinkhole measuring 40 feet in diameter swallowed up a portion of the National Corvette Museum—and several Corvettes—in Bowling Green, Kentucky. Were such acts of God to strike near the pipeline, the extensive network of karst honeycombs and fast-flowing underground streams would make clean-up of spilled NGLs even more difficult than in a typical aquifer. The impact on distilleries, bourbon-lovers began to realize, could be catastrophic. The pipeline's path would also intersect the New Madrid Seismic Zone, which routinely rattles the multistate region and might at any moment spawn an intense earthquake. Compare maps of the Kentucky Holy Land and the Bourbon Trail with geological surveys and the proposed path of the pipeline. What you see is the sheer hubris of this venture.

Divine Energy

The sacredness of the land has long been recognized by the Sisters of Loretto, a 200-year-old community of Catholic nuns and one of four religious orders that occupy the Holy

Land coveted by the pipeline companies. When approached for permission to survey their grounds, the Sisters refused to comply, citing a longstanding sacred trust that binds them to the land. Cistercian monks in the nearby Abbey of Gethsemani, where Thomas Merton is buried, similarly refused. Two other religious orders, the Dominican Sisters of Peace and the Sisters of Charity of Nazareth, confirmed their opposition, and a peaceful protest movement was born. The sisters' ardent defense of their land often evokes the divine spirit as an energizing force that grounds humans in place and inspires moral commitment. "The spirit of God in nature settles our spirit," one Loretto community member explains. "It brings us back to a grounding place, so that our action in the world can really come from a place that's been energized by God" ("Nuns versus Pipeline" 2014).

An energy vision statement, signed by numerous religious leaders from the Holy Land and beyond, sets out a list of beliefs and commitments, including the conviction that the land is "alive with the creative energy of God" and that climate change is caused by humans ("Holy Land" 2014). The Loretto Sisters have used the pipeline controversy to focus attention on the need to transition to clean ENERGY. At an interfaith prayer ritual, surrounded by symbols of RENEWABLE energy, the sisters offer a grateful tribute to the fossil fuels that have brought humans comfort and convenience. These traditional energy sources are also "gifts" of the divine spirit that pervades the land. "We give thanks and welcome a new energy vision" (Runyan 2013).

Spirit of Place

In spring 2014, in a desperate bid to impose the Bluegrass Pipeline on the unwilling citizens of Northern Kentucky, Williams and Boardwalk Pipeline Partners threatened to invoke eminent domain. The bid was unsuccessful. The pipeline was deemed insufficiently public-spirited, because it would neither benefit the local economy nor serve Kentucky's energy needs. Although the companies denied that environmental opposition was a factor in their decision, they suspended their investments in the pipeline on April 28, 2014 (Bruggers 2014).

Eminent domain has tended to regard land as a fungible commodity; it assumes that loss of land can be compensated at fair market value and that just compensation can "make owners whole" (Radin 1993, 136). This understanding of the land and its value is inimical to the spirit in which pipeline opponents, religious and secular alike, mounted their protest. The movement was quickened by a sense of the region's distinctiveness—its utter *non*-fungibility. The Romans referred to this sensibility as *genius loci*: an indwelling presence in the land, the unique tutelary spirit or soul of a place. This spirit merges body with place, forging an intimate bond and sense of belonging. Globalizing forces bring with them a spiritual alienation from the land, a profound forgetfulness regarding the cultural and natural history of our native and adopted places. If connection to place is a psychological necessity, as some have argued, then our species is suffering a kind of collective *de-range*ment (Quinn 2013).

Loss of home entails loss of stories that remind us who and where we are. In response to the seemingly global challenge of modern rootlessness and storylessness, eco-spiritual discourse has swung toward universal storylines and planetary ethics that confer a shared sense of home and global consciousness (Swimme and Tucker 2011). Whether couched as grand narratives that locate us within a "story of the universe" or as ambitious projects like the Earth Charter that identify a common core of ethics for the "whole human family," these projects cultivate planetary—literally, cosmopolitan—rather than localized sensibilities ("What Is the Earth Charter?" 2012). At the same time, the ANTHROPOCENE lens zooms out to reveal one species' unprecedented rise to global dominance. Humans, we now know, are a geological force radically reshaping the planet.

What these visions fail to capture is that we humans are still shaped and defined by geology. Precisely how natural forces shape us varies from region to region, and this variety gives local character and zest to the relationships and livelihoods humans forge in connection with the land. We cannot truly be alienated from the whole human family but only from those whom we have known; similarly, our alienation is not from the entire earth but from particular places. As attention is increasingly paid to the planetary scale of environmental crises and the need for global solutions, the Bluegrass Pipeline is a reminder of the enormous power that still resides in local places and the people who know them well.

See also: COMMUNITY, PIPELINES, RURAL, SPILL, SPILLS.

Statistics

Spencer Morrison

Edward Burtynsky's photograph *Shipbreaking #20*, an IMAGE in his collection *Oil*, depicts a ship demolition yard in Chittagong, Bangladesh, strewn with oil barrels and scrap metal from abandoned ships (see Figure 12). Atop one barrel crouches the photo's sole human figure, a young Bangladeshi ship-breaker splotched with oil; before him lies oil-soaked terrain that descends into a stagnant black pond from which pipes, ladders, and wires haphazardly jut. It might seem counterintuitive to use Burtynsky's photo to discuss statistical reasoning's role in shaping discursive relationships between human subjects and ENERGY REGIMES, since no numbers appear in the photo. Yet by depicting a lone human subject surrounded by oil barrels that are single units in a planetary statistical system (the barrels per day unit so ubiquitous in oil production discourse), *Shipbreaking #20* aesthetically frames a series of tensions inherent to the measuring, quantifying, and encoding operations of statistics themselves. Exposing the carefully concealed status of statistics as only one mode of representing empirical reality among many others (such as images, sounds, and words), Burtynsky's photograph demonstrates statistical reasoning's liminal status between seemingly unclassifiable tactile phenomena and overarching systems of knowledge and classification.

Even in an age of rampant disinformation and spin, statistical representations of energy regimes are thought to be objective, accepted as evidence amid and against fuel discourses that rely on nonnumerical forms of representation. Statistics thoroughly permeate contemporary imaginings of fuel regimes and their ecological consequences. They underlie the claims of politicians who endorse energy independence, of environmentalists who warn of rising ocean levels, and of oil companies proclaiming the economic benefits of pipeline construction. However, *Shipbreaking #20* registers the materiality and tactility of one particular fuel source (oil) by posing its unchecked spread against statistical measuring units (the wasted barrels) whose bounds it exceeds. Such oily profusion defamiliarizes the measuring apparatus by which the substance is so commonly represented in ENERGY debates. Moreover, by foregrounding within this scene a young worker whose future remains indeterminate, Burtynsky's image also evokes how statistics uniquely inflect our imaginings of energy futures, whether of ecological collapse or utopic abundance. The photo decouples the physical phenomenon of oil from its quantification within systems of knowledge, exposing statistical reasoning's role as a nexus between bare particulars and overarching knowledge systems.

A historical subset of political economy, statistics as a discipline has been studied by cultural critics most often in relation to state-based instruments for comprehending and governing national publics; this conception of statistics as both norm-creating and norm-enforcing locates within it the tools of micro-power central to Foucauldian governmentality. Historicizing statistical reasoning specifically in relation to human populations is the primary task of major cultural studies of the field by scholars including Alain Desrosières, Catherine Gallagher, Mary Poovey, and Theodore Porter. Statistics regarding human populations furnish the state with an inventory of a population's energy needs, creating as a byproduct a class of experts whose grasp of statistical measuring tools buttresses their authority in energy debates. Specifically, statistics endow expert claims with an objectivity thought to reside in statistics themselves. In this way, statistics, a term whose very etymology relates to the creation of the state, enables and effects through its measuring procedures the state-based administration of populations (Desrosières 1998, 8).

Yet as the aforementioned cultural critics show, statistical reasoning's vaunted objectivity is itself a discursive construct that conceals a set of cognitive operations; indeed, these cognitive operations subtly shape the very empirical reality that statistics are commonly thought to describe. For instance, as Poovey notes, statistics "embody theoretical assumptions about what should be counted, how one should understand material reality, and how quantification contributes to systematic knowledge about the world" (1998, xii). Statistics register empirical phenomena and give them epistemological shape, ascribing to them an identity within systems of knowledge (while simultaneously denying identity to that which remains statistically unclassified). Moreover, the identities created by statistics privilege specific attributes over others, namely those generalizable attributes that allow equivalencies to be drawn between discrete particulars that become classed within shared statistical categories. These categories of equivalence function as nomenclatures within which data can be added over time, rendering statistical categories "repositories of accumulated knowledge" (Desrosières 1998, 248). And the nomenclatures engendered by statistics serve to transmute worldly materiality into quantifiable abstractions (e.g., "the market," "greenhouse gases," "wattage") that are in turn ideologically inflected in political debates—including debates over energy regimes (Poovey 1998, 15). Scholars often see tensions between the descriptive and prescriptive uses of statistics (i.e., measuring reality vs. endorsing certain policies over others). Attention to the veiled cognitive operations of statistical reasoning shows, however, that the reality "described" statistically is already "prescribed" and shaped by statistical discourse itself.

This attention to the power of statistics to shape both reality and the cognitive structures through which we perceive it leaves open the question of what roles statistical reasoning adopts in cultural discussions of fuel and energy. Statistics has most commonly been studied in relation to human populations, but what about its representations of nonhuman substances and processes related to contemporary fuel regimes? Statistics of this sort provide a shared basis for knowledge around which public discussions of energy and climate issues can swirl (for instance, the Intergovernmental Panel on Climate Change's recent claim of 95 percent certainty that climate change is human-made [17]). However, in a time of information overload and forecasts of ecological collapse, might sheer informational abundance paralyze, rather than galvanize, political will for alternative fuel regimes? While statistics potentially provide a basis for shared public conversation, they also disproportionately empower credentialed "experts," not to mention those who reformat statistical categories in the interests of petro-capital (consider, for instance, recent initiatives to promote "intensity-based" emissions targets rather than absolute ones).[1] A degree of statistical uncertainty over anthropogenic climate change gives life to climate change deniers, who generate their own (non-peer-reviewed) numbers in arguments against GREEN initiatives. The statistical armature of each constituency in energy debates threatens to overwhelm citizens, leaving many befuddled and politically indifferent.

Beyond statistical profusion, however, many of the narratives to which statistics are recruited themselves threaten to instill quietism in citizens. Take, for instance, the oft-cited statistical "tipping point," the precise rise in temperature beyond which disastrous climate change becomes irreversible. The abstracting powers of statistics—the global flows of capital and energy they conjure into being through cognitive operations of equivalence—evoke for everyday citizens a planetary system so vast as to be uncontrollable; system failure consequently appears inevitable. In this way, then, the abstracting operations of statistics both organize knowledge in the present and structure our imaginings of possible energy futures.

Burtynsky's *Shipbreaking #20* evokes two modes of experience antithetical to that of the statistician. Together, these alternative modes illuminate statistical reasoning's powers of abstraction and systemization while also suggesting nonstatistical foundations for cultural narratives of fuel and ecological change. The first emerges from the emotion of empathy. The young ship-breaker in the photo peers impassively at us and compels our acknowledgment and response. This singular subject's oil-splashed body defies the abstracting and sorting processes of statistical reasoning and makes an ethical appeal: to ameliorate his present circumstances and to alter the FUTURE toward which his relative youth gestures. Of

1. For an example in the American context, see George W. Bush's "Statement by the President" of February 12, 2003, published on the official website of The White House.

Figure 12. Edward Burtynsky, "Shipbreaking #20." (Photo © Edward Burtynsky, courtesy of Nicholas Metivier Gallery, Toronto)

course, empathy can also prompt instinctive actions and policies out of line with what the information provided by statistics might suggest is in the best long-run interest of populations, economies, or ecologies. Nevertheless, the relational emotion of empathy produces epistemologies and narratives distinct from those of statistics.

Indeed, our empathic relations with the ship-breaker of Burtynsky's photo spark an awareness of oil's physical particularity. This renewed awareness of unquantifiable materiality is the second mode of experience antithetical to statistical reasoning presented in this image. Oil's sheer viscosity, the congealed lumps it forms upon the oily pond, the blackness of the landscape into which it seeps, all remind us starkly of its empirical qualities, which tend to be abstracted away in contemporary energy discourse that represents the substance exclusively through statistics. The barrel-strewn shipbreaking yard reminds us of these abstracting powers while simultaneously staging for spectators the material effects that statistics conceal. Their aura of objectivity ensures that statistics will remain vital in political and social efforts to develop energy policies sensitive to both planetary ecological needs and the welfare of future generations. Understanding the representational contours of statistical reasoning as statistics intersect with broader cultural narratives entails changing the very planetary futures we imagine to be possible.

See also: affect, Anthropocene, future, photography.

Superhero Comics

Bart Beaty

Fans of superhero comic books divide the history of the genre into at least three periods: the Golden Age, from the first appearance of Superman in *Action Comics* #1 (June 1938) to the Comics Code in autumn 1954; the Silver Age, from the reintroduction of the Flash in *Showcase* #4 (October 1956) to the revision of the Comics Code in 1972; and the Bronze Age, from the mid-1970s to the 1990s. In terms of energy and its relation to this periodization, the most important concern is the distinction between the Golden and Silver Ages. Superheroes created during the Golden Age (many of whom are still tremendously popular today) are, according to Jason Bainbridge (2009), "premodern" because they are unidimensional and adhere to simplistic or now outmoded notions of justice. Examining the subject through another lens, however, would lead us to understand Golden Age characters as not just premodern but pre-energy.

What is notable about characters like Superman and Batman is that, despite their great abilities, their origins and their adventures bear little trace of the logics of ENERGY discovery or scarcity. Of the dozens of superheroes introduced in the late 1930s and early 1940s, none had an energy-based origin. Superman is an alien being on Earth who was born with his powers; Wonder Woman is the daughter of Hippolyta (and later revealed to be the daughter of Zeus), and her abilities are divinely created; Captain Marvel is granted his powers by a benevolent wizard, and they are magical in origin; Captain America is the product of a biochemical experiment; Batman simply exercises a great deal and is not superhuman. Each of these heroes, and the dozens like them created during this period, retains traces of an origin story that may be scientific, divine, or magical but evinces no link to the logics of energy production.

Superheroes of the Silver Age have a similarly diverse set of origins, but many of them bring energy production to the heart of their storytelling. The superheroes created at Marvel Comics under the editorship of Stan Lee developed this idea most fully. These "modern" superheroes, to retain Bainbridge's term, differ from their predecessors insofar as they are imbricated in melodramatic story worlds in which psychological factors play an important role in storytelling. Notions of justice are often complicated by social relations for the modern superhero in ways they are not for the premodern hero. For the modern superhero, the consequences of energy development—particularly NUCLEAR energy—play an important role in the conception of the heroes as more mature or realistic than their predecessors. While Iron Man is the simple product of the technologies of robotics, and while the mutant X-Men are born with their powers (which most often manifest during puberty), other Marvel characters find their origins in the quest to find new forms of energy during the early 1960s. Spider-Man gains his powers when he is bitten by an accidentally irradiated spider at a science exhibition. The Hulk is created when he is exposed to tremendously high doses of gamma radiation during the detonation of a gamma bomb, and the Fantastic Four are transformed when their exploratory spacecraft is inundated with cosmic rays. In the post–World War II context, and during the heightened tensions of the Cold War, stories about nuclear power and its purported effects were increasingly common. The scientization of superhero characters was one element of a new sophistication in superhero storytelling during the 1960s, one that highlighted the exaggerated and fantastical effects of potential new energy sources.

What is most unusual about the energy-based superhero comics of the 1960s is not the simple fact of their origins and powers but how they are represented. As a primarily visual storytelling form, comics are commonly pressed to find interesting ways of presenting the unpresentable. In the origin of the Fantastic Four (November 1961), for example, transformative cosmic rays are presented in two panels as a series of straight-ruled oval shapes that penetrate the spacecraft with a "Rak Tac Tac" sound effect—the rays are somehow both solid (able to generate sounds when they collide with the metal hull of the ship) and not (they pass through the bodies of the crew, transforming their genetic make-up). As energy-based superpowers become more common in the 1960s (Wikipedia's entry for Superhuman Abilities lists no fewer than five subcategories of energy generation, conversion, and manipulation, linking to dozens of characters, and the Superpower wiki (http://powerlisting.wikia.com) identifies twenty distinct energy-based abilities common to superheroes, including energy propulsion, energy generation, and energy absorption), new ways of depicting energy emerged. The leader in this regard was artist Jack Kirby, co-creator of most of Marvel's significant superheroes. By 1966, Kirby routinely depicted ambient nondirected energy as a series of connected and overlapping black circles. As this technique became a visual signature, fans termed it Kirby Krackle—a physical manifestation of raw energy in a defined space. In an article in the fan magazine *Jack Kirby Collector*,

Figure 13. The cover for *Fantastic Four* #50 (May 1966). Art by Jack Kirby with Joe Sinnott. (Marvel Comics)

FIGURE 14. The cover for *Fantastic Four* #72 (March 1968). Art by Jack Kirby with Joe Sinnott. (Marvel Comics)

Shane Foley (2001) notes how Kirby's different strategies for depicting energy are evident in two drawings of the Silver Surfer, an interstellar herald of the planet-devouring Galactus, whose surfboard is constructed by the physicalization of "the power cosmic." The cover for *The Fantastic Four* #50 (May 1966) depicts the third appearance of the character, riding his board toward the reader against a brown background, with a series of motion lines surrounding the character (see Figure 13).

The cover for *Fantastic Four* #72 (March 1968) depicts the character in a similar pose, but here he blasts an energy beam from his left hand while surfing a sea of red-colored Kirby Krackle (see Figure 14). In less than two years, Kirby introduced and consolidated the visualization of fantastical energy sources, laying a foundation for the representation of energy that survives to this day in the pages of superhero comic books.

As continuously published serial narratives of long duration (Superman's adventures have been published uninterruptedly for more than seventy-five years, while the Fantastic Four has appeared monthly for more than fifty), superhero comics offer a strikingly rich archive of popular conceptions of energy generation and consumption. Over time, new creators have sought to recast the origins and powers of superheroes to bring them into closer alignment with contemporary conceptions of energy use. In the 2002 Spider-Man movie, the spider that bites Peter Parker is not irradiated but genetically modified, suggesting a different anxiety about science. Nonetheless, as a visual form, comics—and superhero comics in particular—have placed a greater premium on the visualization of energy than other cultural forms. Scholars seeking to understand the popularization of energy discourses and aesthetics will be well-served by turning to the vast collection of superhero comic books that have circulated for almost eight decades.

See also: EMBODIMENT, FICTION, NUCLEAR, PETROREALISM, RUBBER.

Surveillance

Lynn Badia

Figure 15, produced by the International Energy Agency, maps the flow of the world's total production and consumption of ENERGY (in various states and materials) between myriad origins and destinations. Although the diagram represents sources ranging from water to COAL, they are all converted into a single unit of measurement—the Mtoe, or millions of tons of oil equivalent. This visualization of global energy flows is just one example of how energy topologies (or energyscapes) are mapped and analyzed (Strauss, Rupp, and Love 2012; Appadurai 1990). Energy is analyzed in terms of type, source, capacity, conversion, distribution, etc., across various timespans. Energy flows have been subject to statistical analysis to an extent that rivals analysis of the movement of monies and populations. This information is of vital importance to academic researchers, financial analysts, and industrialists, as well as government institutions and transparency initiatives.

Today the availability of energy data may be taken for granted, but in the mid-twentieth century such information fell within the purview of a military surveillance program dubbed the "War Detection Plan" by the popular press. Indeed, one could argue that the need for comprehensive energy analysis can be traced to this historical conjuncture, in which energy access and energy science were linked to state power. In the final years of World War II, the US State Department commissioned the Engineers Joint Council (EJC), an umbrella organization of civilian engineering societies, to devise a plan for the industrial control of Germany and Japan. In the Yalta and Potsdam Agreements, the Allies called for "the complete disarmament and demilitarization of Germany," and they aimed

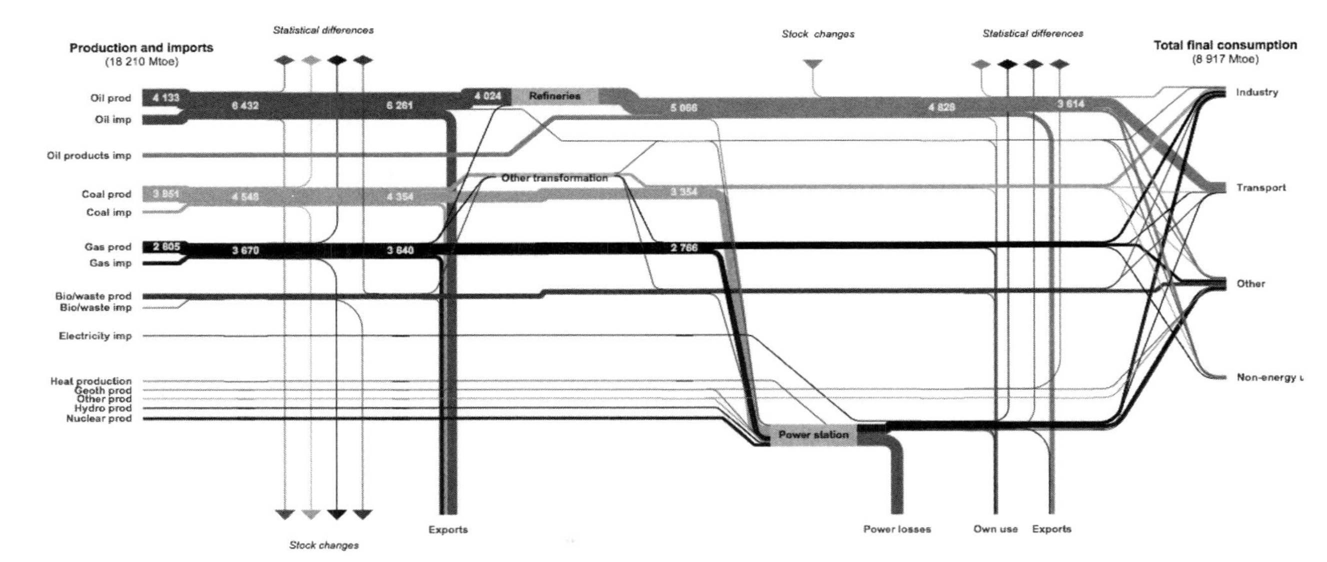

FIGURE 15. Balance of world energy production and consumption, 2013. (International Energy Agency, 2013)

to "eliminate or control all German industry that can be used for military production" (Engineers Joint Council 1945). In response, the EJC assembled an impressive team of engineers guided by the former presidents of the civil, mechanical, electrical, chemical, mining, and metallurgical engineering societies, alongside leaders of the General Electric Company and the Mellon Institute of Industrial Research. All of the reports they produced for the State Department stated the same unifying strategy: to control war potential, the United States would need to control the flow of energy.

When the EJC began work in 1944, comprehensive energy data was not readily available, and their analysis required extensive research and information gathering. Under the EJC, a network of engineers and subcommittees collected new information and consolidated previously separate data streams with ongoing military research in order to track global energy flows and raw materials with new precision. The EJC compiled energy statistics for Germany and Japan and many other European nations by tracking their wartime and peacetime rates of energy use and raw materials consumption; much of this data was published in appendices to EJC policy reports submitted to the State Department (Engineers Joint Council 1945; Engineers Joint Council n.d.; Engineers Joint Council 1946). With well-charted energyscapes in front of them, the EJC could, for the first time, correlate energy rates with specific industrial projects, such as vehicle manufacture or uranium enrichment. If Germany's peacetime electricity rates and steel consumption rose to a certain level, it might indicate that they were producing munitions at a pace to meet wartime (rather than peacetime) need. *LOOK Magazine*'s December 1946 reporting on the War Detection Plan explains, "Early tank production was only a trickle. The [EJC] engineers point out that the changeover to war production took even longer in Germany and Japan in the 1930's, [and] could have been detected accurately years before war struck" ("Peace on Earth" 1946, 33). *LOOK*'s article, "Peace on Earth—Dream or Possibility?" describes a utopian fantasy in which the War Detection Plan could create "permanent peace" through the global surveillance and management of energy and material flows. According to *LOOK*, "any nation changing over to war footing can be instantly spotted through such analysis" (ibid., 21). In this account, world peace could be secured by a team of technical scientists responsible for energy surveillance rather than by diplomats at the United Nations, which, as *LOOK* points out, "although far along in its second year, has found no workable basis on which the nations of the world might lay aside their weapons" (ibid., 25).

In the EJC's work, then, energy surveillance became the means for war detection as well as the control of war potential. For the postwar control of Germany and Japan, the EJC recommended that onsite UN technicians monitor energy and electricity rates to ensure that these nations remained at peacetime levels of production. It is clear from their reports that the EJC was not simply compiling a new database of energy and raw material statistics but instead producing a new apparatus for state control. As they explain in their report on Germany, energy control would "require a minimum of policing" and be "effective for a considerable period after large-scale military occupation of the country has been discontinued" (Engineers Joint Council 1945, 2). In fact, this logic explains why energy was chosen as the master dial for surveilling, managing, and controlling industrial production. Even though raw materials and industrial facilities were also necessary to produce implements of war, these other elements "could be automatically regulated" through their ultimate dependence on energy (ibid., 10). As the EJC states, "It is recommended that a comprehensive postwar control of German industry be based on a system of energy allocation. The fundamental factor is energy, since it is required in one form or another to operate all processing equipment for the transformation of all raw materials into finished goods. Therefore, regulations based on energy are so inclusive that all industrial operations fall within their scope" (ibid., 5–6). This energy management strategy was further shaped by the practical demands of military occupation: "Control of power distribution through allocation of all coal distribution, while fundamentally effective, would be cumbersome, a handicap to peaceful development and would require considerable staff. A rigid control of electric power distribution is therefore important to ensure compliance and regulation of German industry" (ibid., 6). By the time the EJC turned its attention to Japan in 1945, this approach seemed self-evident: "one of the simplest ways to control industrial production is by allocating and limiting electric-power supply" (EJC 1946).

World War II was both the engine that created a new relationship between energy and the state and a devastating demonstration of the power of that relationship. Technically, a strategy of war detection or industrial control became possible as industrialized nations shared increasingly standardized modes of production, which made them susceptible to energy surveillance and management. These nations' abilities to wage war against one another relied on access to expensive technologies and the industrial base to produce them. NUCLEAR science was hardly an exception, as it relied on even more complex and costly industrial installations. For this reason, *LOOK* reports that, "the [EJC] engineers feel that their plan would give advance warning of a 1950-style atomic war just as surely as of a 1940-style air and ground war" ("Peace on Earth" 1946, 21). As the Manhattan Project consolidated links among industry, the state, and energy science, the EJC provided the conceptual means to interpret this complex relationship through the status of energy flows. The creation of the EJC's program (and, consequently, an extensive statistical mapping of energy topologies) thus emerged in the historical moment when nuclear science demonstrated that the status of energy flows was the most salient index of state power.

See also: ENERGOPOLITICS, STATISTICS.

Sustainability

Leerom Medovoi

In recent decades, the vision of a genuinely ecological economics has focused on the principle of sustainability. Capitalism, we are told, should be refashioned as a "sustainable economy" whose growth, in the words of the United Nation's Brundtland Commission Report, "meets the needs of the present without compromising the ability of FUTURE generations to meet their own needs" (Brundtland 1987). Like other liberal political ideals (e.g., democracy, freedom, tolerance), the ideological work performed by sustainability is complex and multivalent. As Joan Martinez-Alier (2009) notes, ecological economics understands the economy not as a system of exchange but rather as a metabolic system of "throughputs." Material and energy flows enter the economy, are digested by it, and exit as waste. From this view, the economy's continued growth depends upon ensuring that inputs can be replenished and that outputs (waste) do not kill the environment that provides the inputs.

Within this metaphorical context of a metabolic economy, two semantic clusters organize uses of the word *sustainability* (Medovoi 2010, 129–32). In the first cluster, *sustainability* indicates sustenance: that which provides nourishment for living existence. This emphasis on maintaining the resources needed for human flourishing appears to champion the cause of life, if *sustainability*'s reigning metaphor is food as life-giving nourishment, as biological fuel. In a different cluster of associations, however, *sustainability* connotes the capacity to absorb harm without succumbing to death, as when one "sustains" a blow to the head. These associations are captured in the ecological notion of RESILIENCE, which, in its lateral transfer from ecological science to sustainability discourse, evokes a living body capable of recovering from injury to itself or its environment. From this perspective, *sustainablity* concerns less the promotion of life than the evasion of death.

Sustainability is thus a profoundly biopolitical project with contradictory tendencies. In movements of the Left and/or the Global South, sustainability may underwrite an affirmative biopolitics, defending human life against ravages of the market by seeking to protect the ecosystems within which human life is embedded. However, sustainability has also emerged as a strategy of capitalist governmentality, a tendency that is my present focus.

In this hegemonic framework, the form of life to be sustained is that of capital ACCUMULATION itself. Here, economic growth stands in for organic growth and species flourishing. The central question becomes, "how can the resilience of capitalism be ensured if the forms of "nourishment" that fuel its growth—including extractive natural resources, fossil fuels, or even that other form of fuel that, in a neoliberal age, we call "human resources"—are depleted?" The instrumentalization of NATURE or humanity as "resources" for sustenance demonstrates that the form of life at stake here is neither human nor biological but instead the abstract life of the economy. On the other side of this metabolism, waste results in the toxification of living conditions (poisoned social relations, damaged natural and human environments) or even the direct inversion of human resources—the excretion of human beings themselves as "waste" or surplus population.

The three pillars of sustainability—social, environmental, and economic—first named at the Rio (or Earth) Summit of 1992, make explicit the kinds of inputs upon which the abstract life of capital depends (Gibson 2006). Pillars are built to support a load—a metaphor that raises questions about how much extraction can be inflicted upon the social, environmental, or economic domains before these pillars collapse. Or in the organic metaphor, how much depletion can be sustained without killing the host?

The pillar of social sustainability emphasizes the need to preserve the social fabric against the forces that erode it. Although this might sound progressive, social sustainability tends to serve as a diluted neoliberal substitute for the concept of social justice (Dillard, Dujon, and King 2009). Instead of naming the ideal of a world where human potential may be realized as fully and universally as possible, the biopoliticized version of social justice (as social sustainability) asks the same question in negative form. That is, not "how do we realize justice fully?" but instead "how much injustice must we remedy in order to avoid fraying the social fabric of human populations to the point of moral, social, or biological death?" Social sustainability assumes that capital must be allowed to draw upon its social host for sustenance but that we are ethically and rationally obliged to keep that depletion within the limits of livability.

Social sustainability is not unrelated to Marx's discussion of the reproduction of labor power—the sustenance of the worker that is the necessary cost in wages offered by capital in order to continue reaping surplus value (Marx 1977, 283–306). To the extent that neoliberalism, as a strategic response to the CRISIS of Fordist accumulation, relies upon what David Harvey calls "accumulation by dispossession"—the direct seizure of wealth—it threatens even that basic level of sustenance needed to return workers back to the workplace (2005, 137–182). This seizure of sustenance entails not merely driving down wages but also diminishing the social wage and the social contract (health care, education, and household support). As neoliberal dispossession drains the government sector and with it the residues of the Fordist social safety net, it imposes mounting levels of debt that amount to borrowing against future sustenance in order to provide for the present. This draws down the population's living energy, its physical and affective reserves, inflicting upon it what Lauren Berlant (2007) calls "slow death." Neoliberalism produces a crisis in social existence to which the discourse of social sustainability seeks to respond, but in reactive form, as a governmental supplement that ascertains the maximum amount of injustice that can be perpetrated without killing the social sources of living energy upon which the abstract life of capital depends.

Environmental sustainability concerns itself with the ecological pressures exerted by economic activity on the nonhuman world. Here too, resources that could nourish ecosystems are instead diverted to fuel capital accumulation. Teresa Brennan (1998) claims that this process constitutes another path for the extraction of surplus value, analogous to that which Marx argues was drawn from the living energy of human labor. The burgeoning economic literature on ecosystem services aims to protect nature by arguing that monetary value is lost when ecosystems are damaged by economic activity and can no longer furnish resources or serve as sinks for waste. In this view, the monetary value of keeping ecosystems intact should be weighed against the gains of the accumulation process that would deplete them. Environmental sustainability proceeds by accepting the premise that maximized accumulation of value is the proper measure of an environment's worth.

The last pillar, economic sustainability, is foundational for the other two. This principle might seem redundant, if taken to mean simply turning a profit indefinitely. But in relation to the social and environmental pillars of sustainability, ongoing profit becomes impossible if social and environmental inputs cannot be reproduced over the long term (or replaced by substitutes if depleted) or if social and environmental outputs (toxicity, climate change) of economic activity destroy the conditions of possibility for the economy's accumulation process.

With regard to energy, sustainability would demand that the fuels necessary for economic activity be coarticulated with the energetics of human power and other natural living activity. But this articulation comes at a high cost. The seeming redundancy of economic sustainability symptomatically reveals the capitalist assumptions that underwrite liberal economic discourse: sustainability in this sense values social and natural living relations primarily for their convertibility into the abstract life of the economy. It asks what other kinds of life must be preserved for the life of capital to continue flourishing, and it applauds itself for having such deep foresight and ethical character. However, this discourse of sustainability circumvents the inverse question of when, and under what conditions, economic life serves as a resource or a means for the concrete life of human beings and other species rather than as a means to their instrumentalization or even demise.

This question has remained alive in discussions of sustainability by way of the discourse of development, as in the 1987 report of the Brundtland Commission, which articulated the combined concept of "sustainable development" (Anand and Sen 2000). That *sustainability* has become an increasingly free-floating term, severed from the aims of development, reflects its gradual delinking from questions of either social or environmental justice.

See also: ENERGOPOLITICS, FUTURE, LIMITS, RESILIENCE.

Tallow

Laurie Shannon

Hamlet, performing his self-styled madman's script, forces his auditors to remember a disturbing truth that is normally repressed: "A man may fish with the worm that hath eat of a king, and eat of the fish that hath fed of that worm" (*Hamlet*, 4.3.27–28).[1] This logic of circulation recalls the Pythagorean doctrine of the transmigration of souls, a view often mocked in early modernity as equivalent to insanity. But Hamlet's line traces no flight by the soul from one body to another. Instead, it joins a traditional Christian perspective on worldly vanities (a fortune's wheel argument) to an insistence on the equivalently gross materiality of all flesh, from worms to kings and back again. The economy of circulation here charts not the routes of individuated souls but rather the dis-individuating paths of recycled ENERGY. "We fat all creatures else to fat us," his speech declaims, "and we fat ourselves for maggots" (4.3.22–23). In Hamlet's recycling vision, fat is fuel—yet fat is (also) us.

To manage the disturbance this view presents between denominators of political or entitled personhood and the commercial metrics of exchange, we designate as "tallow" only the byproduct of nonhumans—"all creatures *else*," in Hamlet's phrase. Tallow is "animal fat (esp. that obtained from . . . about the kidneys of ruminating animals, now chiefly the sheep and ox), separated by melting and clarifying from the membranes, etc., naturally mixed with it" (*OED* 2013, 2). While fat spoils when raw, once processed as tallow it becomes

1. All references to Shakespeare are to *The Complete Works of Shakespeare*, ed. David Bevington (New York: Longman, 1997). The references name scene, act, and lines.

storable and portable, a product used to seal boats, make soaps, dress leather, bind foods (like haggis), and, of course, provide light through combustion.[2] Tallow candlelight ranked beneath that of wax, which was pricier; *Cymbeline* disparages "the smoky light/That's fed with stinking tallow," calling it "base and illustrous" (1.6.109–10). Reeky tallow was later replaced by the oil commandeered by the WHALING industry emerging in Massachusetts; stumping in England for the United States whaling trade in 1785, John Adams vaunted "the fat of the spermaceti whale" as yielding "the clearest and most beautiful flame of any substance that is known in nature" (Adams 1853, 308). Until then, though, tallow was a cheap, readily available staple, the yield of small-scale premodern practices of ANIMAL slaughter that were local and integrated into daily life. Sheep overwhelmingly supplied tallow in a wool-producing economy. Contemporaneous literary contexts, however, persistently defy the official confinement of tallow as something derived from "all creatures else." The most interesting tallow-yielders were people.

In Shakespeare's environs, the ideal deer to kill was one "in grease" or "in prime or pride of grease" (*OED* 1b). The well-fed state of the herd in *As You Like It* provokes their designation as "fat and greasy citizens" (2.1.55). Yet even as deer are measured by their commodifiable fat, their free motion and rightful claim to Arden's woods earn them a name that countervails their commodification: "citizens." Thus, early modern animals resist wholesale reduction to useable matter. Equivocations like this one, however, also work in reverse. Persistent recognition that human matter is fat, oily, grease-laden, meltable, combustible, and consumable erodes tallow's separation of animal fat from human flesh.

The Comedy of Errors assesses Nell in tallow metrics: "she's the kitchen wench, and all grease, and I know not what use to put her to but to make a lamp of her . . . . I warrant her rags and the tallow in them will burn a Poland winter" (3.2.93–99). Falstaff repeatedly blurs personhood and oily substance. Although lean deer were properly "rascals," quibbles make Falstaff a "fat rascal" (i.e., a plump Yorkshire tea-cake); a "fat-kidneyed rascal" (indexing the place from which tallow was drawn); and an "oily rascal" (*The Merry Wives of Windsor*, 2.2.5–6; *1 Henry IV*, 2.2.5, 2.4.521). In *Merry Wives*, Falstaff is a beached whale whose oil might be collected (2.1.61–2); "a barrow of butcher's offal" (3.5.5); and "the fattest" stag "i' the forest" who might "piss his tallow" (expend his fat or energy) in the exertions of "rut-time" (5.5.12–15). These uncongealed metaphors undermine his status as subject. As Wendy Wall specifies, "the play deflates [Falstaff's] bodily pretensions by making him into manageable domestic goods" (Wall 2002, 116–17). Seeking Falstaff, Prince Hal shouts, "call in ribs, call in tallow," and Falstaff enters to vivid insults that reach a climax with "whoreson, obscene, greasy tallow-catch" (Shakespeare 1623, 57).

A *tallow-catch* Falstaff is a container of commodifiable fat. When called "a candle, the better part burned out," Falstaff confirms himself "A wassail candle . . . all tallow"

2. The Worshipful Company of Tallow Chandlers, chartered in 1462, ranks just below the Wax Chandlers in the livery system. According to their website, the present-day Tallow Chandlers "have . . . built up close links with energy company BP, on the basis of *a shared interest in heat and light*" (Worshipful 2010; emphasis added).

(2 *Henry VI*, 1.2.155–58). Noting the contempt that makes a lowly Jonson character "an unsavoury snuff" (i.e., "a tallow candle quickly burning itself out"), Gail Kern Paster excavates the humoral economies enabling the conceit (Paster 2004, 222). The trope of the human body as a combustible candle also had prominent elite precedents. John Foxe's *Book of Martyrs* (1563) recorded Protestant Hugh Latimer's proclamation from the stake, "We shall this day light such a candle by God's grace in England, as (I trust) shall never be put out" (Foxe 2009, 154). Elizabeth herself (decorously adjusting the metaphor to wax, but preserving the logic) claims: "I have . . . been content to be a taper of true virgin wax, to waste myself and spend my life that I might give light and comfort to those that live under me" (Elizabeth I 2000, 347). Both of these self-expending candles are imagined to burn for the public benefit. The trope thus works both ways, representing prodigal waste and public self-sacrifice just as it reveals what official nomenclatures repress about fuel: that we are as combustible as "all creatures else."

The literary apotheosis of the body-candle metaphor comes in Charles Dickens's *Bleak House* with Mr. Krook, who combusts (appropriately enough) in a mercantile setting among the inscrutable commodities of his rag and bottle shop. Soot falls "like black fat," and a "stagnant, sickening oil" leaves a "dark greasy coating on the walls and ceiling"; Krook's death by "Spontaneous Combustion" is "engendered in the corrupted humours of the vicious body itself" (Dickens 1853, 316–20). This event poses a "dreadful mystery" for a coroner (323). Combustibility is no longer a familiar trope reflecting palpable knowledge of the human body's combustible stores of energy.

From the whale oil that lubricated the machines of the Industrial Revolution (retrieved by ships whose journeys recast notions of space and time) to particle physics and nanoengineering (which recast space and time again), Western culture has transitioned to forms of energy whose origins are opaque to ordinary perception, whose material workings are comprehended only by specialists, and whose business operations are shielded and securitized. One result seems clear. In a quite literal sense, visceral knowledge of where energy comes from, or what energy is, has been substantially extinguished.

See also: ABORIGINAL, ANIMAL, FICTION, WHALING.

Texas

Daniel Worden

Texas looms large in contemporary oil culture. The state's first major oil well, Spindletop, came in in 1901 and led to the formation of Gulf Oil and Texaco. Since then, Texas has figured in oil culture as a site of extraction and refining, a center for the multinational corporate oil industry, and an anchor for the oil industry's ideological construction as an innately heroic, individualistic, and deeply American enterprise. This prominence is somewhat odd, considering that Pennsylvania was the site of the modern oil industry's origin in the United States. Alaska, California, Louisiana, and Oklahoma, not to mention the Dakotas, Montana, and the states that share the Appalachian Basin in the Northeast United States, are important sites of oil extraction, yet they loom far less large in our cultural imaginary than Texas, whose pride of place in oil culture is due, in part, to the culture industry. By far the most well-known oil film is *Giant*, an epic 1956 adaptation of Edna Ferber's 1952 novel of the same name, which starred James Dean, Rock Hudson, and Elizabeth Taylor. The television show *Dallas* revised *Giant*'s family melodrama for the age of oil CRISIS, as its thirteen seasons ran from 1978 to 1991. Through international syndication, *Dallas* made oil's association with Texas ubiquitous worldwide; a new production of the show was revived in 2012 on the TNT Network. These cultural artifacts helped to forge the strong association between oil and Texas much more effectively than the less successful films *Louisiana Story* (1948), *Tulsa* (1949), or *Oklahoma Crude* (1973) would do for other sites of extraction and refining. Even *There Will Be Blood*, a 2007 film set during the California oil boom of the early twentieth century, was filmed in Marfa, Texas, where *Giant* was filmed fifty years earlier. The later film's Texas location "was touted by its [Director of Photography] Robert Elswit as a return to authentic filmmaking," invoking the state's long association with the American frontier, an association that extends to its role in the oil industry (LeMenager 2012, 81). What these cultural texts reveal is Texas's ideological function in US oil culture and, by extension, in oil cultures across the globe. Texas has endowed the petroleum industry, its public relations discourse, and its representations in FICTION, television, and film with a rhetoric of heroic, frontier individualism that strives to cast even multinational corporations as wildcat operations. Oil culture is Texas culture.

The Texas oil industry emerged as an individualistic enterprise, and it became the most significant site of oil production in the early twentieth century. As Diana Davids Olien and Roger M. Olien chronicle in their history of the Texas oil industry, "In 1932, the giant East Texas field alone yielded more than the total annual production of most of the other states. Fettered by regulation, in 1940 Texas still produced twice as much oil as California, the next largest producing state and one where production was unlimited. Producing over one-third of the nation's oil, Texas dominated the price of crude oil in national and international markets" (Olien and Olien 2002, vii). Even as oil production ramped up, the Texas oil industry was initially hostile to large oil corporations, with antitrust laws that made it illegal for Standard Oil to operate in the state. After Spindletop came in in 1901, a rumor circulated in Texas newspapers: "these papers printed as fact an incredible account of an alleged Standard Oil project to build, under cover of darkness, a pipeline from the Gulf of Mexico to the Spindletop field. The reported purpose of this secret project was to pump salt water from the Gulf of Mexico into the field, thereby stopping production by Standard's competitors" (Pratt 1980, 821). Over the course of the twentieth century, this fear of and hostility toward large corporations would dissipate, yet the individualistic characters of early oil production in Texas would remain as the face of an increasingly multinational, corporate oil industry. Texas is a site of legitimation for the oil industry, and that legitimation occurs by associating "oilmen" of all kinds with a nostalgic, nationalist, frontier mythology, a mythology that casts the modern technologies and global marketplace through which the petroleum industry conducts its business today as natural extensions of strenuous, heroic individualism.

While Texas oil culture has captured the romance and individualism of the frontier in our cultural imaginary, it also signals the frontier's undoing in ways that often invoke the close ties among oil, environmental damage, and unsustainable, boom-and-bust economics. For example, in Larry McMurtry's 1961 novel *Horseman, Pass By*, adapted in 1963 into *Hud* starring Paul Newman, the town of Thalia is left behind by an oil boom:

> When I was a kid in grade school Thalia had had two picture shows, but in those days the oil-field activity was big, and Thalia was a wild, wet sort of half boom town. Pretty soon the oil production fell off and the oil people took their cars and their dirty scrappy kids down the road to another field . . . . A year or two later one of the shows closed up and they started using the billboards to run advertising for the one that hung on. That night

> there was an old picture playing, *Streets of Laredo*, with Gene Autry and Smiley Burnette. (McMurtry 2002, 35)

In McMurtry's dying Texas town, oil creates not permanent flourishing or stability but obsolescence and a cultural nostalgia for the Wild West, symbolized here by the "old picture" *Streets of Laredo*. Oil promises prosperity, but that wealth is fleeting, temporary.

In another well-known novel about Texas, Cormac McCarthy's 1992 *All the Pretty Horses*, oil is invoked again for its promise of immense wealth, in contrast to an idealized, pastoral life on the frontier. Having left home on horseback to pursue his dream of being a cowboy, John Grady Cole mocks his companion and friend, Rawlins, who begins to yearn for home:

> Wonder what all they're doing back home? Rawlins said.
>
> John Grady leaned and spat. Well, he said, probably they're having the biggest time in the world. Probably struck oil. I'd say they're in town about now pickin out their new cars and all.
>
> Shit, said Rawlins.
>
> They rode. (McCarthy 1992, 36–37)

Cole's mocking reply to Rawlins marks oil's promise as a fantasy—one that does not have the power of the fantasy that he himself is chasing, of living out the frontier myth. If life on the range is idyllic and desirable, oil is false and fleeting, more harmful than heroic.

Today, this divide between the frontier and oil has been bridged, as oil money subsidizes Texas ranches and the neoliberal push to deregulate the oil industry casts multinational oil corporations as beleaguered individuals, hobbled by bureaucracy. As Larry McMurtry notes in a 2012 interview in *Texas Monthly*, oil and gas revenues now keep Texas ranches afloat: "They aren't dependent strictly on cattle. They just wouldn't survive if they were. The King Ranch is a major ranch, but it's not funded by cattle. It has immense oil and gas holdings. And that is very much [the case with] all the ranches" (Silverstein 2012, 76). As Karen R. Merrill notes, this entanglement of oil and ranching would determine the character of the postwar oil industry: "several Texas and Oklahoma oilmen—or men who managed to get rich from oil—helped propel a nostalgia for the nineteenth-century West at the highest levels by building notable western art collections that often included works by Fredric Remington or Charlie Russell as their centerpiece" (Merrill 2012, 202). As in the conclusion of *Giant*, when Bick Benedict's cattle ranch is funded by oil profits, the idyllic vision of ranch life so integral to right-wing politics in the United States is directly subsidized by the oil industry. Oil culture borrows idyllic, wholesome, and individualistic frontier imagery to euphemize an exploitative industry. At the Sid Richardson Museum and the Amon Carter Museum of American Art, endowed by wealthy Texas oilmen in Fort Worth, the ideological conjunction of oil and the frontier is institutionalized.

In a book about how Texas has determined US politics for the past two decades, Gail Collins argues, "Texas frames its political worldview on the ideology of empty places, which holds that virtually any amount of government is too much government" (2013, 206). This "ideology of empty places" also entails a commitment to an ever-expanding frontier and an unwillingness to accept the LIMITS that come with a social and economic system powered by nonrenewable ENERGY. An alternative representation of the oil industry, not as an extension of frontier individualism but rather its negation, might be the only path to loosening Texas's hold on our energy FUTURE.

See also: AMERICA, BOOM, CANADA, CORPORATION, IDENTITY, PETROREALISM, RUBBER.

Textiles

Kirsty Robertson

Every day we dress in oil. At each stage in the process of making, purchasing, and discarding clothing and other textiles, oil is present as a secret companion. Petrochemicals douse the cotton fields; oil powers the complex farm machinery that has replaced hand labor in the fields; it powers the vast looms that weave cotton into fabric. Oil polymers subtend the many synthetic and natural-synthetic hybrid fabrics that make possible a waterproof, stain-resistant, hard-wearing, easily replaced material existence. Underlying the vast transport systems that carry textiles and apparel across the globe is oil. At the end of their lives, as discarded textiles decompose, natural fibers biodegrade while synthetic ones stubbornly resist breaking down. Textiles, particularly synthetics, also shape our understanding of being-in-the-world: they are literally the material of the current moment, an archive of its labor processes and an infolded, fluid record of what is considered important.

Philosopher Michael Serres uses textiles (among other examples) to describe the nature of time, connection, and communication in the contemporary world. Serres writes that "tissue, textile and fabric provide excellent models of knowledge, excellent quasi-abstract objects, primal varieties: the world is a mass of laundry" (Serres 1998, 100–01; see also Connor 2002, 2009). For Serres, the pliable forms of textiles can be seen (and felt) as "metaphorical matter," an interstitial space always in-between, open to the senses. In short, textiles are a milieu through which to understand the world: "veil, canvas, tissue, chiffon, fabric, goatskin and sheepskin . . . all the forms of planes or twists in space, bodily envelopes or writing supports, able to flutter like a curtain, neither liquid nor solid, to be sure, but participating in both conditions. Pliable, tearable, stretchable . . . topological" (Serres 1994, 45). In extending Serres's metaphor, one might look to the way that fabric folds in on itself, and the way that knitting and weaving link and unravel themselves, in order to think through the tangled yet fluid systems of power and communication that underlie capital in the contemporary moment.

Serres was not alone in using textiles to theorize a wider project. Gilles Deleuze and Félix Guattari also looked to textiles—felt and patchwork on the one hand, woven textiles on the other—to illustrate their theory of smooth and striated space (Deleuze and Guattari 1988, 474). With no center, felt and patchwork are an expansive, infinite, smooth space, while, by contrast, the loom delimits woven fabric. Woven fabric is a striated space that can only expand within the boundaries allowed by warp and weft. Striated space is arboreal, organized, regulated, while smooth space is rhizomatic, multiplying, and mobilizing (Munro n.d.). Without duality, the smooth and the striated are simultaneous states of being, two parts of a continuum.

Both theories are useful, but in mobilizing textiles as metaphor, they overlook the literal and epistemological facticity of textiles: textiles are the most material of materials. They might be the fabric of the world, the fabric of our lives, but they are also just fabric, grown from the muck of cotton fields, shaved from the backs of muddy sheep, dried from the skins of animals we eat, boiled from the primordial murk of fossil fuels. Nothing seems further removed from oil than a clean piece of white cotton, after the greenwashing of capitalism bleaches that cotton clean of its petrochemical past. In the material world, textiles are smooth space and striated, communicator and silencer.

The rise of synthetic fabrics was part of the larger mid-twentieth-century effort to turn wartime inventions to profit-generating peacetime consumables. Polyester and polyamide fabrics were manufactured with ethylene and propylene byproducts from distilling and refining crude oil. Several companies and laboratories were working toward this process when DuPont first successfully synthesized polyester fabrics in 1936 (Ndiaye 2006). After World War II, DuPont manufactured nylon for commercial markets, where it and other synthetics quickly became immensely popular, reaching the height of the fashion world in the 1950s, before succumbing to the oil CRISIS in the 1970s (Robertson forthcoming).

Synthetic textiles come in and out of fashion. While Nylon, Rayon, and Lycra persist, other once-popular brands, such as Dacron, have long since fallen from grace. In the popular imagination, synthetic textiles now tend to be associated with shiny polyester suits: sheen and cheapness. They are considered at best kitschy, at worst downright tacky. But in actual fact, a significant percentage of clothing remains synthetic; the new synthetics hide their glisten. They are more about function, less about fashion. Jeans with a bit of stretch, waterproof jackets and boots, stain-resistant work pants appear somehow clean, new, untouched by the sheen of petroleum. I have talked to many people wearing Lycra exercise gear, fleece jackets, or Gore-Tex who insist they only wear natural fibers. This naturalization of synthetics reflects the operation by which oil is rendered invisible. It seems hard to imagine that what keeps one dry is the repellent relationship between oil and water.

If there is a continuum between oil and textile (two forms of striation? two forms of oily smoothness?), it has now come full circle. While synthetic textile production once used petroleum by-products, synthetics are now increasingly made from the DETRITUS of consumer products. Since the discovery of the process to break down PET PLASTICS (primarily water bottles) into tiny chips that could be further shredded, spun into filaments, and pressed into cloth, recycled fleece has become an increasingly marketable way of dealing with the intractable problem of plastic pollution. Making fabric from garbage is presented as a GREEN solution, a way of diverting plastic from the landfill and repurposing it. It looks textile but is actually plastic.

Companies large and small have become involved in this process, from Patagonia to plastic bottle producer and PET manufacturer Invista (formerly a DuPont subsidiary acquired by the Koch Brothers). The popularity of this process is demonstrated by a February 2014 headline, "Pharrell Is Making Jeans out of Recycled Ocean Plastic" (Richmond 2014). The singer and producer, known for his hit "Happy" (can't nothing bring him down), partnered with denim designer label G-Star RAW to create "bionic yarn," a cotton/plastic hard-wearing hybrid thread to be woven into high-end denim. This partnership—called "RAW for the Oceans" and announced during New York Fashion Week at the American Museum of Natural History—was made possible by plastic pollution ("Pharrell Williams Announces" 2014).

The problem, as numerous studies have documented, is that recycled plastic fabrics disaggregate, shedding particles in washing cycles that end up in waterways and eventually in the ocean, contributing to the very problem that Pharrell's RAW jeans seek to solve (or at least profit from). The smooth textile sheds its striated partner in microfilaments that then bioaccumulate, likely finding their way into invertebrates, fish, and eventually back into the food chain (Browne et. al. 2011).

The metaphor of textiles—as a pliable whole, woven or patchworked from several components—depends on their integrity, on their not shedding filaments that return, invisibly, to the ocean's primordial deep water, where they join in the slow-moving dance of a giant gyre of garbage, mostly imperceptible to the human eye.[1] In garbage gyres, both textile and metaphor disintegrate. (What is the point of a fluid and pliable metaphor if it shreds into a million destructive pieces?) As Pharrell himself notes, it is really only clean plastic bottles that are useful ("Pharrell Williams Announces" 2014). The trash from the ocean is but one component—most of the plastic used in RAW jeans comes from land-based recyclers. Pharrell's removal of plastic bottles is a microscopic drop in a macro problem.

Textiles, as metaphor, as material, have the kind of expansiveness that Serres theorizes as an intermediary in-betweenness and Deleuze and Guattari describe as smooth space.

1. There are currently five gyres or garbage vortexes in the world's oceans composed of plastics, chemical sludge, and other debris trapped in ocean currents. Despite their enormous size, the gyres are not immediately visible, as many of the particles are tiny or microscopic.

Plastic fleece demonstrates something even more remarkable: it is a felted nonwoven textile along the lines of Deleuze and Guattari's smooth space, yet also striated by the passage of oil through it. It is a Möbius strip of the material consequences of contemporary human life. The relationship between oil and textiles, expansive and ubiquitous, tends toward invisible dependency. Textiles are more than the sum of their parts, metaphorical, material, and otherwise.

See also: CHINA, DETRITUS, EMBODIMENT, NUCLEAR, PLASTICS, PLASTIGLOMERATE.

Unobtainium

Crystal Bartolovich

According to the Wiki for James Cameron's 2009 blockbuster *Avatar*, unobtanium "is a highly valuable mineral found on the moon Pandora" that "humans mine . . . to save the Earth from its energy crisis" ("Unobtainium" 2016). Against this narrowly human-interested "one thing needful" approach to saving a planet, the film insists—reassuringly—that "Nature" can do the saving: Eywa—the local Nature goddess—sends forth battalions of conveniently powerful creatures to drive the imperialistic humans out. More interesting than either of these dubious salvation narratives is a moment in the film that implicitly calls both into question. Disgusted with the mining operation that threatens the flora and fauna of Pandora, a scientist denounces a corporate executive's evil ways. Unabashed, he reminds her: "This is why we're here. *Unobtainium*. Because this little gray rock sells for twenty million a kilo. That's the only reason. It's what pays for the whole party. It's what pays for your science." Though the film gives viewers an out (allowing them the pleasure of "going native" with the hero), the executive's challenge does not. As the CEO thrusts the "little gray rock" toward the camera, viewers too are confronted with issues of cost—and complicity. The CEO asserts that agency—and responsibility—are distributed ("this is why *we're* here"); the scientist, however outraged, does not inhabit a zone of purity outside the military-industrial complex that funds her lab.

Thanks are due to Christian Thorne and Henry Turner.

For readers today, allusions to "distributed agency" probably bring to mind Bruno Latour's "human/nonhuman assemblages" (2011). But debates about agency have a much longer, more varied history, including a prominent role in the emergence of the concept of class, whose purpose is to theorize collective human struggle (D. Harvey 2010). Class analysis insists that so long as costs and benefits are unequally shared, the problems of hierarchy, collective responsibility, and struggle cannot be abandoned—nor, as we shall see, can the distinction between unobtainium and utopia raised by *Avatar*.

Unobtainium is an engineering term, around since the 1950s, to describe elusive materials (or, less often, ideas or conditions) that, were they available, would solve some immediate technological problem. From Cold War aeronautics and rocketry labs, the term migrated to sci-fi novels, whence, presumably, it entered Cameron's vocabulary. Its usefulness is manifest: unobtainium inhabits the gap between the actually existing and the desired. It is close kin to Utopia (eu/ou-topia)—the "good place" that is (at least for now) "no place." Unobtainium differs from UTOPIA in that utopia aspires to total social transformation while unobtainium seeks to solve a specific problem (Adorno 1988)—the energy source, metal, or technique needed to realize a particular goal: a fix for Earth's energy crisis, or development of the motion-capture 3D cameras necessary to film *Avatar* according to Cameron's vision. With their focus directed intensively on a specific agenda, pursuers of unobtainium can ignore or discount the totality of relations and effects produced by their quest. There is no way to combat this sanctioned ignorance without examining particular unobtainium projects within the totality of relations in which they participate, with an eye to determining uneven benefit and cost, in order to overcome them—exactly what the Marxian concept of class undertakes.

The more typical, mainstream approach is evident in a recent Shell advertising campaign that proffers the fantasy of a solution to an unobtainium problem while keeping the totality of corporate-privileging structures intact. In large letters written on a blackboard, the ad urges, "Say no to no." It continues: "Isn't it high time someone got negative about negativity? Yes it is." Then it informs us, "The world is full of things that, according to nay-sayers, should have never happened . . . . And yet . . . continents have been found . . . men have played golf on the moon . . . [and] straw is being turned into biofuel to power cars." The ad concludes by proclaiming Shell the purveyor of "real energy solutions for the real world": a champion of positive transformative powers. Something, the ad promises, will (as in *Avatar*) "save the Earth from its energy crisis." Just trust Shell's R & D!

While in real-world engineering parlance *unobtainium* is often a joke—not unlike conservative dismissal of *utopia* as unrealistic dreaming—Shell insists that solutions to unobtainium problems have been found in rocketry and the like. *Avatar*, though, emphasizes what Shell does not: that with the pursuit of unobtainium come costs that have never been equally borne (just as the benefits have not been equally shared). To the extent that *Avatar* stages a dialectic between desire and cost, it opens up a true utopian (totalizing) moment that complicates its facile ecological message. When the CEO reminds the scientist of her complicity, he may be attempting to offset his own responsibility, but he nonetheless gestures toward the systemic relations on which the pursuit of unobtainium depends. Viewers can see that its costs are unevenly borne when Home Tree ignites.

Then comes the hard part: Do these viewers—when inhabitants of the Global North or privileged sectors of the South—recognize themselves as systemic beneficiaries of unobtainium pursuits and therefore as responsible for global inequality and planetary destruction as they grumble about gasoline prices or scramble to get by on decreasing incomes?[1] Usually not. That is why it is possible to drive to the multiplex, eat a food court meal whose ingredients have been transported hundreds (even thousands) of miles, while basking in the climate-controlled comfort provided by a petro-fueled electrical grid, *identify with the Na'vi*, and then drive home, flip on the lights, take a snack from the fridge, and play a video game. The misrecognition of such viewers is not unlike that of the film's head scientist, who does not (at first) let herself think too hard about where the dosh for her lab comes from because her heart is in the right place. Unfortunately, costs do not work that way; they accrue no matter how we feel about them, or whether we think about them at all.

Knowledge of global unevenness and planetary destruction and one's rôle in it is a necessary (but insufficient) step in the process of transforming unjust structures, which requires the assumption of strong responsibility of the class kind. Until the little band of human dissidents joins the Na'vi in anti-imperial struggle—having recognized that reform of the oppressive apparatus will make no difference whatsoever—they remain part of the problem. Since the urgent task for social justice and continued planetary existence is for the humans most implicated collectively (the ones who drive cars, travel by jet, use massive amounts of fossil fuels to light their cities, produce their commodities, etc.) to take responsibility for global warming and act to stop it, we must continue to emphasize the distributed agency of human collectives.[2]

Political theorist Jane Bennett understands agency differently, as distributed among human/nonhuman assemblages; she challenges the commonplace notion that "humans are *special*" (Bennett 2010, 36). The only alternative version of agency that she recognizes would blame individual CEOs for things like dispossessing the Na'vi. But such models of distributed agency fail to recognize that some humans have indeed been special as collective agents in creating our uneven global predicament by successfully diverting human (and nonhuman) relations to serve their own interests. Bennett's take fits all too tidily into the long-standing human practice of distributing agency (and therefore responsibility) for oil SPILLS to O-rings (e.g., BP) or global warming to trees (e.g., Ronald Reagan; see Radford 2004) when it suits their purposes—that is, evading human responsibility. Bennett recognizes this problem but not the crucial importance of considering uneven relations

1. Lauren Berlant (2011) describes as "cruel optimism" the subjective experience of making choices against our own flourishing. I am more interested in the choices we make against other people's flourishing, but Berlant's characterization of "impasse" works at both levels.

2. For an assessment of the still massive differential in human consumption between the Global North and South, see Worldwatch Institute (2013).

among human collectivities to each other and NATURE, in order to undermine this ruse of nonresponsibility.

Understanding distributed agency in terms of human collectivities—as Marxism does—continues to matter, then, because when theorists like Bennett displace intentionality into a mode in which humans and human agency are deprivileged, they not only mischaracterize the situation but, more important, provide an alibi for human inaction. Citing Derridean "messianicity," Bennett insists that "to be alive is to be *waiting* for someone or something"—an attribute she extends, animistically, to all matter (2010, 32). This view seeks to mortify human "exceptionalism" (ibid., 34), but it also participates in the widespread fantasy that something will, say, "save the Earth from its energy crisis." Humans can go on waiting alongside everything else. Bennett tries to evade this problem by returning human intentionality (though she does not call it that) in the form of ostensible individual choices to participate (or not) in human/nonhuman assemblages. It is hard see how such choices are possible, though, if our agency and responsibility are as weak as she suggests ("strong responsibility seem[s] to me to be empirically false"; 2010, 37). Assemblages, in her account, choose us (as individuals) as much as we choose them. In them, nonhumans assume responsibility for the planet's fate equally with humans (ibid., 37). Why, then, CHANGE what we do?

To extricate ourselves from this impasse, we must continue to assess the totality of relations among humans assembled into oppressing-oppressed collectivities and their implications for the planet, locally and globally. This moment of critique cannot be bypassed in the quest for social justice or planetary survival. Only utopian (totalizing), dialectical critique provides an antidote to the persistent unobtainium fantasies that something will free human beneficiaries (and their excluded others) from the hard collective choices and struggles necessary to systemically transformative redress.

See also: CHANGE, ENERGOPOLITICS, FICTION, GUILT, RESILIENCE, RESOURCE CURSE, SPILLS.

Urban Ecology

Allan Stoekl

What does urban ecology have to do with ENERGY, specifically with the fuels that power our lives?

Urban ecology proposes the study of *all* life in cities, considered not as supremely autonomous human artifacts but rather as ecologies: sites of the interdependence of living things, flora and fauna of all types (Douglas et al. 2011). Urban life is the aggregate lives of all beings and their modes of survival and flourishing. Human life is certainly a function of urban ecology, but so is rodent life, bird life, earthworm life, canine life, and bedbug life.

As David Owen points out in *Green Metropolis* (2010), human life in cities is actually more "sustainable" than in RURAL regions, since people in cities, especially big cities, drive less, walk more, use public transport, live in apartment buildings that are inherently more energy efficient that stand-alone homes, and so on. No matter how grimy, the city is GREEN; we can spend months in a city without seeing a tree, but we will still be more green than rural Vermonters who must drive to their CSA pickup points (and everywhere else). The city may be dependent on vast peri-urban areas that provide food and fuel, but it is in the city that these inputs are concentrated and make the flourishing of human life possible. City space is human space, lived in its greatest cultural intensity and energy efficiency. But human space to the max is, perhaps paradoxically, also the concentrated space of all life, and all life forms.

Understanding ecologies of organisms in cities is inseparable from understanding their ENERGY REGIMES. How animals, plants, and people flourish, survive, or die out in urban contexts is dependent on the types of food (fuel) available to them—their energy sources. People are dependent not just on fossil-fuel inputs, which are inherently limited (finite), but above all on SOLAR energy inputs mediated through living things, fellow beings in the urban ecology: plants and animals. A densely populated city with appropriate INFRASTRUCTURE (active transport, urban agriculture) will enable people to consume less, thereby encouraging life and human settlement. It is impossible to extricate human energetic needs—the very basis of human existence—from those of all the other living things (edible by humans or not) within the urban environment. We are all consumers of energy, and the end of energy consumption is hardly the triumph of the human (as a species or as a metaphysical category) in isolation.

Leonard Dubkin, one of the great twentieth-century urban ecologists—unemployed, he roamed the streets of Chicago, observing what was invisible to everyone else—discovered that attention to *all* life in the city prevents one from privileging the human perspective. After observing a polyphemus moth discovered in a neighbor's yard, he writes:

> When one saw a caterpillar, I thought, one visualized the butterfly it would become; but the caterpillar had a life of its own, it did not go through its days with the expectation of someday becoming a butterfly . . . . In our language the word "end" has two meanings: it means conclusion or termination, and it also means the object or purpose. When we say the end of the caterpillar's life is the butterfly, we imply that this is the aim or purpose of the caterpillar. But does "end" have two meanings in nature? Is it not just as likely that the end of the caterpillar—that is, the butterfly—is merely the termination of the insect's life, and that it exists only so there will be another generation of caterpillars? (Dubkin 1947, 198–99)

This consideration of the double end—the endpoint (or terminus) as distinguished from the goal (or telos)—is, as Montaigne would remind us,[1] the crux of the problem: Does NATURE in the city exist to further human lives—is it merely a means, with its end being our triumph—or is the end humans provide merely an arbitrary termination, not a goal, and the larger flourishing of all life, in all its forms, the real goal (ultimately beyond our mastery and comprehension)?

Urban ecologists have recently posed the question of the very presence of any given species (perhaps including, we might add, our own) in cities. In "Living Cities: Toward a Politics of Conviviality" (2008), Steve Hinchliffe and Sarah Whatmore imagine an urban field populated by humans and animals, but instead of populations defined by simple absence and presence, even the very awareness of communities (on the part of humans and other animals) is characterized by "experimental activities [that] involve elements of not knowing, of the unknown and the unknowable" (111). The presence of an urban species is not simply a given; it is tied to ecologies of place, and to speculation—scenarios, fictions—concerning the movements of species. They cite as an example the black red-

1. "La mort est le bout, non le but de la vie" ("Death is the end, not the goal, of life") (Montaigne 2001, 1633).

start, a rare bird in Britain that inhabits disturbed ecologies (abandoned factories, bombed out ruins, etc.). The bird's "presence in the topography is difficult to ascertain in practical terms" (112). How can a species whose presence is in large part a matter of speculation and possibility/probability be offered legal protection? "Likely presence" is the term coined to address this challenge, and developers in Birmingham are starting to "act as if there was presence" (113), setting up intricate nesting boxes and plantings and installing green roofs.

The urban is a larger scenario of protection, interaction, and RISK (and also human aggression; witness the fate of crow colonies in urban habitats); human and nonhuman agencies interact, and the relations between biology and politics become extremely complex.[2] As Hinchliffe and Whatmore put it,

> Cities are inhabited with and against the grain of urban design; . . . inhabitants are not static beings but entangled in complex processes of becoming; . . . and attempts to engage with urban heterogeneity require realignments of people and things in ways that are responsive to uncertainties, indeterminacies, materialities and passions, charm and magic. (2008, 116)

Among those entanglements are the relations of humans to their energy sources. People have long lived in cities to make life easier, more efficient. And all living organisms operate under the same imperative: to use energy in the most efficient and sustainable manner possible. It is literally a question of life and death for all. People and nonhuman organisms will thus always exist together in the city; the imperative of SUSTAINABILITY will make human dependence on other urban biological systems—flora and fauna of all types—that much clearer.

But it will also become clearer that in order to coexist with other biological systems we will never simply be able to dominate them. They may have visibility, agency, but not necessarily as subordinates; their ends cannot be provided by us, with the totality of their existence remaining invisible. To really see them is to see their autonomy, irreducibility, profound difference, and even their indifference to human goals. They make visible to us, as Dubkin argues, not just their presence but our own contingency. We are not their end or the end-all of all creation.

Thus, following Hinchliffe and Whatmore, we can ask: Is our very presence in the city such a certainty? Might we be, like any other species—like the enigmatic redstarts—determined by "uncertainties, indeterminacies, materialities and passions, charm and magic"? Might the passing of the (all too finite) fossil fuel era lead us to a clearer recognition not only of the NECESSITY of other energy regimes but finally, and along with that,

2. See Penn State Cooperative Extension, College of Agricultural Sciences, "Managing Urban Crow Roosts: In Pennsylvania and the Northeast" (http://www.extension.psu.edu). This rather exhaustive document on urban crow colonies lists any number of methods of "harassment" (sic) to be used by humans against corvid adversaries: pyrotechnics (i.e., fireworks), Methyl-anthranalite, effigies, and, presumably as a last resort, "lethal methods."

of our urban indeterminacy, our profound complicity with other species that can never serve as simple means to our end? To see the human in the midst of the animal—or vice versa—is to see the energy of the urban not in fossil fuels but in the renewable energy of all animals' unknowable and indefinable movement. All this in the city, that incubator, that teeming gut,[3] of what we like to call (post)human culture.

See also: ANIMAL, RURAL, SUSTAINABILITY, UNOBTAINIUM.

3. "Our" seemingly autonomous bodies are ecologies in themselves, incapable of functioning without billions of microbes in our guts and other places. Human bodies are a kind of urban ecology in miniature, an ecology linking human biological needs and desires to those of myriad other creatures.

Utopia

Philipp Lehmann

Whether as optimally employed labor or an infinite supply of power, ENERGY is often central to utopian designs for FUTURE societies. Conversely, utopian impulses have been particularly strong during times of actual or predicted energy transitions, inspiring designs for reorganizing socioeconomic, political, and environmental conditions with the help of new sources and unprecedented quantities of energy.

The advent of NUCLEAR power after the Second World War presents the classic example in this story of utopian ideas about energy. Just a few years after atomic energy's destructive force was demonstrated in Hiroshima and Nagasaki, nuclear power inspired numerous utopian visions of human futures transformed by scientific, clean, and cheap energy. This was true not only in the United States and RUSSIA but even in Japan, where faith in the civilian uses of nuclear energy was scarcely affected by the wartime experience of nuclear destruction. In the United States, Project Plowshare envisioned using nuclear explosions to geoengineer a more perfect planet, while the Atoms for Peace program aspired to ease Cold War tensions through peaceful exploitation of fissionable material (Kaufman 2013; Kirsch 2005; Krige 2006).

RENEWABLE energy sources also inspired utopian visions. In the 1920s, a German engineer was already envisioning a network of enormous wind towers—each higher than the Eiffel Tower—that could satisfy the energy needs of entire countries (Honnef 1932; Heymann 1995). Even the sun, volcanoes, and the tides inspired designs for industrial exploitation.

In the early twentieth century, hydropower arguably sparked the most spectacular ideas about the future of energy. Ever since the long-range transmission of ELECTRICITY had become industrially and economically viable around the turn of the twentieth century, "white coal" inspired designs of ever-larger hydropower stations. These projects addressed common anxieties about dependence on nonrenewable energy, an enduring motive since the British economist William Jevons warned in 1865 of the "probable exhaustion of our coal-mines." In the first decade of the twentieth century, journals and magazines devoted to the exploitation of hydropower emerged across Europe. Alpine states made the most of their opportune geography by building reservoirs and connecting new hydroelectric power plants to GRIDS.

Far removed from the wooden watermills of the preindustrial era, hydropower presented itself as a technology at the forefront of progress: the production sites, with impressive DAMS and oversized penstocks and turbines, were feats of modern engineering. Backers advertised "white coal" as clean, scientific, inexhaustible, and—because of the small staff required to run power stations—free of the danger of organized labor and social strife. The atmosphere was conducive to visions of what the full use of the world's hydropower potential might accomplish. It even inspired projects to expand that potential. This was the central objective of Herman Sörgel, a German architect who proposed a radical solution to interwar European conflicts: Atlantropa (see Figure 16).

Atlantropa

Atlantropa was many things at once: a precocious attempt at European unification, a geopolitical master plan to preserve the ascendancy of the Occident, and a futuristic vision of land reclamation and climate engineering on a continental scale (Gall 1998, 2006; Voigt 1998). These projected changes, however, were dependent on Atlantropa's foundational feature, which became almost synonymous with its name: it was designed to be the largest hydropower station the world had ever seen. Sörgel's idea was to isolate the Mediterranean Sea with a giant dam at the Strait of Gibraltar and smaller dams at the Suez Canal and Gallipoli. Evaporation would slowly lower the water level of the new "Mediterranean Lake," expose new land, and lead to a buildup of potential energy across the Gibraltar dam. Hydropower generators could transform the potential energy into enormous amounts of electricity—initial calculations estimated about 120 GW, or a thousand-fold increase over the Walchensee Power Plant in Bavaria, one of the largest hydroelectric power stations in Europe at the time.

"This enormous energy," Sörgel wrote in explaining his neoimperialist vision, would "transform large areas of North Africa into thriving plantations, tap the raw products of a large part of the world, and cultivate the emerging new land in the Mediterranean" (Sörgel 1931, 216). Beyond these plans for environmental transformation and economic exploitation, Sörgel believed that the availability of almost limitless energy could overcome political divisions within Europe and halt the decline that Oswald Spengler (2006) and his devotees feared. Atlantropa was to herald a new society of the Occident, a new order built on technocratic principles and the primacy of hydropower.

Plans for Atlantropa stirred the imaginations of modernist architects in Weimar Germany and prompted several newspaper articles in the early 1930s. Most contemporary critics doubted the political feasibility of European cooperation but did not question the project's technological foundations. During the Third Reich, Sörgel had a harder time garnering attention, although not for lack of trying. The Nazi leadership had little patience for a project that spoke openly about a unified European cultural sphere and that focused its utopian energies on North Africa rather than Central Asia, the prime location of German colonial plans during the Third Reich. After the Second World War, Sörgel solicited support from political leaders, public intellectuals, and the newly founded United Nations. Following his death in 1952, the project slipped into oblivion—a turn of events that had as much to do with the passing of Atlantropa's visionary architect as with the fact that proposals for a unified Europe seemed increasingly implausible with the rapid intensification of the Cold War.

DESERTEC

Atlantropa's disappearance from the public eye did not forestall the development and construction of large hydropower projects around the world. In fact, nine out of the ten largest hydroelectric power plants were built in the last fifty years. The world's largest power station was completed in 2012 at the Three Gorges Dam in CHINA, with one-fifth the capacity of Sörgel's projected Gibraltar works. Despite—or maybe because of—the ongoing expansion of the world's hydroelectric capacity, "white coal" had already lost some of its utopian appeal by the mid-twentieth century. Initially, oil and nuclear power offered fresher visions of personal mobility and cheap, scientific energy. But the promise soon dissipated after the oil crises of the 1970s and the Chernobyl DISASTER in 1986. Enthusiasm for oil and nuclear power waned further with growing awareness of the environmental costs of the fossil fuel–based industrial economy and the inherent hazards of atomic energy, as demonstrated most recently by the Fukushima CATASTROPHE.

Plans for large-scale renewable energy production have lately made a comeback. One of the most ambitious examples is the DESERTEC project, which mirrors Atlantropa's geographical focus and its attempt to exploit economically marginal desert lands. A foundation supported by a European consortium of public and private institutions, DESERTEC aims to cover large swaths of the Sahara with high-yield solar-thermal collectors, which would feed into an enormous energy grid spanning North Africa, the MIDDLE EAST, and Europe. Wind turbines, photovoltaic cells, biomass and geothermal generators, and hydropower stations would supply additional energy to the decentralized grid.

As with Atlantropa before it, DESERTEC has lofty goals. According to its official mission statement, the project "demonstrates how to combat climate change, ensure a reliable energy supply and promote security and development by generating sustainable power

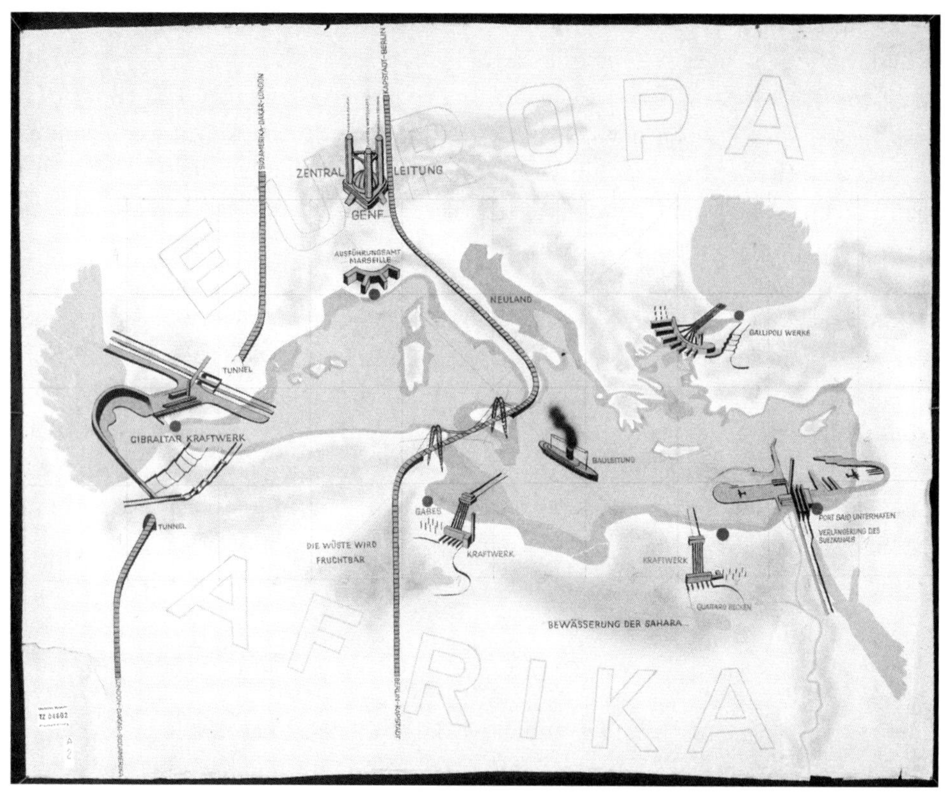

FIGURE 16. Map of Atlantropa with power stations, transcontinental railway lines, and recaplined areas of land. (Photo Deutsches Museum, Munich)

from the sites where renewable sources of energy are most abundant" ("DESERTEC Foundation" n.d.). Similar to previous large-scale energy projects, DESERTEC evinces a strong belief in the ascendancy and forward march of technology. The project will have to overcome both complex political and economic obstacles and the technological challenges of constructing a massive electricity grid, partly underwater.

Still, an emissions-free energy supply for Europe, the Middle East, and North Africa is an appealing prospect for a world threatened by the consequences of anthropogenic global warming. And while the odd wind turbine can provoke the vitriol of nearby residents, DESERTEC's great scale might rally people behind a vast and ambitious renewable energy project. It might even endow renewable energy with a dose of the "technological sublime" (D. E. Nye 1994) of large hydroelectric power plants like the Walchensee Power Plant, the Hoover Dam, or the envisioned Atlantropa works. The big questions for the future of DESERTEC will be, first, whether the managers and engineers can overcome the utopian trap and find practical ways toward realization and, second, whether the project can avoid the dystopian side effects that have so often accompanied the implementation of previous large-scale energy visions.

See also: ARCHITECTURE, DAMS, INFRASTRUCTURE, NETWORKS, RENEWABLE, UNOBTAINIUM.

Venezuela

Donald V. Kingsbury

The absolute centrality of oil in Venezuela has produced profound anxieties about vulnerability, dependency, and modernity. Here the "RESOURCE CURSE"—in which abundance of a highly valued commodity like oil is said to doom a country to inequality and maldevelopment—is not merely an academic concept but the topic of street corner debate. The symbolic and material substance of daily life, oil forms the "transcendent and unifying agent of the nation," in Fernando Coronil's account: "Venezuela was seen as having two bodies, a political body made up of its citizens and a natural body made up of its rich subsoil" (1997, 5). The relation of these "two bodies" forms an internal, antagonistic frontier of the political subject: Venezuela is defined by oil but also by the national project to transcend the status of "oil country." The dialectic of this political subject is ultimately tragic and self-defeating. It lacks faith in its ability not to "drown . . . in the devil's excrement," as one Venezuelan statesman famously concluded (Pérez Alonso 1976). This constituent bad faith of the subject—its two bodies drowning in oil—is fostered by what we might think of as the coloniality of oil.

From its beginnings in the Caracazo—a series of popular uprisings against neoliberal structural adjustment policies and government crackdowns that left thousands dead in February 1989—the Bolivarian Revolution has been driven by subjects that refuse this coloniality. Coloniality is a global logic in which race and geography determine access to resources, labor, and status. It divides the world into "developed" and "developing" camps while enforcing a common sense acceptance that the latter will always lag behind the former economically, politically, and culturally. In terms of oil, coloniality charts which states are defined by oil and which states simply use it, for better or worse. The Bolivarian Process, then, is a revolution in subjectivity, defined by new forms of collective life opposed to the modern/colonial world system's culture of submission.

The Coloniality of Oil

Venezuela has long been an oil country. Indeed, in all modern states, oil contours the economy, the geography of town and country, the reasons for and technologies of warfare, the limits and prerogatives of sovereignty, and the exploitation and desecration of environments and their inhabitants—but unevenly so. In Venezuela, this unevenness is best understood through Aníbal Quijano's notion of coloniality, a logic of domination, distinct but also arising from colonization proper. It "determine[s] the social geography of capitalism"; it underlies "all forms of the control of subjectivity, culture, and especially of knowledge and the production of knowledge" (2003, 208–9). Coloniality naturalizes structures of domination and submission, forming and placing subjects through binaries and exclusions: modern/nonmodern, developed/developing, European/non-European. While modernity is soaked through with oil, its effects—the ways in which it distributes costs and benefits, and the degree to which the relation to oil is seen as positive or negative—are unevenly patterned. This uneven pattern is the coloniality of oil.

North Atlantic oil countries were defined by open ROADS, suburban sprawl, and the middle-class consumerism that emerged in the mid-twentieth century. According to oil's colonial logic they remain so, despite (for example) the United States' and CANADA's increasing dependence on extraction industries (and financial speculation) tied to tar sands and FRACKING (Mitchell 2011, 259). When citizens of the Global North raise alarms over oil, they tend to worry that the convenience, freedom, and wealth of oil-driven industrialization have poisoned the planet and should not be expanded elsewhere. Such concerns reinforce the modern/colonial hierarchy under the guise of ecoconsciousness, dividing the world into "good" (or at least piously guilty) oil countries and "bad" (irresponsible, short-sighted, polluting) ones. In this "post-petroleum" developmentalism, the North is the agent of Enlightenment, the South an ill-disciplined pupil.

As a producer-state, Venezuela also illustrates a fatalistic dimension of the coloniality of oil that defines its very being as nonmodern. Oil is said to be a "curse" that fosters corruption and a lack of accountability among elected officials, fostering deficits in productive and economic INFRASTRUCTURE (Karl 1997). Venezuela's "petro-state" has been described as particularly overbearing—"magical," in Coronil's terms—in the sense that, as a rent-seeking and distributing entity, it transforms NATURE into the trappings of modernity at no cost to the population. It is not access to capital or labor that determines wealth and status in Venezuela but rather one's proximity to the state (Coronil 1997, 4). In a more Weberian register, the rentier capitalism of the twentieth-century Venezuelan petro-state impeded the formation of a domestic bourgeoisie, an autonomous civil society, and the culture of self-creation necessary for modern citizens.

The coloniality of oil in Venezuela internalizes concerns surrounding maldevelopment and the petro-state. As early as 1936, Arturo Úslar Pietri warned of the need to "*sembrar el petróleo*" (sow the oil), lest Venezuela suffer the economic and moral degradation he saw as "inevitable" for a society that was less a country than a "parasite" feeding off a particular industry. As long as Venezuela is defined by oil, it cannot hope to define itself as modern.

However, the call to "sow the oil" has historically carried a deeper resonance than the economic call to diversify the economy, as it has been articulated within a larger civilizational discourse regarding Venezuela's IDENTITY as a developing nation. Perhaps the most telling expression of this discourse has been the policy—up to and during the dictatorship of Marcos Pérez Jiménez (1948–1958)—of importing skilled labor from Europe to modernize the oil industry. Venezuelans, it was held, lacked the technical competence and the WORK ethic to do so themselves. As Miguel Tinker Salas notes, "foreigners assumed the characteristic role of modernizers, confronted by a 'backward' labor force that had to be transformed" (2009, 95). The habits of modernity, in other words, had to be imported to Venezuela through foreign agents.

Revolution against Modernity?

Some critics of the Bolivarian Revolution have dismissed it as a mere variation on this familiar petro-state: developmentalist, but ineptly and unsustainably so. Speaking of late Venezuelan president Hugo Chávez, Mexican neoliberal strategist Jorge Castañeda concluded, "Chávez is not Castro; he is Perón with oil" (2006, 38). By situating Chávez and the Bolivarian Revolution within a familiar spectrum of politics in Latin America, Castañeda denies that anything new is afoot. He claims, rather, that when Chávez—who stands in for the entire Bolivarian Revolution here—speaks of "socialism for the twentieth century," he is merely attempting to purchase elements—modernity, socialism, democracy—for which others toiled and sacrificed. What is more, for Castañeda whatever happens in Venezuela is unexportable. Oil is the precondition for social CHANGE in Venezuela, as it has been since long before the election of Hugo Chávez.

And yet, the tone of most responses to the Bolivarian Revolution has been more hostile than dismissive. Venezuela poses a threat to the modern/colonial world system not because of potential disruptions to the flow of oil or because of its drive to democratize consumption and to eradicate poverty. Rather, the Bolivarian Revolution is most threatening to global order as a precedent: that is, as a successful experiment in de-linking from the modern/colonial world system, driven by the grassroots energies of new political subjectivities.

This generative politics can be characterized in calls to deepen the state's reliance on protagonistic citizenship in the construction of the communal state. Protagonistic citizenship rests on a participatory, autonomous, inclusive, and creative subject explicitly opposed to liberal and representative models of democracy. Protagonism is expressed in multiple forms: the decentering of state authority to neighborhood-level communal councils and Bolivarian missions (grassroots social welfare, cultural, and economic programs), the occupation of factories, and the outing of corrupt bureaucrats. The red thread uniting these activities is an expression of autonomous social power that opens space into which the Bolivarian state moves—not the reverse. The state, in this sense, is the residue of resistance.

In other words, the Bolivarian Revolution has restaged the question of how constituent and constituted power—of autonomous social NETWORKS and state institutions—can and should relate to one another. Rather than worrying over its status as an oil country, or answering the impossible riddle of how best to "catch up" with Europe, it instead insists that its identity is based in the expression and generation of political subjectivities—the nonwhite, the purposefully nonmodern, the informal, the poor, the multitude—all of which were covered over and excluded throughout the twentieth century. And in so doing, it breaks from the coloniality of oil.

See also: CANADA, ENERGOPOLITICS, RESOURCE CURSE.

Whaling

D. Graham Burnett

Animal fats have served human beings as sources of ENERGY ever since humans merited the name. Crammed in the gullet, a little marbling or caul afforded early hominids the same kilocalories that such comestibles afforded any other creature equipped to function as a carnivore. And we can assume that whenever those restless hominids mastered the runaway oxidation reaction known as fire, they likely noticed that the white bits of their roasted meat flamed up impressively. Control over these little grease fires presumably followed, in the form of TALLOW lamps and candles.

There is no ANIMAL on Earth, and never has been, with as much fat as a mature blue whale (*Balaenoptera musculus*). All of the big cetaceans, equipped for mammalian life in cold oceans, pack on a considerable layer of blubber beneath their relatively thin skin. The history of whaling is the history of the pursuit of this layer, which nineteenth-century American whalemen took to calling the "blanket."

When did whaling start? Interpreted broadly, whaling is probably about as old as humanity—which is to say, prehistoric peoples who spent time near the seas would have had occasion to witness cetacean strandings. Who can doubt that a clutch of semi-starved hunter-gatherers would have gleefully thumped the remaining life out of such expiring creatures: stupefying, destiny-changing windfalls of meat, bone, and oil. Indeed, strong circumstantial evidence indicates that some of those hungry clans may have organized a more proactive form of scrounging. There were once gray whales in the Atlantic Ocean. These animals (which now only exist in the Pacific) have a penchant for shallow water lagoons where they calve and mate. Although not yet confirmed by archaeological evidence, it seems probable that early humans helped to exterminate this archaic Atlantic population: the animals would have been relatively easy (and irresistible) targets, lolling in mudflats at low tide, or rolling in light surf. A single kill could have supported a medium-sized coastal settlement for an entire winter.

Later textual sources (from Scandinavia and elsewhere) show how this sort of windfall could galvanize a subsistence COMMUNITY. Strandings in the medieval period could still produce a charivari chaos at the margins of Europe: everyone in the surrounding countryside streamed to the coast armed to cut out some blubber, and, if necessary, cut their way through their neighbors to do so. Nor did such seafaring peoples content themselves with waiting for the rare good fortune of a stranding. Even before the formal, specialized pursuit of large whales solidified into a seasonal enterprise (which had certainly occurred in the Bay of Biscay by the eleventh century, and quite possibly earlier in Northeastern Asia, or among Pacific or Caribbean island peoples), the spearing or droguing of animals encountered in the course of other activities—like sealing or fishing—would have been relatively common. Such wounded animals might later wash up on the beach, particularly weakened or juvenile specimens.

Not until the nineteenth century, however, did the hunt for these great marine kegs of oil become a global, open-ocean, cosmopolitan affair. This is the era of "traditional" whaling, an open-boat chase that ended with cold steel harpoons and lances. This highly specialized, labor intensive expropriation of natural resources reached its apogee with Yankee whalers' pursuit of sperm and right whales in the 1840s and 1850s. Among its primary commercial products were lamp oil and spermaceti wax for fine candles (Burnett 2007). In fact, in much of nineteenth-century Europe and AMERICA, the finest domestic illumination was provided by whale products. This held true until the watershed mobilization of liquid fossil fuels in the late 1850s, a discovery that rapidly undermined the economics of the traditional whaling industry (Davis, Gallman, and Gleiter 1997).

But there was more to come. Around 1870, enterprising Scandinavians developed new technologies that catalyzed a new kind of whaling—one that could take species (like the larger, faster blue and fin whales) previously beyond the reach of even the most intrepid harpooner. The key was explosive cannons and steam- or diesel-driven catcher vessels, together with onboard air compressors with hypodermic attachments that could inflate the carcass of a dead rorqual, insuring that these dense and muscular animals did not go to the bottom before they could be dragged away for the flensers. Processing in modern whaling was industrialized as well, since whales thus secured were "tried out" as never before: rather than merely peeling the blanket layer, and setting the rest of the animal adrift to feed the sharks, gargantuan boiler technologies let whalemen wring a lucrative additional measure of oil from the meat of the carcasses themselves (Tønnessen and Johnson 1982). This system of hyper-intense harvesting—supplemented after 1923 by mobile factory vessels that could roam the open ocean, processing thousands of animals for months at a time—wrought staggering destruction on the world's whales for nearly a century.

Significantly, however, modern whaling companies did not sell their barrels of whale oil for either heat or light. The highest quality product went largely into margarine for human consumption (never an important consumer item, whale meat represented a marginal aspect of modern whaling). Lower grades of oil, together with ground and dried meat-meal, supplemented the animal feeds essential to intensive production of supermarket chicken and beef—a macabre meat-economy that is still incompletely understood. Still lower quality output was saponified (i.e., made into soap; the byproduct of that process, glycerin, could be used in explosives). A variety of other industrial uses—including cosmetics, lubricants, and fertilizers—waxed and waned in significance across the fifty years that saw giant factory ships roam the high seas, grinding up and melting down the majority of the cetacean biomass of the global ocean (Burnett 2013).

A deep dive into the history of human utilization of cetaceans is not for the faint of heart or stomach. The archives seethe with an appalling chronicle of smoking entrails and mephitic wastes: factory vessels afloat in small seas of their own fetid excretion; men lost on deck under giant fetuses blasted from rot-swollen bodies; beaches so deep in a froth of flocculent, wind-churned organic DETRITUS as to be impassable even to draft animals. And from these scenes emerged such a queer array of emollients, combustibles, and consumer knickknacks: watch oil, coach-whips, transmission fluid, leather dressings, growth hormones, cattle feed. The bones of the giant whales killed in California, burned to rank CHARCOAL, whitened sugar on the plantations of Hawaii. English schoolboys, quite unknowing, spread a thickness of pulped Bryde's whale on their muffins. Russian silver foxes, locked in vast, stinking, Soviet fur farms, chomped on pellets of cetacean jerky, sourced in the icy waters of the Antarctic convergence.

In chapter 65 of *Moby-Dick*, the bottomless Bible of all things whale, Herman Melville depicts the second mate, Stubb, sitting down to a meal of fresh whale, even as the carcass lashed to the Pequod still seethes with sharks, slashing each other as they gobble at the offal. All the unholy circuits of immolation and consumption close upon this scene, where the hungry mate scarfs down a portion of the same creature that illuminates the repast. Here is Melville on this strange incest of eye and tongue: "That mortal man should feed upon the creature that feeds his lamp, and, like Stubb, eat him by his own light, as you may say; this seems so outlandish a thing" (1956, 238). And so it is, though perhaps not more so than the other scene of cyclical fueling that stands near the center of the book: the IMAGE of the whalemen feeding the fire under the try pots with the oil soaked bits of crispy whale skin they fish from the bubbling cauldrons themselves—a cannibal efficiency.

In these ways, and others, the great whales fueled the cultures that consumed them.

See also: ANIMAL, FICTION, KEROSENE, TALLOW.

Wood

Vin Nardizzi

"There's wood enough within": projected from offstage, this response to Prospero's summoning in *The Tempest* launches Caliban into literary history (Shakespeare 1999, 1.2.315). Its emphasis on adequacy indicates that the slave has completed his work. Stemming from this sense of closure, its disgruntled tone suggests an insubordination later elaborated in Caliban's plan to murder Prospero and burn his books. Such acts of defiance have made Caliban, as Jonathan Goldberg says, "a byword for anticolonial riposte" (2004, ix). But what of the wood? This question may seem slight when weighed against empire and resistance to it, but Caliban uncovers the indispensability of wood as the primary ENERGY source fueling subsistence and manufacture in the preindustrial era.[1] Moreover, the response encodes a fantasy of plenty articulated during a time of shortage in England. This resonance has fallen off our cultural radar because, unlike Shakespeare and his contemporaries, most of us in the Global North no longer live in the "age of wood."[2] Were we to substitute "oil" for "wood" in Caliban's first line, we would more readily comprehend its reference to a

1. In this epoch, wood was also a primary building material (M. Williams 2003).

2. Pearson (2006) shows that petroleum cars can be retrofitted to run on wood. Youngs (1982) argues that the age of wood has not ended, and wood's ubiquity as a source of energy in the Global South bears this proposition out. The matter of energy simultaneity is outside the scope of this piece.

necessary energy source that can cause environmental devastation. Thoughts about supply, source, and price may have crossed the minds of *The Tempest*'s earliest audience members when Caliban offers this accounting of the island's energy security.

The "age of wood" is an epochal designation probably unfamiliar to literary scholars. In environmental history, it names a period stretching from prehistory to the second half of the eighteenth century, when COAL generally replaced CHARCOAL (an energy source plucked from the ashes of cone-shaped piles of lumber that had been charred) in industrial iron making and fuelwood in homes, where it heated food and consumer alike. Sometimes dubbed "the wooden age" (Warde 2006, 6), this ligneous era bursts the strictures of traditional nomenclature for Anglo-American literary periods, outstripping epochs retrospectively parceled into temporal units (the [long] eighteenth century) or labeled for cultural movements (the Renaissance), monarchies (the age of Elizabeth), or position in relation to other periods (the Middle Ages and early American). To apprehend the sweep of the age of wood, we could do worse than to reflect on the life span of two of the planet's most mature organisms: Methuselah, a bristlecone pine in California, and Old Tjikko, a Norway spruce in Sweden. Dendrochronological research has determined that these trees are roughly 4,800 and 9,550 years old, respectively ("Swedes" 2008). They are colossal measuring sticks for approximating the age of wood's breathtaking temporal reach.

The counterpart of the era's mindboggling temporal coordinates is its geographic range. In a discussion of colonial Brazil, Shawn William Miller observes, "prior to 1800 one had almost no place to go but the forest to obtain a practical source of heat" (2000, 3). Supporting Miller's thesis are case studies of wood dependency in other preindustrial locales—colonial AMERICA (Perlin 2005), Easter Island (Diamond 2005), England (Nef 1966), Germany (Warde 2006), Japan (Totman 1989), and the Venetian Republic (Appuhn 2009)—as well as comparative accounts that start with the despoiling of woodlands in the ancient and the early modern worlds (Richards 2003; M. Williams 2003). Both approaches to environmental history have begun to chart the global forest that once was.

Given the temporal and geographic magnitude of the wooden age and the diverse expertise that its study entails, how do we bring into focus the grain of literatures dating from this era? Robert Pogue Harrison (1993) offers a model; his survey of an array of literatures demonstrates the transhistorical force the forest has exerted on the imagination. My sense, however, is that "wooden-age literature" tends to represent spectacular employments of this energy source, from the funeral pyres of ancient epic to public burnings of presumed heretics, and eschews its routine uses in hearth and home. When these mundane practices come into view, special circumstances frame their inclusion. Robinson Crusoe remarks that he "found it absolutely necessary to provide a place to make fire in, and fewel to burn" (Defoe 1985, 80). Although he is a meticulous chronicler of everyday life, we never see him search for either on the island. Instead, he mentions these matters in connection with other events: his illness (106, 108), his firing of pottery and first baking of bread (132–34), and his discovery of a cave (182–83). Are his energy sources, despite being "absolutely necessary" to survival, paradoxically not significant enough for literary representation?

Are they too prosaic to be described in their own right? Undertaking a parallel project to Harrison's that would trace the presence of fuelwood and charcoal might prove well-nigh impossible.

Caliban's riposte is no exception to this representational rule. Yet it does not exhaust the presence of wood in the play: Prospero tells Miranda that they "cannot miss" Caliban because he "does make our fire" and "fetch in our wood" (1.2.312–13); Caliban throws down a bundle of wood at the start of one scene (2.2), and Ferdinand, the play's mock slave, hauls a log onstage in the next (3.1). This log is a synecdoche for the "some thousands" that he must "pile . . . up" (3.1.10). Why might this energy source have such stage prominence? It may well be, as *There Will Be Blood* (2007) and *Avatar* (2009) suggest, that energy sources—oil and "UNOBTAINIUM," respectively—rise to the level of detailed representation during times of energy insecurity. I have argued elsewhere that Shakespeare's wooden Os—his playhouses—are uniquely self-reflexive spaces for meditating on wood: its expense, indispensability, scarcity, and centrality to dreaming (Nardizzi 2013). *The Tempest* offers both an imaginative record of an unprecedented wood scarcity gripping Shakespeare's England and a complex response to energy insecurity. Prices for this staple were increasing when *The Tempest* was first performed (M. Williams 2003, 170), and polemics describing an unremedied shortage predicted ecopolitical collapse. In a pamphlet contemporaneous with the play, Arthur Standish articulates the potential fallout: "no wood no Kingdome" (1611, 2). How might audience members affected by the shortage of wood have apprehended its abundance in *The Tempest*, and the fact that Prospero ships off to Europe without any of it? *The Tempest*'s depiction of "wooden slavery" may thus have stoked colonialist desire for restocking a depleted resource (perhaps through plantations in Ireland or the New World) long before Caliban spoke on behalf of anticolonial resistance (Shakespeare 1999, 3.1.62). From our vantage, it also emblematizes a historical tendency to take essential energy sources for granted and simultaneously to mobilize them in the exercise of power, as Prospero does.

See also: ENERGY REGIMES, FICTION, PETROREALISM, UNOBTAINIUM.

Work 1

Susan Turcot

Turning Down the Noise for a While: Ten Days Drawing in an Oil Camp

How can one make sense of a place as vast and inaccessible as CANADA's tar sands?

I enter an oil company HQ. It has taken a year to find a way through the fortress wall to access a shift worker live-in camp. Having learned portraiture as a way to spend time with shift workers, I'm aware that a portrait artist gains access to workers in a way that a researcher or journalist might not.

Most of the twenty-four people I drew needed to talk, to articulate their pasts, presents, or futures, sometimes with great intensity. Observing is about recognizing subtlety, discerning light from dark. The process moves you inward, expands and contracts time while unfolding a unifying space between observer and observed. In this 24/7 transitory environment, relationships to labor, the environment, and one's self can be tenuous. Being drawn means being listened to, being seen, and being felt—the antithesis of how people generally experience this work environment.

People say there are not enough oooo's in smooth to describe the onsite manager who leads tours for VIP guests visiting the largest oil project in the world. He is a Pentecostal minister, well-suited to delivering a good news story. We arrive in the extraction area, where up to forty thousand workers live transiently.

Artifacts, things that could link us to a past dating back thousands of years, lay at rest with their stories in the earth until the soil and all the life growing from it was declared mere overburden in the way of the oil below. Day and night, the diggers claw away at this vast space, a raped and haunted landscape left in their wake.

While pop music plays nonstop, I set up two chairs in the corner of the camp foyer, near a security desk, a counter with coffee and donuts, vending machines, and workers checking in or out with their bags. This space for drawing allows more fragile beings to emerge from behind the masks of the laborers, surviving their long shifts fueled by adrenaline.

Human beings transmogrify into vehicle bodies with solitary heads visible at their windows. Crisscrossing a landscape devoid of plant and ANIMAL life, they move past toxic ponds, sulfur mountains, and layers of earth separated by chemicals and by other machines. They stop only just before the next twelve-hour shift begins.

Wearing an identity badge that says "First Impressions Manager," a smiling woman greets me. She shows us around the camp that sleeps up to three thousand and whose design was borrowed from a prison. We visit a room for playing cards, where poker winners are sent each year to Las Vegas to play for bigger stakes. Outside there is a racing track for miniature battery-operated vehicles and an overgrown sports court. She mentioned she was a gardener, and I suggested planting vegetables. But she said everything is poisoned up here and you would not plant, fish, swim, or forage anywhere. I ask, "Did anyone ever ask you about second impressions?"

An absence of value permeates everything in this barren zone, right up to our front doors. Nobody belongs here now. Each job carried out by each person amounts to removing something, extracting more memory, life, and health from the landscape. Yet few workers in this giant mine wonder what was here before. It's best to shut down, shut off, be used by the system, and hope the FUTURE price of waking up is not too high.

> This is not a place, it's all sex drugs and alcohol, it's not worth it. The guys who make $2000 a week and have nothing left by Monday will be here another thirty years. You do not ask too many questions; you would not be here if you did. You have to be careful not to want bigger and better things, I work three days then three nights—I'm here for that. (Man from Quebec)

Most conversations here are about places far from here. Thinking about the denial of a place or the history of a place, I remember this idea from Lucy Lippard: "For many displacement is the factor that defines a colonized or expropriated place. Even if we can locate ourselves we haven't necessarily examined our place or our actual relationship to that place" (1998, 9).

> At work there is terrible back-stabbing, I just keep my head down and don't say anything. The stuff you see on site . . . you just have to carry on . . . . There's a lot of people you don't like—you don't say anything—carry on. The company screws you, my wife pushes, pushes too much—she wants an even bigger house now, she wants more and more. (Chinese man from Edmonton)

In the absence of a settled environment and grounded relationships, notions of place swell in the minds of shift workers living in camps, with ideas of belonging in some other place cast like a lifeline into the denuded landscape.

"I have never sat still for this long. I do not like it, it makes me think."
ST: "That you do not want to be here?"
"Yes"
ST: "If you could go back in time would you have stayed in Lebanon?"
"Yes."

Most shift workers spend more time living in camps than they do at home. At camp, each private space must be cleaned every other day; it's part of the security regime. Guards check the workers on their way into the canteen (no bags or hats allowed), and then it's an all-you-can-eat buffet, comfort, and SURVEILLANCE.

> I clean the camp rooms seven days on, then I have seven off where I have nothing to do but hang around the camp, as flights home are too expensive. Most of the rooms I clean are disgusting, with spit in the sink, sweaty clothes on the floor, flicked snot on the ceiling and walls. I am up and down like the other women my age here [about 50], but I find it hard to stop working; I've been working since I was 14. (Woman from Cape Breton)

A sense of self is also a resource that can be exhausted, through lack of sleep, support, privacy, and connection to family, friends, and place. What is the cost of recovering a human?

> I come from a very quiet place, very quiet parents. I hear the men in the adjacent rooms here. I hear them in the toilet. I have to speak to my girlfriend under the covers. It's no good up here, relationships break down up here. (Digger/operator from PEI off work after an aquifer was pierced)

Ground water routes are largely unmapped in Canada, and contamination travels far from areas of excavation, FRACKING, or steam-assisted drainage. Trying to stop high-pressure leaks from punctured aquifers can take years. Regulations for water used by the industry suit profit models, not the safety of drinking water hundreds of miles away.

> I would not work on a fly-in fly-out site where people are exhausted and disoriented from the long flights from home and where more accidents happen. [Shows me pictures of toppled cranes.] In my first week I was sprayed by benzene coming out of a pipe and no one had bothered to report it. These very young engineers have little practical experience or sense of responsibility. I found open valves that had the potential to explode an area the size of several football fields. (BC man writing safety report)

> Companies keep you on with the promise of work even when they don't have it or punish you if you quit. I refuse to be a foreman—this requires taking responsibility for all the workers when some are overworked, under-slept, on drugs. Some are assholes on the job because they know the union won't fire them. There was a mentally unstable guy who disappeared and they found him hitching down a heavy hauler road. (Acadian from Cape Breton)

The operation here is so vast that that jobs and money seem endless. But the money comes at great cost. Once they arrive, workers are drawn into complex, anonymous patterns of shift work that can leave them feeling bereft of autonomy and any sense of common purpose. It's difficult to imagine the oil companies will ever take any genuine interest in enabling anything or anyone who gets in the way of their numbing drive for profit.

See also: BOOM, CORPORATION, EXHAUSTION, IDENTITY, NECESSITY, TEXAS.

Work 2

Stevphen Shukaitis

Zerowork Training?

"One of the problems," my friend and comrade Ben said, "is that while we've been quite good at celebrating the refusal of work, we never had anything like zerowork training."

This statement struck me as strange, and not merely because of the context—a meeting of the editorial collective for Autonomedia, a long-running Brooklyn-based autonomist publisher. After a decade of involvement with the project, Ben was moving on. In the autonomist equivalent of an exit meeting, Ben declared his exit from a collective whose stated goal was to exit from work itself, to "substruct the planetary work machine," in the words of p.m. (2011).

But the substance of Ben's comment seemed absurd. What exactly would zerowork training be? Could you train someone to refuse work? Would the sign of mastery be the refusal of the training itself?

Ultimately, Ben's insight seems strangely true. Simply declaring one's desire to abolish work does not magically produce the skills and organizational capacities to make that happen. Even as a provocation or utopian demand (Weeks 2011), the refusal of work is a practice and a relation to the political, not an incantation.

The Pleasures (and Mostly Sorrows) of Work

Midway through *The Pleasures and Sorrows of Work*, Alain de Botton identifies as the most remarkable feature of modern work practices the idea that work should make us happy. The notion that "we should seek to work even in the absence of financial imperative" (2009, 106) is a far cry from notions of work as NECESSITY, drudgery, or even punishment. De Botton raises a telling point about the contemporary celebration of work as a cultural good and value to be cherished beyond its remuneration. This celebration arguably ties together the valorization of artistic and cultural work with attempts to impose unpaid or poorly paid work on the recipients of social benefits. While these forms of work differ greatly in their conditions and cultural prestige, they share an underlying assumption of an intrinsic "good" of work that exceeds external rewards.

This widespread valorization of work reveals something interesting about the fueling of culture. Why would work be celebrated at the same time that flexibility in production has been achieved through mass outsourcing and the devaluing of labor? Perhaps work is valued as a good in itself—an unlimited resource—precisely because of the LIMITS of material resources that have become impossible to ignore.

In *The Human Motor* (1992), Anson Rabinbach shows how debates in physics about the conservation of energy underpinned the European science of work and the eventual rise of management and industrial relations—in other words, the social technologies for the intensified extraction of labor. Rabinbach argues that energy conservation became the Continental answer to a Darwinian-Spencerian vision of society propelled by conflict and struggle. Although T. H. Huxley popularized thermodynamics as a metaphor for capitalist superiority (grounded in scientific materialism and Lamarckian biology), such ideas also pointed toward "an equilibrium of economic expansion and social justice" (Rabinbach 1992, 179). This equilibrium was imagined to stand above social classes and political imperatives, as the domain of expert state planners and scientists who could use the study of efficiency to transform society into an industrial enterprise that would maximize productivity and social justice. Few would embrace this notion today, even if there is something appealing in the notion that efficiency could reduce work and improve conditions rather then intensify exploitation.

The Limits to Immaterial Labor

When the Club of Rome published *The Limits to Growth* in 1972, the notion that natural resources were finite, and that limits to their exploitation needed to be taken into account, came as something of a shock. As David Harvey (2005) argues repeatedly, capital does not solve its problems so much as it moves them around. The recognition of limits to growth and finite resources has been paired with continued attempts to reduce the cost of labor, or the value of work, in varied and vicious ways, including attacking unions, dismantling the welfare state, reorganizing global production, and financialization. In the realm of management, worker discontent has met with responses ranging from humanization and job enrichment to teamwork and "fun at work." The emergence of the digital economy and network culture has been accompanied by the massive explosion of free work, where tasks are outsourced as bite-sized parcels that break down the labor process along with the wages. Free labor has become an infinitely exploitable resource of the sort that natural resources once were.

Immaterial and cultural forms of work have been hailed by management and business theorists as a superior source of value production, while also being celebrated in contemporary post-autonomist thought as the basis of a new form of communism: a curious conjuncture indeed! Perhaps the demand for meaningful work means one thing in the context of rejecting the factory line and the drudgeries of industrial labor and quite another when it becomes the ideological apparatus that renders us into an infinite reservoir of workers willing to work because we believe in what we are doing. This subjective attachment to work beyond its external rewards developed largely among culture workers, but it has drastic consequences when generalized beyond this sector.

Franco "Bifo" Berardi departs from the post-autonomist celebration of immaterial labor to focus on its costs: namely EXHAUSTION, depression, and the breakdown of possibilities for social recomposition that have been the focus of autonomist analyses of class composition. He argues that this transformation of work practices actively prevents the emergence of new political movements or forms of social antagonism capable of radically transforming the present. This analysis may seem overly pessimistic, especially given the outbreak of new forms of social movements during the past few years. But Bifo argues for an active withdrawal from labor, returning to the autonomist notion of the refusal of work but in expanded forms that create exodus from the economic sphere altogether.

Work! Work! It's the Sound of the Police

It seems that work, as Nietzsche argues (1997), is the best policeman. It governs social life even when its role in adding productive value seems to slip away and we become "dead men working" (Fleming and Cederstrom 2012). In an era of biopolitical production, where the policing function of work tends to police all of life, it might seem that the refusal of work is the refusal of life itself. Such notions lead to rather dismal conclusions about the possibility of autonomy and social recomposition, but one can think about work in other ways. Stefano Harney points to the black radical tradition, which takes up the problem of refusal in a situation where one's very being is defined and captured as work. For Harney this refusal "is the dimension of original exodus; this is the practice of fugitivity . . . the escape that goes nowhere but remains escape" (2008).

In his new introduction to Paul Lafargue's classic *The Right to be Lazy*, Bernard Marszalek hints at another important insight: that the opposite of work is neither leisure or idleness but instead "autonomous and collective activity—ludic activity—that develops our unique humanity and grounds our perspective of reversing perspective" (Lafargue 2011, 19). From a compositional approach, refusing work is not doing nothing but developing the capacities, organization, and collective becomings that enable and sustain these ludic activities and social wealth. In short, it is the zerowork training my friend Ben asked for. In this pedagogy of learning not to labor, refusal is not individual but socialized. Learning not to labor conjoins the refusal of work with a re-fusing of the social energies of refusal back to the continued affective existence of other forms of life and ways of being together—as practice and embodied critique.

See also: AFFECT, EMBODIMENT, EXHAUSTION, LIMITS, UTOPIA.

FIGURE 17. "Oil/Lie." (Image by Pedro Reyes)

Afterword

Imre Szeman

It is always possible to dismiss even the most threatening problems with the suggestion that something will turn up.

—E. F. SCHUMACHER, *Small Is Beautiful*

No mark survives this place, you too will yield
to unmemory. Give everything you are
in three-day pieces. Watch the gypsy iron
move, follow its commands.
Tend the rusted steel like a shepherd.

—MATHEW HENDERSON, "The Tank"

The contributions to *Fueling Culture* offer ample evidence of the multiple and varied ways in which energy has figured and transfigured human life. By highlighting the key role of fossil fuels in shaping the experience and reality of the modern, these essays provide an important and much-needed corrective to our understanding of the forces that shape societies, organize geopolitics, and, perhaps most surprisingly, animate cultural and intellectual life. Unprecedented access to massive and ever-increasing amounts of cheap energy from fossil fuels—first coal, later oil and gas—is rarely identified as a constitutive element in the narrative of modernity, which tends to be told as a story organized around a more familiar cast of characters, such as the expansion of individual and social freedoms, class struggle, challenges to cultural norms and expectations, and industrial and technological progress. While attending to energy has the potential to recast many of the dominant periodizing accounts of our present, *Fueling Culture* does not aim to offer an etiology of the modern or to nominate energy as history's deus ex machina. Instead, our intent is to demonstrate the necessity of including energy in all of our investigations of the past and present, and—perhaps most important—in our projections of the shape of futures to come.

The impact and import of fossil fuels in modernity have long been invisible—one of the many cultural phenomena that the essays in this book seek to explain. But once we finally have energy on the brain, we cannot help but recognize that it was there all along, hidden in plain sight, a lumbering presence in need of an episteme to bring it to the fore. For instance, the landscape of oil derricks that opens Douglas Sirk's *Written on the Wind*

(1956) now generates critical notice and curiosity, just as the offhand remark of Rock Hudson's character, Mitch Wayne, about relocating from TEXAS to Iran figures the geopolitics of energy into the landscape of melodrama. Set just a few years after the 1953 coup d'état that ousted the nationalist government of Mohammad Mosaddegh, the film intimates that Iran's oil fields are once again a safe site for US professionals to seek their fortune. In the 1945 novel by Robert Wilder upon which *Written on the Wind* is based, the source of the Hadley family's fortune is tobacco; the shift to oil as an obvious marker of near limitless wealth and an origin point of capitalist ACCUMULATION speaks to a symbolic function that oil still possesses—despoiled nature, masculine power, technological prowess, environmental mastery, youthful extravagance, and inevitable conflict.

Greater attention to fossil fuels in literature and culture generates changes in how we name and frame energy. This is true both for texts in which its presence is obvious—for instance, films about the social forces brought into existence by oil, ranging from George Stevens's *Giant* (1956) to Paul Thomas Anderson's *There Will Be Blood* (2007)—as well as for texts in which energy's absence needs to be thought through and challenged. To be sure, Jack Kerouac's *On the Road* (1957)—the classic American novel about the AUTOMOBILE's promise to realize desires for freedom and mobility—meditates on many of the subjective limits and failings of the modern. Yet it seems crucially oblivious to the constitutive relationships among its eponymous "road," the burgeoning US oil industry, and the vast infrastructures being brought into existence via the Federal-Aid Highway Act of 1956 (see Yaeger 2011). Revisiting classic texts and challenging the premises of newer ones from the perspective of oil and energy cuts to the heart of tired, boringly familiar, and utterly hypostasized tropes and figures, making them strange, disturbing, and materially animate to the present.

A focus on energy and especially fossil fuels as an interpretive strategy might well be seen as constituting little more than the application of old methods to new themes—a new sub-theme within the broader, already well-established practices of environmental criticism.[1] Such a perspective fails to grasp what makes a focus on energy so unnervingly powerful. As this book demonstrates, our culture is *fueled*. We are creatures of fossil fuels and the petrocultures that they have enabled us to inhabit. The energy of COAL and oil has allowed each of us to become more powerful than our forebears could ever have imagined. Greater attention to energy does not simply flesh out environmental studies or the practice of eco-criticism, though it certainly enables us to address the challenges of global warming and environmental crisis with greater insight and understanding. It also generates a broader, fuller and (it has to be said) *truer* sense of the operations of contemporary politics and society than would otherwise be the case. As the contributions to *Fueling Culture* ar-

1. This sense of energy criticism would be in accord with Lawrence Buell's broader assessment of the impact of environmental concern in literary studies, which he claimed in 2005 had "thus far, not changed literary studies or environmental studies so much as it has been increasingly absorbed therein. Its durability so far rests on its having introduced a fresh topic or perspective or archive rather than in distinctive methods of inquiry" (130).

gue, if we are to properly grasp the dynamics and forces shaping the texts, contexts, and imaginaries of cultural and social life, energy has to become a part of the critical vocabulary of every subject or topic under investigation.

In this sense, *Fueling Culture* makes the case for, forecasts, and proposes methodological innovations in the exploration of culture, society, and politics beyond the purview of the environment. This critical method lies in the deep connections that the essays in this volume establish between energy and human capacity or capability. There are significant links between the forms of energy that we use and depend on and the shape and character of our values, ethics, political practices, and sociopolitical imaginaries. John Urry argues, for instance, that "car culture has developed into a dominant culture generating major discourses of what constitutes the good life and what is necessary to be a mobile citizen in the twentieth century" (2007, 117). The beasts of the fossil fuel era have reshaped our expectations of what it is to be human by transforming both the landscapes we inhabit and the space and time of everyday life. In many parts of the world, one becomes an adult when one gains access to the mobility offered by a driver's license. Dipesh Chakrabarty goes even further, claiming in his influential account of the ANTHROPOCENE that "the mansion of modern freedoms stands on an ever-expanding base of fossil-fuel use. Most of our freedoms so far have been energy-intensive" (2009, 208). The hard work of explicating and understanding the full significance of the connections between cars and the good life, between fossil fuels and the freedoms we associate with modernity, has only just begun. *Fueling Culture* highlights how our values, verities, and capacities have been engendered by fossilized sunlight—liquid forces made up of condensed time, an uncanny historical anomaly we have learned to greet with a shrug whenever we encounter it at the gas pump.

The contributions to this volume demand not only the inclusion of energy in our critical narratives, or the production of new critical methodologies, but also the reimagining of our political sensibilities and orientations. In *Carbon Nation*, historian Bob Johnson (see EMBODIMENT) insists that modern life and the depths of the modern self are the direct outcome of the "deceptively deep ecological revolution" (2014, 3) that accompanies the use of fossil fuels. The discovery of oil and the refashioning of human social life around it played an essential role in legitimating the narrative, emergent in the West since the seventeenth century, of a continuously "improving" society.[2] The dominant political philosophy of the fossil fuel era has been liberalism, a theory of society that functions by misrecognizing our temporary push beyond Malthusian constraints as a function of social struggle and Enlightenment maturity rather than the unrepeatable good fortune of stumbling upon

2. One of the grand tricks of the modern has been to reimagine all of life and experience as shaped by endless expansion and accumulation, which has in turn sedimented ideals of growth and images of frontiers to be conquered deeply into our social imaginaries. For an elaboration of the role played by access to cheap fossil fuels in establishing growth as a social life, see Johnson (2014, 3–40) and Mitchell (2011, 109–43). For a discussion of growth as a value in literary and cultural production, see Szeman (2011).

nonrenewable resource plentitude. If liberal democracy and its attendant freedoms (such as they are) necessitate an abundance of energy, how are we going to invite everyone (more than 11.2 billion people by the end of the century, according to the UN) into the mansion of modern freedoms when we no longer have fossil fuels as the foundation to keep the structure standing? If the energy provided by fossil fuels makes the modern possible, what does this mean for the vast majority of the planet's denizens that have yet to use these resources with the same intensity as North Americans but who desperately desire to inhabit fully the social and personal freedoms and capacities of oil modernity?

When fossil fuels are figured as an essential aspect of the political, the practices and imaginaries that have guided liberalism are drawn into question. Liberalism is a meliorative political philosophy that imagines gradual improvements to social life against the backdrop of a history unfolding without end. When it comes to the environment, the limits of such "trickle down" meliorism are obvious enough: we are already facing environmental crisis and things have to be changed radically, and *now*. If liberalism appears unable, or unwilling, to address global warming and other environmental problems, it is not only because of its lack of speed but also because its desires for improvement come with durable, if hard to detect, limits on what or how far it can reform. For instance, one thing that could never be addressed within liberal capitalism is differences in wages or wealth: "abolish poverty" is okay, "abolish wealth" is not. The dependence of liberalism on dirty energy to fuel its meliorative system constitutes an even more intractable limit or blind spot than the unjust mechanisms of its economics. As its social improvements require the energy of oil, liberalism can only imagine that there will be ever more of the stuff, even if it is, by definition, a limited resource. And because what fuels liberal programs and policies is a principal cause of environmental damage and destruction, liberalism gets stuck, unable to grasp how helping the environment is also a way of hurting it. Recognition of the true role that energy has played in fashioning the political—its figuration, beliefs, and self-imaginings—suggests that we need a new, more radical politics, one more aware of the consequences of the liberal fantasia within which we appear to be stuck.

This brings me to my final point. Looming in the background of any discussion of oil and energy are *finitude* and *the future*. If there were an endless amount of oil, we would not have to worry about its disappearance, at least from the perspective of our ability to retain, sustain, and even expand the individual and social capacities we moderns have come to associate with our lives. And were our dominant source of energy merely in limited supply—which as an ancient, organic compound, it necessarily is—analyses of it would already constitute an eschatology: the end of a specific way of life, an oil modernity that we have come to accept as our own, but which we would need to somehow refashion around other sources of energy. However, the fact that our use of oil also causes significant and ever-increasing damage to the environment introduces yet another, more profound eschatology: not only the end of modernity but of human life as such. Since burning oil damages the environment on a significant scale and with growing force, we have to worry about its very use, not just its limited availability. There is a contradiction at the heart of the project of modernity: the source of energy so essential to its operations is also a material threat to its existence. How to figure this contradiction lies at the center of those theories and philosophies that have begun to grapple meaningfully with oil and energy.

In some cases, the intellectual project to render oil visible and nameable has been motivated by desires to push back these approaching end times and to return to the open horizon of the FUTURE that has been one of the signal features of modernity. Such projects want to retain the shape and form of oil modernity and the familiarity of the present social and political landscape, pushing back fears of finitude by appearing to name the problem. The fatal limit of these modes of inquiry is that, in their struggle to make oil newly visible, they ultimately seek to preserve and shore up all of the social practices and beliefs that arose in conjunction with oil, when its significance was still invisible. Put bluntly: the current way of dealing with our energy crisis is to state the problem in a way that affirms the necessity of the liberal status quo, as in the recent announcement by the G7 to end fossil fuel use by 2100 (the end of oil, but certainly not the end of the G7 and its values and verities!) (see Connolly 2015). This form of visuality imagines that it has the capacity to bring what Timothy Morton (2013) calls "hyperobjects," such as global warming, into the existing space of scientific and economic calculations, thereby domesticating a threat to comprehension and pushing oil back into the comfortable space of our political and social blind spot. Nicholas Mirzoeff has described such forms of conceptualization as "Anthropocene visuality," which "allows us to move on, to see nothing and keep circulating commodities, despite the destruction of the biosphere. We do so less out of venal convenience, as some might suggest, than out of a modernist conviction that 'the authorities' will restore everything to order in the end" (2014, 217).

In an effort to move beyond the limits of liberalism, Anthropocene visuality, and a capitalism ready to sacrifice everything for its own survival, the most challenging and productive investigations of energy have attempted to confront head-on the energy contradictions of modernity by making its epistemologies conceptually visible in a manner that might generate a forceful and transformative political intervention into the basic configurations of oil society. Our global society became what it is through the use and abuse of fossil fuels and now faces the twinned prospect of their decreasing supply and increasing harm. To date, our collective response to this situation amounts to little more than hoping that "something will turn up"—that is, when we permit ourselves to think about it at all. The provocative thought pieces in *Fueling Culture* help us understand why we do this, why it is a problem, and what we might do instead, in the hope that we might yet free ourselves from tending the rusted steel of oil fields that care not for our survival and will not remember us when we are gone.

ACKNOWLEDGMENTS

Any project of this scale and duration will bring any number of challenges and changes for its editors—this one perhaps more than most.

We are grateful to our contributors for their patience and to Richard Morrison, Tom Lay, and Fredric Nachbaur of Fordham University Press for their support of *Fueling Culture* throughout a time of transition for the Press. We are also indebted to Matt Flisfeder and Sean O'Brien for their assistance in managing the details and workflow of this massive project.

Imre Szeman wishes to acknowledge the insights and encouragement of his amazing colleagues in the Petrocultures Research Group at the University of Alberta, especially Sheena Wilson, Mike O'Driscoll, and Mark Simpson. At Alberta, he has been fortunate to work with one of the best groups of graduate students anywhere, including such superstars-in-the-making as Lynn Badia, Brent Bellamy, Sarah Blacker, Adam Carlson, Jeff Diamanti, Dan Harvey, David Janzen, Jordan Kinder, Sean O'Brien, Sina Rahmani, and Valerie Savard. He thanks members of the After Oil research partnership for their role in the conceptualization of this book—Darin Barney, Ruth Beer, Dominic Boyer, Stephanie LeMenager, Janet Stewart, and the amazing Graeme Macdonald. Andrew Pendakis and Justin Sully are critical interlocutors anyone would love to have in their lives. For her energy, enthusiasm and smarts, he thanks Jennifer Wenzel for being a model co-editor and colleague. And he thanks Eva-Lynn Jagoe for keeping his head buzzing with ideas and for filling his life with excitement and possibility.

Jennifer Wenzel wishes to acknowledge the encouragement and helpful suggestions for the Introduction offered by audiences at Austin College, the CUNY Graduate Center, the University of Kansas, New York University, the University of the Witwatersrand, the University of York, and particularly the attendees of the June 2014 Marxist Literary Group Institute on Culture and Society in Banff. Mike Schoenfeldt and Jane Johnson at the University of Michigan's Department of English Language and Literature provided travel support at a key moment. She is profoundly grateful to Patsy Yaeger for the invitation to join the endeavor that became *Fueling Culture* and to Imre Szeman for seeing it through. Work on the Introduction and the rest of the book would have become simply impossible without the circle of friends of Patsy—Michael Awkward, Carol Bardenstein, Marjorie Levinson, Brenda Marshall, Anita Norich, Yopie Prins, and Valerie Traub—as well as Daniel Braun,

Christi Merrill, Megan Sweeney, and Gillian White. As ever, but particularly during the most difficult periods of working on this project, Joey Slaughter has been a true companion, incisive first reader, and indispensable voice of reason.

Finally, this book is dedicated to the memory of Fordham's Editorial Director Helen Tartar and our co-editor Patricia Yaeger, both of whom died before the project was completed—Helen suddenly, the victim of an automobile accident in March 2014; Patsy in July 2014, after a brave struggle with ovarian cancer that lasted just over a year. Helen died doing what she loved—knitting, on her way from a tai chi retreat to the Fordham booth at an academic conference in Colorado. Patsy was brimming with interesting questions and big ideas until the end. They were each strikingly elegant, luminous women one would spot instantly across a crowded room even if they hadn't been so tall. Energy metaphors come readily to mind for both, and their premature loss has taught us how great ideas are grounded in, but ultimately not limited by, the infrastructure of the human body. Although she was not well enough to read a word of what we eventually received from our contributors, Patsy was the driving force behind our successive conceptualizations of *Fueling Culture*—its animating spirit. At a very deep level, Helen immediately understood our idea for the book when we three met with her at MLA in January 2013. This project is much the poorer for not having the benefit of their scintillating intelligence as it came together as a book, but it would not exist at all without them. We are also deeply grateful to Helen's husband, Bud Bynack, and Patsy's husband, Rich Miller, for their grace and generosity in helping to continue the work that was so important to these extraordinary women—bright lights both, gone too soon.

WORKS CITED

The 180 with Jim Brown. 2014. *CBC*. Radio One, Calgary. January 31.

Abramsky, Kolya. 2010. *Sparking a Worldwide Energy Revolution: Social Struggles in the Transition to a Post-Petrol World*. Oakland, CA: AK Press.

Achenbach, Joel. 2010. "The 21st Century Grid: Can We Fix the Infrastructure That Powers Our Lives?" *National Geographic Magazine*, July. http://ngm.nationalgeographic.com/print/2010/07/power-grid/achenbach-text.

Acland, Charles. 2006. *Residual Media*. Minneapolis: University of Minnesota Press.

"Activistas y Artistas Critican la Relacion de Centros Culturales Ingleses con la Petrolera BP." 2010. *terc3ra*. January 7. http://www.tercerainformacion.es/spip.php?article16502.

Adams, Carol J. 2010. *The Sexual Politics of Meat*. New York: Continuum.

Adams, John. 1853. *The Works of John Adams, Second President of the United States*. Edited by Charles Francis Adams. Vol. 8. Boston: Little Brown.

Adorno, Theodor. 1988. "Something's Missing: A Discussion between Ernst Bloch and Theodor W. Adorno on the Contradictions of Utopian Longing." In *The Utopian Function of Art and Literature: Selected Essays*, by Ernst Bloch, translated by Jack Zipes and Frank Mecklenburg, 1–17. Cambridge, MA: MIT Press.

Aeschylus. 1979. *The Oresteia: Agamemnon; the Libation Bearers; the Eumenides*. Edited by W. B. Stanford. Translated by Robert Fagles. New York: Penguin.

Agamben, Giorgio. 1999. *Potentialities*. Edited by Daniel Heller-Roazen. Translated by Daniel Heller-Roazen. Stanford, CA: Stanford University Press.

"The Age of Wire." 1884. *The Electrical World*, March 29.

Ahmed, Akbar Shahid. 2015. "Solar Panels Could Save Patients in Gaza's Hospitals, Thanks to a New Fundraising Campaign." *World Post*, May 21. http://www.huffingtonpost.com/2015/05/21/gaza-hospitals-solar-power_n_7338188.html.

Ahn, Daniel. 2011. "Improving Energy Market Regulation: Domestic and International Issues." *Council on Foreign Relations*. http://www.relooney.com/NS4053-Energy/CFR-Energy_1.pdf.

Alaimo, Stacy. 2010. *Bodily Natures: Science, Environment, and the Material Self*. Bloomington: Indiana University Press.

Alatout, Samer, and Chelsea Schelly. 2010. "Rural Electrification as a 'Bioterritorial' Technology: Redefining Space, Citizenship and Power during the New Deal." *Radical History Review* 107: 127–38.

Alberta Energy Regulator. 2013. "What Is Unconventional Oil and Gas?" https://www.aer.ca/about-aer/spotlight-on/unconventional-regulatory-framework/what-is-unconventional-oil-and-gas.

Alberta's Industrial Heartland Association. "Who We Are." n.d. http://industrialheartland.com/.

Alfred, Taiaiake. 2009. "Colonialism and State Dependency." *Journal of Aboriginal Health* 5 (2): 42–60.

American Psychiatric Association. 2000. *Diagnostic and Statistical Manual of Mental Disorders: DSM-IV-TR*. Washington, DC: American Psychiatric Association.

Amit, Vered. 2010. "Community as 'Good to Think With': The Productiveness of Strategic Ambiguities." *Anthropologica* 52 (2): 357–75.
Anand, Sudhir, and Amartya Sen. 2000. "Human Development and Economic Sustainability." *World Development* 28 (2): 2029–49.
Anders, Günther. 2008. *Hiroshima Est Partout*. Paris: Seuil.
Anderson, Ben. 2010. "Security and the Future: Anticipating the Event of Terror." *Geoforum* 41 (2): 227–35.
Anderson, Daniel Gustav. 2012a. "Accumulating-Capital, Accumulating-Carbon, and the Very Big Vulnerable Body: An Object of Responsibility for Ecocriticism." *Public Knowledge* 3 (2). http://pkjournal.org/?page_id=1698.
———. 2012b. "Critical Bioregionalist Method in Dune: A Position Paper." In *The Bioregional Imagination: Literature, Ecology, and Place*, edited by Tom Lynch, Cheryll Glotfelty, and Karla Armbruster, 226–44. Athens: University of Georgia Press.
———. 2012c. "Natura Naturans and the Organic Ecocritic: Toward a Green Theory of Temporality." *Journal of Ecocriticism* 4 (2): 34–47.
Anderson, Jon. 2004. "The Ties That Bind? Self- and Place-Identity in Environmental Direct Action." *Ethics, Place and Environment* 7 (1): 45–57.
Anderson, Perry. 1983. *In the Tracks of Historical Materialism*. London: Verso.
———. 2013. "Homeland." *New Left Review* 81: 5–32.
Appadurai, Arjun. 1986. *The Social Life of Things: Commodities in Cultural Perspective*. Cambridge: Cambridge University Press.
———. 1990. "Disjuncture and Difference in the Global Cultural Economy." *Theory, Culture & Society* 7 (295): 295–310.
Apple. 2012. "Apple Facilities: Environmental Footprint Report." https://www.apple.com/jp/environment/pdf/Apple_Facilities_Report_2013.pdf.
Appuhn, Karl. 2009. *A Forest on the Sea: Environmental Expertise in Renaissance Venice*. Baltimore, MD: Johns Hopkins University Press.
Apter, Andrew. 2005. *The Pan-African Nation: Oil and the Spectacle of Culture in Nigeria*. Chicago: University of Chicago Press.
Arab, Paula. 2011. "Language Tars Debate over Alberta Oilsands." *Calgary Herald*, April 21, A14.
Aradau, Claudia, and Rens van Munster. 2011. *Politics of Catastrophe: Genealogies of the Unknown*. London: Routledge.
Archer, W.J., ed. 1922. *Mexican Petroleum*. New York: Pan American Petroleum and Transport Company.
Arendt, Hannah. 1958. *The Human Condition*. Chicago: University of Chicago Press.
Aristotle. 1982. *Poetics*. Translated by James Hutton. New York: Norton.
———. 1984. *The Complete Works of Aristotle: The Revised Oxford Translation*. Edited by Jonathan Barnes. Volume 2. Princeton, NJ: Princeton University Press.
———. 2004. *De Anima (On the Soul)*. Translated by Hugh Lawson-Tancred. New York: Penguin.
Associated Press. 2013. "NM County Ordinance Bans Oil, Gas Development." *Associated Press*, May 1. http://fuelfix.com/blog/2013/05/01/nm-county-ordinance-bans-oil-gas-development/.
Atkinson, Lucy. 2013. "Smart Shoppers? Using QR Codes and 'Green' Smartphone Apps to Mobilize Sustainable Consumption in the Retail Environment." *International Journal of Consumer Studies* 37: 387–93.
"Atomic Energy Act of 1954, as Amended in NUREG-0980." 2013. *United States Nuclear Regulatory Commission*. October 21. http://pbadupws.nrc.gov/docs/ML1327/ML13274A489.pdf.
Attel, Kevin. 2009. "Potentiality, Actuality, Constituent Power." *Diacritics* 39 (3): 35–53.
Austen, Jane. 1995. [1816]. *Emma*. Oxford: Oxford University Press.
Auty, Richard. 1993. *Sustaining Development in Mineral Economies: The Resource Curse Thesis*. London: Routledge.
Avatar. 2009. Directed by James Cameron. Twentieth Century Fox.
Bachelard, Gaston. 1994. *The Poetics of Space: The Classic Look at How We Experience Intimate Places*. Translated by Maria Jolas. Boston: Beacon.

Bacigalupi, Paolo. 2009. *The Windup Girl*. New York: Night Shade Books.
Back to the Future. 1985. Directed by Robert Zemeckis. Universal Pictures.
Badiou, Alain. 2012. *The Rebirth of History: Times of Riots and Uprisings*. Translated by G. Elliott. London: Verso.
Bailey, Ian, Rob Hopkins, and Geoff Wilson. 2010. "Some Things Old, Some Things New: The Spatial Representations and Politics of Change of the Peak Oil Relocalisation Movement." *Geoforum* 41: 595–605.
Bainbridge, Jason. 2009. "Worlds within Worlds: The Role of Superheroes in the Marvel and DC Universes." In *The Contemporary Comic Book Superhero*, edited by Angela Ndalianis, 64–85. New York: Routledge.
Baker, Peter, and John Schwartz. 2013. "Obama Pushes Plan to Build Roads and Bridges." *New York Times*, March 30. http://www.nytimes.com/2013/03/30/us/politics/obama-promotes-ambitious-plan-to-overhaul-nations-infrastructure.html?_r=0.
Bakhtin, Mikhail. 1994. "Social heteroglossia." In *The Bakhtin Reader*, edited by Pam Morris, 73–80. London: Edward Arnold.
Balakrishnan, Gopal. 2009. "Speculations on the Stationary State." *New Left Review* 59: 5–26.
Baldwin, James Mark. 1960. "Innervation." In *Dictionary of Philosophy and Psychology*. Gloucester: Peter Smith. https://archive.org/details/dictionaryphilooobaldgoog.
Ballard, J. G. 1974. *Concrete Island*. New York: Farrar, Straus and Giroux.
Balousha, Hazem. 2014. "As Power Cuts Continue, Gaza Turns to Solar Energy." *Al-Monitor*, January 22. http://www.al-monitor.com/pulse/originals/2014/01/solar-energy-gaza-electricity-outages.html.
Barad, Karen. 2007. *Meeting the Universe Halfway: Quantum Physics and the Entanglement of Matter and Meaning*. Durham, NC: Duke University Press.
Barber, Daniel A. 2013. "The World Solar Energy Project, c. 1954." *Grey Room* 51: 64–93.
Barrera, Jorge. 2013. "PM Harper Believes Idle No More Movement Creating 'Negative Public Reaction,' Say Confidential Notes." *APTN National News*, January 25. http://aptn.ca/news/2013/01/25/pm-harper-believes-idle-no-more-movement-creating-negative-public-reaction-say-confidential-notes/.
Barrett, Ross, and Daniel Worden, eds. 2014. *Oil Culture*. Minneapolis: University of Minnesota Press.
Barry, Andrew. 2001. *Political Machines: Governing a Technological Society*. London: Athlone.
———. 2009. "Visible Invisibility." In *New Geographies 2: Landscapes of Energy*, edited by Rania Ghosn, 67–74. Cambridge, MA: Harvard University Press.
Barth, Lawrence. 1996. "Immemorial Visibilities: Seeing the City's Difference." *Environment and Planning A* 28: 471–93.
Barthez, Paul-Joseph. 1772. *Oratio Academica de Principio Vitali Hominis*. Montpellier, France: Rochard.
Bataille, Georges. 1991. *The Accursed Share: An Essay on General Economy*. Vol 1. *Consumption*. Translated by Robert Hurley. New York: Zone Books.
Bateson, Gregory. 1972. *Steps toward an Ecology of Mind*. New York: Ballantine.
Baudrillard, Jean. 1998. "When Bataille Attacked the Metaphysical Principle of Economy." In *Bataille: A Critical Reader*, edited by Fred Botting and Scott Wilson, 191–95. London: Blackwell.
Bauman, Zygmunt. 2001. *Community: Seeking Safety in an Insecure World*. Cambridge: Polity.
"Beijing Slashes Car Sales Quota in Anti-pollution Drive." 2013. *Reuters*. November 5. http://www.reuters.com/article/2013/11/05/us-china-cars-idUSBRE9A40AP20131105.
Benjamin, Walter. 1977. *Illuminations*. Edited by Hannah Arendt. Glasgow: Fontanta/William Collins Sons.
———. 1997. "One-Way Street." In *Selected Writings*. Vol. 1. Translated by Edmund Jephcott, 444–88. Cambridge, MA: Harvard University Press.
———. 2001. "Surrealism." In *Selected Writings*. Vol. 2. Translated by Edmund Jephcott, 207–221. Cambridge, MA: Harvard University Press.

———. 2002. *The Arcades Project*. Translated by Rolf Tiedemann, Howard Eiland, and Kevin McLaughlin. New York: Belknap.
———. 2003. "On Some Motifs in Baudelaire." In *Selected Writings*. Vol. 4. Translated by Harry Zohn, 313–55. Cambridge, MA: Harvard University Press.
Bennett, Dean. 2013a. "Alberta Buys *New York Times* Ad to Make Its Case for Keystone XL Pipeline." *Edmonton Journal*, March 18.
———. 2013b. "Two Senior Alberta Politicians Euro-Bound to Head Off Anti-Oil Sands Resolution." *Financial Post*, September 25. http://business.financialpost.com/news/energy/two-senior-alberta-politicians-euro-bound-to-head-off-anti-oil-sands-resolution.
Bennett, Jane. 2010. *Vibrant Matter: A Political Ecology of Things*. Durham, NC: Duke University Press.
Berger, John. 1991. "Why Look at Animals?" In *About Looking*, 3–30. Vintage International Edition. New York: Vintage-Random.
Bergson, Henri. 1992. "Creative Evolution." In *Art in Theory: 1900–1990*, edited by Charles Harrison and Paul Wood, 140–43. Oxford: Blackwell.
Berland, Jody. 2009. *North of Empire: Essays on the Cultural Technologies of Space*. Durham, NC: Duke University Press.
Berlant, Lauren. 2006. "Cruel Optimism." *differences: A Journal of Feminist Cultural Studies* 17 (3): 20–36.
———. 2007. "Slow death (sovereignty, obesity, lateral agency)." *Critical Inquiry* 33 (4): 754–80.
———. 2011. *Cruel Optimism*. Durham, NC: Duke University Press.
The Big Fix. 2012. Directed by Joshua Tickell and Rebecca Harrell Tickell. Big Picture Ranch.
Bird, Louis. 2007. *The Spirit Lives in the Mind: Omushkego Stories, Lives and Dreams*. Edited by Susan Elaine Gray. Montreal: McGill-Queens University Press.
"The Birth of Bourbon." n.d. *Heavenly Hills Distillery Bourbon Heritage Center*. http://www.bourbonheritagecenter.com/history/bourbon-pioneers/.
Black, Brian. 2012. *Crude Reality: Petroleum in World History*. Lanham, MD: Rowman & Littlefield.
Blais, Jacqueline. 2007. "Vonnegut 'Still Had Hope in His Heart.'" *USA Today*, April 13. http://usatoday30.usatoday.com/life/people/2007-04-12-kurt-vonnegut-appreciation_N.htm.
Blake, William. 1893. *The Poems of William Blake*. London: Lawrence & Bullen.
Blanchot, Maurice. 1986. *The Writing of the Disaster*. Translated by Ann Smock. Lincoln: University of Nebraska Press.
Bloch, Ernst. *The Principle of Hope*. Cambridge, MA: MIT Press, 1986.
Blondin, George. 2006. *Trail of the Spirit: The Mysteries of Medicine Power Revealed*. Edmonton: NeWest Press.
Blum, Andrew. 2012. *Tubes: A Journey to the Center of the Internet*. New York: Ecco.
Blumenbach, Johann Friedrich. (1781) 1971. *Über den Bildungstrieb und das Zeugungsgeschäfte*. Göttingen, Germany: Dieterich. Reprint, Stuttgart, Germany: Fischer.
———. (1787) 1971. *De niso formativo et generationis negotio nuperae: observationes*. Göttingen, Germany: Dieterich. Reprint, Stuttgart, Germany: Fischer.
Boetzkes, Amanda, and Andrew Pendakis. 2013. "Visions of Eternity: Plastic and the Ontology of Oil." *e-flux journal* 47. http://www.e-flux.com/journal/visions-of-eternity-plastic-and-the-ontology-of-oil/.
Böhm, Steffen, Campbell Jones, Chris Land, and Matthew Paterson. 2006. "Introduction: Impossibilities of Automobility." In *Against Automobility*, edited by Steffen Böhm, Campbell Jones, Chris Land, and Matthew Paterson, 3–16. New York: Wiley.
Bond, Patrick, and Trevor Ngwane. 2010. "Community Resistance to Energy Privatisation in South Africa." In *Sparking a Worldwide Energy Revolution: Social Struggles in the Transition to a Post-Petrol World*, edited by Kolya Abromsky, 197–207. Oakland, CA: AK Press.
Bookchin, Murray. 2004. *Post-Scarcity Anarchism*. Oakland, CA: AK Press.
Borasi, Giovanna, and Mirko Zardini, eds. 2007. *Sorry, Out of Gas: Architecture's Response to the 1973 Oil Crisis*. Montreal: Canadian Center for Architecture.

Bougen, Philip, and Pat O'Malley. 2008. "Bureaucracy, Imagination and US Domestic Security Policy." *Security Journal* 22 (2): 101–18.
Bougrine, Hassan. 2006. "Oil: Profits of the Chain Keepers." *International Journal of Political Economy* 35 (2): 35–53.
Boyer, Dominic. 2011. "Energopolitics and the Anthropology of Energy." *Anthropology Newsletter* (May): 5–7.
———. 2013. *The Life Informatic: Newsmaking in the Digital Era*. Ithaca, NY: Cornell University Press.
———. 2014. "Energopower: An Introduction." *Anthropological Quarterly* 8 (2): 309–34.
Boyer, Dominic, and Imre Szeman. 2014. "The Rise of Energy Humanities: Breaking the Impasse." *University Affairs*, February 12. http://www.universityaffairs.ca/opinion/in-my-opinion/the-rise-of-energy-humanities/.
Boyle, Godfrey, and Peter Harper. 1976. *Radical Technology: Food, Shelter, Tools, Materials, Energy*. New York: Pantheon.
Bozak, Nadia. 2012. *The Cinematic Footprints: Lights, Camera, Natural Resources*. New Brunswick, NJ: Rutgers University Press.
Bradsher, Keith. 2013. "Suntech Unit Declares Bankruptcy." *New York Times*, March 20. http://www.nytimes.com/2013/03/21/business/energy-environment/suntech-declares-bankruptcy-china-says.html.
Braidotti, Rosi. 2013. *The Posthuman*. Cambridge: Polity.
Brand, Russell. 2013. "Russell Brand on Revolution: 'We No Longer Have the Luxury of Tradition.'" *New Statesman*, October 24. http://www.newstatesman.com/politics/2013/10/russell-brand-on-revolution.
Brannen, Peter. 2013. "Headstone for an Apocalypse." *New York Times*, August 16. http://www.nytimes.com/2013/08/17/opinion/headstone-for-an-apocalypse.html.
Breaking the Waves. 1996. Directed by Lars von Trier. Twentieth Century Fox.
Brecht, Bertolt. 1977. *Brecht on Theatre*. Edited and translated by John Willett. New York: Hill and Want.
Brennan, Shane. 2016. "Making Data Sustainable: Backup Culture and Risk Perception." In *Sustainable Media: Critical Approaches to Media and Environment*, edited by Nicole Starosielski and Janet Walker, 56–76. New York: Routledge.
Brennan, Teresa. 1998. "Why the Time Is Out of Joint: Marx's Political Economy without the Subject." *South Atlantic Quarterly* 97 (2): 263–80.
———. 2004. *The Transmission of Affect*. Ithaca, NY: Cornell University Press.
Brontë, Emily. 1847. *Wuthering Heights*. London: Thomas Cautley Newby.
Brooks, Max. 2006. *World War Z*. New York: Three Rivers Press.
Browne, Mark Anthony, Phillip Crump, Stewart J. Nivens, Emma Teuten, Andrew Tonkin, Tamara Galloway, and Richard Thompson. 2011. "Accumulation of Microplastic on Shorelines Worldwide: Sources and Sinks." *Environmental Science and Technology* 45 (21): 9175–79.
Bruggers, James. 2014. "Developers Suspend Investment in Bluegrass Pipeline." *Courier-Journal*, April 28. http://www.courier-journal.com/story/watchdog-earth/2014/04/28/bluegrass-pipeline-suspension/8402379/.
Brundtland, Gro Harlem, and the World Commission on Environment and Development. 1987. *Our Common Future: Report of the World Commission on Environment and Development*. Oxford: Oxford University Press.
Buell, Frederick. 2003. *From Apocalypse to Way of Life: Environmental Crisis in the American Century*. New York: Routledge.
———. 2012. "A Short History of Oil Cultures: Or, the Marriage of Catastrophe and Exuberance." *Journal of American Studies* 46 (2): 273–93.
———. 2013. "Post-Apocalypse: A New U.S. Cultural Dominant." *Frame* 26 (1): 9–24.
Buell, Lawrence. 1995. *Environmental Imagination: Thoreau, Nature Writing, and the Formation of American Culture*. Cambridge, MA: Harvard University Press.

———. 2005. *The Future of Environmental Criticism*. Oxford: Blackwell.
———. 2009. *Writing for an Endangered World: Literature, Culture, and Environment in the U.S. and Beyond*. Cambridge, MA: Harvard University Press.
Burke, Edmund. 2009. "The Big Story: Human History, Energy Regimes, and the Environment." In *The Environment and World History*, edited by Edmund Burke III and Kenneth Pomeranz, 33–53. Berkeley: University of California Press.
Burnett, D. Graham. 2007. *Trying Leviathan: The Nineteenth-Century New York Court Case That Put the Whale on Trial and Challenged the Order of Nature*. Princeton, NJ: Princeton University Press.
———. 2013. *The Sounding of the Whale: Science and Cetaceans in the Twentieth Century*. Chicago: University of Chicago Press.
Burroughs, Edgar Rice. 2010. *Tarzan of the Apes*. Oxford: Oxford University Press.
Burt, Jonathan. 2001. "The Illumination of the Animal Kingdom: The Role of Light and Electricity in Animal Representation." *Society and Animals* 9 (3): 213–28.
Burtynsky, Edward. 2007. *Alberta Oil Sands #6*. Photograph. Nicholas Metivier Gallery, Toronto, and Paul Kuhn Gallery, Calgary.
Bush, George W. 2003. "Statement by the President." *The White House: President George W. Bush*. The White House. February 12. https://georgewbush-whitehouse.archives.gov/news/releases/2003/02/20030212.html.
———. 2004. "Fourth State of the Union Address." *The American Presidency Project*, January 20. http://www.presidency.ucsb.edu/ws/index.php?pid=29646.
Butler, Judith. 1990. *Gender Trouble*. New York: Routledge.
———. 1997. *The Psychic Life of Power: Theories in Subjection*. Stanford, CA: Stanford University Press.
Bütschli, Otto. 1901. *Mechanismus und Vitalismus*. Leipzig, Germany: Engelmann.
Campelo, Adriana, Robert Aitken, and Juergen Gnoth. "Visual Rhetoric and Ethics in Marketing of Destinations." *Journal of Travel Research* 50 (1): 3–14.
Campion-Smith, Bruce. 2013. "Idle No More: Spence Urged by Fellow Chiefs to Abandon Her Fast." *Toronto Star*, January 18. https://www.thestar.com/news/canada/2013/01/18/idle_no_more_spence_urged_by_fellow_chiefs_to_abandon_her_fast.html.
Canadian Association of Petroleum Producers. 2010a. "Alberta Is Energy." *Canadian Association of Petroleum Producers*, April 7. http://www.capp.ca/media/news-releases/alberta-is-energy.
———. 2010b. "Canada's Oil Sands—Come See for Yourself." February 13. http://www.capp.ca/media/commentary/environmental-performance-and-communication-equals-earned-reputation-for-canadas-oil-gas-industry.
Canadian Press. 2012. "Western Premiers to Talk Environment, Energy and Tom Mulcair." *CBC News*, May 27. http://www.cbc.ca/news/politics/western-premiers-to-talk-environment-energy-and-tom-mulcair-1.1297995.
———. 2013. "Oil and Gas Ad Campaign Cost Feds $40M at Home and Abroad." *CBC News*, November 27. http://www.cbc.ca/news/politics/oil-and-gas-ad-campaign-cost-feds-40m-at-home-and-abroad-1.2442844.
Canguilhem, Georges. 2008. *Knowledge of Life: (Forms of Living)*. Translated by Stefanos Geroulanos and Daniela Ginsburg. New York: Fordham University Press.
Caplan-Bricker, Nora. 2013. "This Is What Happens When a Pipeline Bursts in Your Town." *New Republic*, November 18. https://newrepublic.com/article/115624/exxon-oil-spill-arkansas-2013-how-pipeline-burst-mayflower.
Cariou, Warren. 2012. "Tarhands: A Messy Manifesto." *Imaginations* 3 (2): 17–34.
Carlson, Adam. 2013. "Petrorealism on The Oil Road." Introduction to Mika Minio-Paluello's unpublished talk "Revolution, Oil, Climate Change: Fossil Fuels in North Africa and the Arab Uprisings." Humanities Centre, University of Alberta, October 18.
Carlton, Larry. 1986. "The BP Blues." *Last Nite*. MCA Records.
Carlyle, Ryan. 2013. "Taking Pictures on an Offshore Oil Rig Is Serious Business." *PetaPixel*, June 11. http://petapixel.com/2013/06/11/taking-pictures-on-an-offshore-oil-rig-is-serious-business/.

Carlyle, Thomas. 1889. *Works*. 30 vols. London: Chapman and Hall.
Carey, James. 1983. "Technology and Ideology: The Case of the Telegraph." *Prospects* 8: 303–25.
Carson, Rachel. 1962. *Silent Spring*. New York: Houghton Mifflin.
Carus, Carl Gustav. 1838–1840. *System der Physiologie umfassend das Allgemeine der physiologischen Geschichte der Menschheit, die des Menschen und die der einzelnen organischen Systeme im Menschen, für Naturforscher und Aerzte*. 2 vols. Dresden and Leipzig: Gerhard Fleischer.
Castañeda, Jorge G. 2006. "Latin America's Left Turn." *Foreign Affairs* 85 (3): 28–43.
CBC News. 2013. "Colleges, Universities Get Mandate Letters from Province." *CBC News*, March 13. http://www.cbc.ca/news/canada/edmonton/colleges-universities-get-mandate-letters-from-province-1.1314033.
———. 2008. "Low Voter Turnout in Alberta Election Being Questioned." *CBC News*, March 5. http://www.cbc.ca/news/canada/edmonton/low-voter-turnout-in-alberta-election-being-questioned-1.761174.
Chakrabarty, Dipesh. 2009. "The Climate of History: Four Theses." *Critical Inquiry* 35: 197–222.
———. 2012. "Postcolonial Studies and the Challenge of Climate Change." *New Literary History* 43 (1): 25–42.
———. 2014. "Climate and Capital: On Conjoined Histories." *Critical Inquiry* 41: 1–23.
Chari, Sharad. 2013. "Detritus in Durban: Polluted Environs and the Biopolitics of Refusal." In *Imperial Debris: On Ruins and Ruination*, edited by Ann Laura Stole, 131–61. Durham, NC: Duke University Press.
———. 2015. "African Extraction, Indian Ocean Critique." *South Atlantic Quarterly* 114 (1): 83–100.
Chazan, Guy. 2011. "U.K. Gets Big Shale Find: Discovery by Cuadrilla Resources May Rival Vast, New U.S. Gas Properties." *Wall Street Journal*, September 22. http://www.wsj.com/articles/SB10001424053111904563904576584904139100880.
Cheadle, Bruce. 2013. "Canada's Natural Resources Ads Light on Facts, Heavy on Patriotism." *Financial Post*, February 13. http://business.financialpost.com/news/energy/canadas-natural-resources-ad-campaign-light-on-facts-heavy-on-patriotism.
Chen, Mel. 2012. *Animacies: Biopolitics, Racial Mattering, and Queer Affect*. Durham, NC: Duke University Press.
"China's Soviet-style Suburbia Heralds Environmental Pain." 2013. *Bloomberg News*, November 7. http://www.bloomberg.com/news/2013-11-06/china-s-soviet-style-suburbia-heralds-environmental-pain.html.
The China Syndrome. 1979. Directed by James Bridges. Columbia Pictures.
Clark, Nigel. 2011. *Inhuman Nature: Sociable Life on a Dynamic Planet*. Los Angeles: Sage.
Clarke, Lee. 1990. "Oil-Spill Fantasies." *Atlantic Monthly*, November, 65–77.
Clarke, Tony. 2008. *Tar Sands Showdown: Canada and the Politics of Oil in an Age of Climate Change*. Toronto: Lorimer.
Clausewitz, Carl von. *On War*. Trans. Michael Howard and Peter Paret. Princeton: Princeton University Press, 1976.
Clements, Marie. 2003. *Burning Vision*. Vancouver: Talonbooks.
———. 2005. *The Unnatural and Accidental Women*. Vancouver: Talonbooks.
Cloke, Paul. 2006. "Conceptualizing Rurality." In *Handbook of Rural Studies*, edited by Terry Marsden, Patrick H. Mooney, and Paul Cloke, 18–28. London: Sage.
Coffman, Keither. 2013. "Colorado an Energy Battleground as Towns Ban Fracking." *Reuters*, November 6. http://www.reuters.com/article/us-usa-fracking-colorado-idUSBRE9A50QT20131106.
Colborn, Theo, Dianne Dumanoski, and John Peterson Myers. 1996. *Our Stolen Future: Are We Threatening Our Fertility, Intelligence, and Survival? A Scientific Detective Story*. New York: Dutton.
Cole, David R. 2013. *Traffic Jams: Analysing Everyday Life through the Immanent Materialism of Deleuze & Guattari*. Brooklyn: Punctum Books.
Cole, Raymond J., and Paul C. Kernan. 1996. "Life-Cycle Energy Use in Office Buildings." *Building and Environment* 31 (4): 307–17.

Coll, Steve. 2012. *Private Empire: ExxonMobil and American Power*. New York: Penguin.
Collier, Paul. 2007. *The Bottom Billion*. Oxford: Oxford University Press.
Collier, Stephen J., and Andrew Lakoff. 2008. "The Vulnerability of Vital Systems: How 'Critical Infrastructure' Became a Security Problem." In *Securing "the Homeland": Critical Infrastructure, Risk and (In)Security*, edited by Myriam Dunn and Kristian Soby Kristensen, 17–39. New York: Routledge.
Collins, Gail. 2013. *As Texas Goes . . . : How the Lone Star State Hijacked the American Agenda*. New York: Liveright.
Colomar-Garcia, Marta, and Xingjian Zhao. 2011. "Tapping Latin America's Lithium." *Latin Business Chronicle*, April 13. http://latintrade.com/tapping-latin-americas-lithium/.
Connery, Christopher. 2010. "Sea Power." *PMLA* 125 (3): 685–92.
Connolly, Kate. 2015. "G7 Leaders Agree to Phase Out Fossil Fuel Use by End of Century." *Guardian*, June 8. http://www.theguardian.com/world/2015/jun/08/g7-leaders-agree-phase-out-fossil-fuel-use-end-of-century.
Connor, Steven. 2002. "Michel Serres's Milieux." Conference Paper. *Brazilian Association for Comparative Literature*, July 23–26. http://www.stevenconnor.com/milieux/.
———. 2009. "Michael Serres: The Hard and the Soft." Conference Paper. *Modern Studies*, November 26. http://stevenconnor.com/hardsoft/hardsoft.pdf.
Contagion. 2011. Directed by Steven Soderbergh. Warner Bros.
Cooper, Melinda. 2010. "Turbulent Worlds: Financial Markets and Environmental Crisis." *Theory, Culture & Society* 27 (2–3): 167–90.
Coopersmith, Jennifer. 2010. *Energy, the Subtle Concept: The Discovery of Feynman's Blocks from Leibniz to Einstein*. Oxford: Oxford University Press.
Corcoran, Patricia, Charles Moore, and Kelly Jazvac. 2014. "An Anthropogenic Marker Horizon in the Future Rock Record." *GSA Today*, June. http://www.geosociety.org/gsatoday/archive/24/6/article/i1052-5173-24-6-4.htm.
"Coroner's Inquest." 1803. *Times*, September 29. Issue 5827, 2.
Coronil, Fernando. 1997. *The Magical State: Nature, Money, and Modernity in Venezuela*. Chicago: University of Chicago Press.
———. 2012. "The Future in Question: History and Utopia in Latin America (1989–2010)." In *Business as Usual: The Roots of the Global Financial Meltdown*, edited by Craig Calhoun and Georgi Derluguian, 231–64. New York: New York University Press.
Credit Suisse. 2012. "Global Wealth Report 2012." October. https://publications.credit-suisse.com/tasks/render/file/index.cfm?fileid=88EE6EC8-83E8-EB92-9D5F39D5F5CD01F4.
Costa, James T. 2009. Introduction to *Charles Darwin, The Annotated Origin: A Facsimile of the First Edition of On the Origin of Species*, ix–xx. Cambridge, MA: Harvard University Press.
Creed, Gerald W. 2006. "Reconsidering Community." In *The Seductions of Community: Emancipations, Oppressions, Quandaries*, edited by Gerald W. Creed, 23–48. Santa Fe, NM: School of American Research Press.
Crutzen, Paul J. 2002. "Geology of Mankind." *Nature* 3: 23.
Crutzen, Paul J., and Eugene F. Stoermer. 2000. "The 'Anthropocene.'" *IBGP Newsletter*, 17–18.
Cubbit, Sean. 2005. *Eco Media*. Amsterdam: Rodopi B.V.
———. 2013. "Integral Waste." Conference Paper. *Transmediale*, February 1.
Curley, Bob. 2010. "DSM-V Draft Includes Major Changes to Addictive Disease Classifications." *DrugFree.org*, February 12. http://www.drugfree.org/join-together/dsm-v-draft-includes-major-changes-to-addictive-disease-classifications/.
Dahlgren, Peter. 2009. *Media and Political Engagement: Citizens, Communication, and Democracy*. New York: Cambridge University Press.
Darmstadter, Joel, and Hans H. Landsberg. 1976. "The Economic Background." In *The Oil Crisis*, edited by Raymond Vernon, 15–38. New York: W. W. Norton.
Darwin, Charles. 2009. *The Annotated Origin: A Facsimile of the First Edition of* On the Origin of Species. [1859.] Cambridge, MA: Harvard University Press.

Davidson, Debra J., and Mike Gismondi. 2011. *Challenging Legitimacy at the Precipice of Energy Calamity*. New York: Springer.
Davis, Lance E., Robert E. Gallman, and Karin Gleiter. 1997. *In Pursuit of Leviathan: Technology, Institutions, Productivity, and Profits in American Whaling, 1816–1906*. Chicago: University of Chicago Press.
Davis, Mike. 2010. "Who Will Build the Ark?" *New Left Review* 61: 29–46.
Debeir, Jean-Claude, Jean-Paul Deléage, and Daniel Hémery. 1991. *In the Servitude of Power: Energy and Civilization through the Ages*. Translated by John Barzman. London: Zed Books.
de Botton, Alain. 2009. *The Pleasures and Sorrows of Work*. London: Vintage Books.
Deckard, Sharae. 2012. "Editorial: Reading the World-Ecology." Special Issue on Global and Postcolonial Ecologies. *Green Letters: Studies in Ecocriticism* 16 (1): 5–14.
Defoe, Daniel. 1985. *Robinson Crusoe*, edited by Angus Ross. London: Penguin.
DeLanda, Manuel. 1995. "Uniformity and Variability: An Essay in the Philosophy of Matter." Conference Paper. *Doors of Perception 3: On Matter*, November 7–11.
Deleuze, Gilles. 1986. *Foucault*, translated by S. Hand. Minneapolis: University of Minnesota Press.
———. 1989. *Cinema 2: The Time-Image*. London: Continuum.
Deleuze, Gilles, and Felix Guattari. 1983. *Anti-Oedipus: Capitalism and Schizophrenia*. Translated by Mark Seem, Helen R. Lane, and Robert Hurley. Minneapolis: University of Minnesota Press.
———. 1987. "The Geology of Morals." *A Thousand Plateaus: Capitalism and Schizophrenia*, translated by Brian Massumi, 48–50. Minneapolis: University of Minnesota Press.
———. 1988. *A Thousand Plateaus: Capitalism and Schizophrenia*. Translated by Brian Massumi. London: Athlon Press.
Deng, Chao. 2013. "Newest Pollution Concern: 'Ugly' Sperm." *Wall Street Journal/China*, November 13. http://blogs.wsj.com/chinarealtime/2013/11/07/chinas-newest-pollution-concern-ugly-sperm/.
Dening, Greg. 2003. Afterword to *Islands in History and Representation*, edited by Rod Edmond and Vanessa Smith, 203–6. London: Routledge.
Denning, Michael. 2010. "Wageless Life." *New Left Review* 66: 79–97.
Derrida, Jacques. 1991. "'Eating Well,' or the Calculation of the Subject: An Interview with Jacques Derrida." In *Who Comes after the Subject?*, edited by Eduardo Cadava, Peter Connor, and Jean-Luc Nancy, 96–119. New York: Routledge.
DESERTEC Foundation. 2012. "Global Mission of the DESERTEC Foundation." *DESERTEC Knowledge Platform*. http://knowledge.desertec.org/wiki/index.php5/Global_mission_of_the_DESERTEC_Foundation.
Desrosières, Alain. 1998. *The Politics of Large Numbers: A History of Statistical Reasoning*. Translated by Camille Naish. Cambridge, MA: Harvard University Press.
Devereux, Cecily. 2014. "Surface Tensions: Looking through Petroculture's Images of Women." Conference Paper. *Institute on Culture and Society*, June 13.
Devon Canada Corporation. 2004. *Devon Beaufort Sea Exploration Drilling Program*. Calgary, Alberta: National Energy Board.
Dewan, Shaila. 2008. "At Plant in Coal Ash Spill, Toxic Deposits By the Ton." *New York Times*, December 29. http://www.nytimes.com/2008/12/30/us/30sludge.html.
Di Leo, Jeffrey R., ed. 2013. Critical Climate [special issue]. *Symploke* 21 (1–2).
Diamanti, Jeff. 2014. "Thoughts on Marxism and Energy for MLG-ICS 2014." *The Analogous City*, June 15. http://www.analogouscity.com/2014/06/thoughts-on-marxism-and-energy-for-mlg.html.
Diamond, Cora. 1988. "Losing Your Concepts." *Ethics* 98 (2): 255–77.
Diamond, Jared. 1987. "The Worst Mistake in the History of the Human Race: Agriculture." *Discover Magazine*, May. http://discovermagazine.com/1987/may/02-the-worst-mistake-in-the-history-of-the-human-race.
———. 2005. *Collapse: How Societies Choose to Fail or Succeed*. New York: Viking.
Dickens, Charles. 1853. *Bleak House*. London: Bradbury & Evans.

Dickinson, Adam. 2013. *The Polymers*. Toronto: House of Anansi Press.
Dilger, Daniel Eran. 2013. "Apple's Reno iCloud Data Center Taps AT&T's Latest DWDM Tech." *Apple Insider*, April 13. http://appleinsider.com/articles/13/04/03/apples-reno-data-center-to-serve-icloud-users-at-light-speed-with-fiber-optic-tech.
Dillard, Jesse F., Veronica Dujon, and Mary C. King. 2009. *Understanding the Social Dimension of Sustainability*. Vol. 17. London: Taylor & Francis.
Dirty Energy. 2012. Directed by Bryan D. Hopkins. Flood Films and Media.
Dirty Oil. 2009. Directed by Leslie Iwerks. Leslie Iwerks Productions.
Dittmer, Jason. 2010. *Popular Culture, Geopolitics and Identity*. Lanham, MD: Rowman & Littlefield.
Dodd, Susan. 2012. *The Ocean Ranger: Remaking the Promise of Oil*. Halifax: Fernwood.
Dolin, Eric Jay. 2007. *Leviathan*. New York: Norton.
Doris, Stacy, and Lisa Robertson. 2012. "From 'The Perfume Recordist.'" In *I'll Drown My Book: Conceptual Writing by Women*, edited by Caroline Bergvall, Laynie Browne, Teresa Carmody, and Venessa Place, 235–47. Los Angeles: Les Figues Press.
Dorow, Sara. 2015. "Gendering Energy Extraction in Fort McMurray." In *Alberta Oil and the Decline of Democracy in Canada*, edited by Meenal Shrivastava and Lorna Stefanick, 275–92. Athabasca, Alberta: Athabasca University Press.
Dorow, Sara, and Sara O'Shaughnessy. 2013. "Fort McMurray, Wood Buffalo, and the Oil/Tar Sands: Revisiting the Sociology of 'Community.'" *Canadian Journal of Sociology* 38 (2): 121–40.
Dorsey, Kurkpatrick. 2013. *Whales Nations: Environmental Diplomacy on the High Seas*. Seattle: University of Washington Press.
Douglas, Ian, David Goode, Mike Houck, and Rusong Wang, eds. 2011. *The Routledge Handbook of Urban Ecology*. London: Routledge.
Doyle, Arthur Conan. 2008. *The Lost World*. [1912.] Oxford: Oxford University Press.
Dr. Strangelove or: How I Learned to Stop Worrying and Love the Bomb. 1964. Directed by Stanley Kubrick. Columbia Pictures.
Dubkin, Leonard. 1947. *Enchanted Streets*. Boston: Little Brown.
Duchamp, Marcel. 1973. "Apropos of Readymades." In *The Essential Writings of Marcel Duchamp*, edited by Michel Sanouillet and Elmer Peterson, 141–42. London: Thames and Hudson.
Duffy, Enda. 2009. *The Speed Handbook: Velocity, Pleasure, Modernism*. Durham, NC: Duke University Press.
Duggan, Jennifer. 2013. "China Hit by Another Airpocalypse as Air Pollution Cancer Link Confirmed." *Guardian*, October 24. http://www.theguardian.com/environment/chinas-choice/2013/oct/24/china-airpocalypse-harbin-air-pollution-cancer.
Durham Peters, John. 2001. *Speaking into the Air: A History of the Idea of Communication*. Chicago: University of Chicago Press.
Dworkin, Craig. 2005. "Fact." *Chain* 12: 73.
Eastenders. 1985–. Created by Tony Holland and Julia Smith. British Broadcasting Company.
Edmond, Rod, and Vanessa Smith. 2003. "Introduction." In *Islands in History and Representation*, edited by Rod Edmond and Vanessa Smith, 1–18. London: Routledge.
Effendi, Rena. 2010. *Pipe Dreams*. Amsterdam: Schilt Publishing.
Ehrenburg, Ilya. 1929. *The Life of the Automobile*. Translated by Joachim Neugroschel. London: Serpent's Tail.
Eisenhower, Dwight D. 1953. "Address Before the General Assembly of the United Nations on Peaceful Uses of Atomic Energy, New York City." *The American Presidency Project*, December 8. http://www.presidency.ucsb.edu/ws/?pid=9774.
Elections Alberta. 2012. *Candidate Summary of Results (General Elections 1905–2012)*. December 21. http://www.elections.ab.ca/reports/statistics/candidate-summary-of-results-general-elections/.
"*The Electric Company* Theme Lyrics." Lyrics on Demand. http://www.lyricsondemand.com/tvthemes/theelectriccompanylyrics.html.
Eliot, T.S. 1975 [1923]. "*Ulysses*, Order, and Myth." In *Selected Prose of T. S. Eliot*, edited by Frank Kermode, 175–78. New York: Harvest Books.

Elizabeth I. 2000. *Collected Works*. Edited by Leah Marcus, Janel Mueller, and Mary Beth Rose. Chicago: University of Chicago Press.
Ellis, Blake. 2013. "North Dakota Sees Surge in Homeless Population." *CNNMoney*, December 17. http://money.cnn.com/2013/12/17/pf/north-dakota-homeless/index.html?iid=EL.
———. 2014. "How North Dakota's Economy Doubled in 11 Years." *CNNMoney*, July 14. http://money.cnn.com/2014/06/11/news/economy/north-dakota-economy/index.html?iid=SF_PF_Lead.
Ellis, Erle C. 2011. "Anthropogenic Transformation of the Terrestrial Biosphere." *Philosophical Transactions of the Royal Society A* 369: 1010–35.
Ellul, Jacques. 1964. *The Technological Society*. Translated by John Wilkinson. New York: Knopf.
Elsheshtawy, Yasser. 2012. "The Production of Culture: Abu Dhabi's Urban Strategies." In *Cities, Cultural Policy and Governance*, edited by H. Anheier and Y. R. Isar, 133–44. London: Sage.
Emmott, Stephen. 2013. *Ten Billion*. Harmondsworth, England: Penguin. Kindle edition.
Endnotes. 2010. "Misery and Debt: On the Logic and History of Surplus Populations and Surplus Capital." *Endnotes* 2. https://endnotes.org.uk/en/endnotes-misery-and-debt.
"Energiebranche: RWE-Chef sieht niedrigere Gewinne als 'neue Normalität.'" 2014. *Der Spiegel*, April 16. http://www.spiegel.de/wirtschaft/unternehmen/rwe-chef-terium-stimmt-aktionaere-auf-karge-zeiten-ein-a-964856.html.
Engineers Joint Council. 1945. "Industrial Disarmament of Aggressor States (Germany): Report of the National Engineers Committee of Engineers Joint Council." *National Engineers Committee*, September. New York: The Council. https://babel.hathitrust.org/cgi/pt?id=mdp.39015080043949;view=1up;seq=3.
———. 1946. "Report of Power Task Committee—Japan." *National Engineers Committee*, May 13. National Archives, College Park.
———. n.d. "Report on the Industrial Disarmament of Japan: Submitted to the Secretaries of State, War, & Navy." *National Engineers Committee*. National Archives, College Park.
Epstein, Mitch. 2009. *American Power*. Göttingen, Germany: Steidl.
Eriksen, Lars. 2011. "Bjarke Ingels Designs Incinerator that Doubles as Ski Slope in Copenhagen." *Guardian*, July 3. http://www.theguardian.com/environment/2011/jul/03/bjarke-ingels-incinerator-ski-slope.
Errázuriz, Tomás. 2010. "El asalto de los motorizados. El transporte moderno y la crisis del tránsito públic en Santiago, 1900–1927." *Historia* 43: 357–411.
Evans, Brad and Julian Reid. 2014. *Resilient Life: The Art of Living Dangerously*. Cambridge: Polity.
Facebook. 2014. "Facebook Prineville Data Center Building 2 & Cold Storage." https://www.facebook.com/PrinevilleDataCenter.
Facebook Luleå. 2014. "Starting Work on Luleå Building 2." *Facebook*, March 7. https://www.facebook.com/notes/luleå-data-center/starting-work-on-luleå-building-2/587772454624780/.
Farley, Edward. 1990. *Good Evil: Interpreting a Human Condition*. Minneapolis, MN: Fortress.
Featherstone, Mike. 2005. "Introduction." In *Automobilities*, edited by Mike Featherstone, Nigel Thrift, and John Urry, 1–24. London: Sage.
Fehrenbacher, Katie. 2013. "Apple Makes Progress on Its Solar-Powered Data Center in Reno, But (of Course) It's Controversial." *Gigaom*, December 17. https://gigaom.com/2013/12/17/apple-makes-progress-on-its-solar-powered-data-center-in-reno-but-of-course-its-controversial/.
Ferber, Edna. 2003. [1952]. *Giant*. New York: Harper Perennial.
Fernandes, Sujatha. 2011. "Urbanizing the San Juan Fiesta: Civil Society and Cultural Identity in the Barrios of Caracas." In *Ethnographies of Neoliberalism*, edited by C. Greenhouse, 96–111. Philadelphia: University of Pennsylvania Press.
Ferris, Joshua. 2010. *The Unnamed*. New York: Little, Brown.
Ferry, Luc. 1995. *The New Ecological Order*. Translated by Carol Volk. Chicago: University of Chicago Press.

Feynman, Richard P. 1995. *Essentials of Physics Explained by Its Most Brilliant Teacher*. Reading, PA: Perseus.

Fifth Estate. 2014. "Federal Programs and Research Facilities That Have Been Shut Down or Had Their Funding Reduced," January 10. http://www.cbc.ca/fifth/blog/federal-programs-and-research-facilities-that-have-been-shut-down-or-had-th.

Fiil-Flynn, Maj, with the Soweto Electricity Crisis Committee. 2001. "The Electricity Crisis in Soweto." *Municipal Services Committee Occasional Papers No. 4*, August. http://www.municipalservicesproject.org/sites/municipalservicesproject.org/files/publications/OccasionalPaper4_Fiil-Flynn%20_The_Electricity_Crisis_in_Soweto_Aug2001.pdf.

Fine, Ben and Zavareh Rustomjee. 1996. *The Political Economy of South Africa: From Minerals-Energy Complex to Industrialisation*. Boulder, Colorado: Westview Press.

"First Nation Teen Told Not to Wear 'Got Land?' Shirt at School." 2014. *CBC News: Saskatchewan*. CBC. January 14. http://www.cbc.ca/news/canada/saskatchewan/first-nation-teen-told-not-to-wear-got-land-shirt-at-school-1.2497009.

Fleming, Peter, and Carl Cederstrom. 2012. *Dead Man Working*. Winchester, UK: Zero Books.

Flink, James. 1988. *The Automobile Age*. Cambridge, MA: MIT Press.

Foley, Shane. 2001. "Kracklin' Kirby: Tracing the Advent of Kirby Krackle." *Jack Kirby Collector*, 68–71.

Folke, Carl, Steve Carpenter, Brian Walker, Marten Scheffer, Thomas Elmquist, Lance Gunderson, and Crawford Stanley. 2004. "Regime Shifts, Resilience, and Biodiversity in Ecosystem Management." *Annual Review of Ecology, Evolution, and Systematics* 35: 557–81.

The Forgotten Space. 2010. Directed by Noel Burch and Allan Sekula. Wildart Film.

Forsyth, Frederick. 1974. *The Dogs of War*. New York: Viking Press.

Fort McMoney. 2013. Directed by David Dufresne. National Film Board of Canada and TOXA.

Forty, Adrian. 1986. *Objects of Desire: Design and Society, 1750–1890*. London: Thames and Hudson.

Foster, John Bellamy. 2013. "Marx and the Rift in the Universal Metabolism of Nature." *Monthly Review* 65 (7): 1–19.

Foucault, Michel. 1973. *The Order of Things*. New York: Pantheon.

———. 1977. "Intellectuals and Power." In *Language, Counter-Memory, Practice: Selected Essays and Interviews*, edited by Donald F. Bouchard, 205–17. Ithaca, NY: Cornell University Press.

———. 1978. *The History of Sexuality*. Vol. 1. *An Introduction*. New York: Random House.

———. 2008. *The Birth of Biopolitics: Lectures at the College de France 1978–1979*. Edited by Michel Senellart. Translated by Graham Burchell. New York: Picador.

Fouquet, Roger, and Peter J. G. Pearson. 1998. "A Thousand Years of Energy Use in the United Kingdom." *Energy Journal* 19 (4): 1–41.

Fox, Stephen. 2010. "DSM-V, Healthcare Reform Will Fuel Major Changes in Addiction Psychiatry." *Medscape*, December 6. http://www.medscape.com/viewarticle/733649.

Foxe, John. 2009. *Foxe's Book of Martyrs: Select Narratives*. Edited by John King. Oxford: Oxford University Press.

Francis, Pope. 2013. "Apostolic Exhortation." *Holy See*, November 24. https://w2.vatican.va/content/francesco/en/apost_exhortations/documents/papa-francesco_esortazione-ap_20131124_evangelii-gaudium.html.

Freese, Barbara. 2003. *Coal: A Human History*. New York: Penguin.

French, Desiree (personal communication). 2016. Public Affairs Specialist. *US Federal Railroad Administration*, June 9.

French, Hilary. 2000. *Vanishing Borders: Protecting the Planet in the Age of Globalization*. New York: Norton.

Freudenburg, William R., and Robert Gramling. 1994. *Oil in Troubled Waters: Perceptions, Politics, and the Battle Over Offshore Drilling*. Albany: SUNY Press.

Friedman, Thomas L. 2006. "The First Law of Petropolitics." *Foreign Policy* 154: 28–36.

Frosch, Dan. 2013. "An Energy Source with Some Special Problems." *International Herald Tribune*, August 12. https://www.questia.com/newspaper/1P2-36306956/an-energy-source-with-some-special-problems-amid.

Frumhoff, Peter, Richard Heede, and Naomi Oreskes. 2015. "The Climate Responsibilities of Industrial Carbon Producers." *Climatic Change* 132 (2): 157–171.
Fuentes Rabé, Arturo. 1923. *Tierra del Fuego*. 2 vols. Valdivia, Chile: Imprenta Central e Lambert.
Fyodorov, Evgeny. 1939. *Scientific Work of Our Polar Expedition*. Moscow: Foreign Languages Publishing House.
Gagliano, Monica. 2012. "Green Symphonies: A Call for Studies on Acoustic Communication in Plants." *Behavioral Ecology*. http://beheco.oxfordjournals.org/content/early/2012/11/24/beheco.ars206.
Gahagan, Kayla. 2014. "Oil Boom Jars Small-town North Dakota." *Aljazeera America*, May 3. http://america.aljazeera.com/articles/2014/5/3/oil-boom-changingthelandscapeofsmalltownnorthdakota.html.
Gaita, Raimond, ed. 2010. *Gaza: Morality, Law & Politics*. Perth: UWA Publishing.
Gall, Alexander. 1998. *Das Atlantropa-Projekt: Die Geschichte einer gescheiterten Vision. Herman Sörgel und die Absenkung des Mittelmeers, vols*. Frankfurt: Campus.
———. 2006. "Atlantropa: A Technological Vision of a United Europe." In *Networking Europe: Transnational Infrastructures and the Shaping of Europe, 1850–2000*. Edited by Erik van der Vleuten and Arne Kaijser, 99–128. Sagamore Beach, MA: Science History Publications.
García Pérez, J. D. 2003. "Early Socio-Political and Environmental Consequences of the Prestige Oil Spill in Galicia." *Disasters* 27 (3): 207–23.
Genosko, Gary. 2013. *When Technocultures Collide*. Waterloo, Ontario: Wilfrid Laurier University Press.
Gere, Charlie. 2011. "Sunbeam." *Chris Meigh Andrews*, May 6. http://www.meigh-andrews.com/archives/2569.
Gessner, David. 2011. *The Tarball Chronicles: A Journey Beyond the Oiled Pelican and into the Heart of the Gulf Oil Spill*. Minneapolis: Milkweed Editions.
Ghosh, Amitav. 1992a. "Petrofiction: The Oil Encounter and the Novel." In *Incendiary Circumstances*, 138–51. Boston: Houghton Mifflin.
———. 1992b. "Petrofiction: The Oil Encounter and the Novel." *New Republic*, 29–34.
Giant. 2003. [1956.] Directed by George Stevens. Warner Brothers. DVD.
Gibson, Robert B. 2006. "Beyond the Pillars: Sustainability Assessment as a Framework for Effective Integration of Social, Economic and Ecological Considerations in Significant Decision-Making." *Journal of Environmental Assessment Policy and Management* 8 (3): 259–80.
Gillis, John R. 2013. "The Blue Humanities." *Humanities: The Magazine of the National Endowment for the Humanities* 34 (3). http://www.neh.gov/humanities/2013/mayjune/feature/the-blue-humanities.
Gilmore, Paul. 2009. *Aesthetic Materialism: Electricity and American Romanticism*. Stanford, CA: Stanford University Press.
Gitelman, Lisa. 2006. *Always Already New: Media, History and the Data of Culture*. Cambridge, MA: MIT Press.
Giucci, Guillermo. 2012. *The Cultural Life of the Automobile: Roads to Modernity*. Translated by Anne Mayagoitia and Debra Nagao. Austin: University of Texas Press.
Gladwell, Valerie F., Daniel K. Brown, Carly Wood, Gavin R. Sandercock, and Jo L. Barton. 2013. "The Great Outdoors: How a Green Exercise Environment Can Benefit All." *Extreme Physiology & Medicine* 2: 3.
Gleick, James. 1999. *Faster: The Acceleration of Just About Everything*. New York: Pantheon.
Godzilla (Gojira). 1954. Directed by Ishirô Honda. Toho Film.
Goldberg, Jonathan. 2004. *Tempest in the Caribbean*. Minneapolis: University of Minnesota Press.
Goldenberg, Suzanne. 2013a. "Children Given Lifelong Ban on Talking about Fracking." *Guardian*, August 5. http://www.theguardian.com/environment/2013/aug/05/children-ban-talking-about-fracking.
———. 2013b. "Obama Campaign Launches Plan to Shame Climate Sceptics in Congress." *Guardian*, April 25. http://www.theguardian.com/environment/2013/apr/25/obama-for-america-shame-climate-sceptics.

———. 2013c. "A Texan Tragedy: Ample Oil, No Water." *Guardian*, August 11. http://www.theguardian.com/environment/2013/aug/11/texas-tragedy-ample-oil-no-water.
———. 2014. "Just 90 Companies are Responsible for Two-Thirds of Greenhouse Gas Emissions." *Guardian*, January 20. https://www.theguardian.com/environment/2013/nov/20/90-companies-man-made-global-warming-emissions-climate-change.
Gore, Al. 2009. Appearance on *Saturday Night Live*. November 23. NBC Studios.
Gosden, Emily. 2013. "Water Firms Raise Fears over Shale Gas Fracking." *Daily Telegraph*, July 19. http://www.telegraph.co.uk/finance/newsbysector/utilities/10189331/Water-firms-raise-fears-over-shale-gas-fracking.html.
Gould, Deborah G. 2012. "Political Despair." In *Politics and the Emotions: The Affective Turn in Contemporary Political Studies*, edited by Simon Thompson and Paul Hoggett, 95–114. New York: Continuum.
Government of Alberta. 2010. "Alberta's Oil Sands: About." June 24. http://oilsands.alberta.ca/about.html.
———. 2013a. "Economic Summit Participant Guide." February.
———. 2013b. "Highlights of the Alberta Economy 2013." June. http://brookschamber.ab.ca/wp-content/uploads/2013/05/Highlights-of-the-Alberta-Economy-2013.pdf.
———. 2014. "Alberta Education's Curriculum Development Prototyping Partners." March 12. https://open.alberta.ca/dataset/alberta-education-s-curriculum-development-prototyping-partners/resource/8172f07a-b817–460d-boff-2b2ac8df5157.
Government of Canada. 2013. "Financial Security—Family Income: Regions." September 7. http://well-being.esdc.gc.ca/misme-iowb/.3ndic.1t.4r@-eng.jsp?iid=21#M_3.
Graeber, David. 2012. "Afterword: The Apocalypse of Objects—Degradation, Redemption and Transcendence in the World of Consumer Goods." In *Economies of Recycling: The Global Transformation of Materials, Values, and Social Relations*, edited by Catherine Alexander and Josh Reno, 277–90. London: Zed Books.
Gray, John. 2013. "Are We Done For?" *Guardian*, July 6: 6.
Greenpeace. 2001. "A Decade of Dirty Tricks." *Greenpeace UK*, July 31. http://www.greenpeace.org.uk/media/reports/a-decade-of-dirty-tricks.
———. 2010. "Koch Industries: Secretly Funding the Climate Denial Machine." *Greenpeace USA*. http://www.greenpeace.org/usa/research/koch-industries-secretly-fund/.
Groenwegen, Peter. 2001. "Thomas Carlyle, 'the Dismal Science,' and the Contemporary Political Economy of Slavery." *History of Economics Review* 34: 74–94.
Gross, Daniel. 2015. "Israeli Solar Warms Up." *Slate*, August 7. http://www.slate.com/articles/business/the_juice/2015/08/ketura_solar_energy_field_israeli_is_finally_embracing_renewable_power.html.
Gruley, Bryan. 2012. "The Man Who Bought North Dakota." *Bloomberg Businessweek Magazine*, January 19. http://www.businessweek.com/magazine/the-man-who-bought-north-dakota-01192012.html.
Guattari, Felix. 1995. *Chaosmosis: An Ethico-Aesthetic Paradigm*. Translated by Paul Bains and Julian Penafis. Bloomington: Indiana University Press.
Guattari, Felix, and Toni Negri. 1990. *Communists Like Us: New Spaces of Liberty, New Lines of Alliance*. Translated by Michael Ryan. New York: Semiotext(e).
Guay, Justin. 2012. "Eight19 and the 'Un-Grid.'" *Compass: Pointing the Way to a Clean Energy Future*, March 30. http://blogs.sierraclub.org/compass/2012/03/eight19-and-the-un-grid.html.
Guillory, John. 2010. "Genesis of the Media Concept." *Critical Inquiry* 36 (2): 321–62.
Gumbrecht, Hans Ulrich. 2004. *Production of Presence: What Meaning Cannot Convey*. Stanford, CA: Stanford University Press.
H2Oil. 2009. Directed by Sheldon Walsh. Loaded Pictures. DVD.
Habila, Helon. 2011. *Oil on Water: A Novel*. New York: W.W. Norton.
Hahnemann, Samuel. 2001. *Gesammelte kleine Schriften*. Edited by Josef M. Schmidt and Daniel Kaiser. Heidelberg, Germany: Haug.
Hall, Stuart. 1996. "Who Needs Identity?" In *Questions of Cultural Identity*, 1–17. London: Sage.

Hall, Stuart, and Martin Jacques. 1983. *The Politics of Thatcherism*. London: Lawrence and Wishart.
Hansen, Mark. 2006. "Media Theory." *Theory, Culture & Society* 23 (2–3): 297–305.
Hansen, Mark, and W. J. T. Mitchell. 2010. Introduction to *Critical Terms for Media Studies*, vii–xxiii. Chicago: University of Chicago Press.
Hansen, Miriam. 1999. "Benjamin and Cinema: Not a One-Way Street." *Critical Inquiry* 25: 306–45.
———. 2012. *Cinema and Experience: Siegfried Kracauer, Walter Benjamin, and Theodor W. Adorno*. Berkeley: University of California Press.
Haraway, Donna. 1991. "A Cyborg Manifesto: Science, Technology, and Socialist-Feminism in the Late Twentieth Century." In *Simians, Cyborgs and Women: The Reinvention of Nature*, 149–81. New York: Routledge.
Hardt, Michael. 2010. "Two Faces of the Apocalypse: Letter from Copenhagen." *Polygraph* 22: 265–74.
Harman, P. M. 1982. *Energy, Force, and Matter: The Conceptual Development of Nineteenth-Century Physics*. Cambridge: Cambridge University Press.
Harney, Stefano. 2008. "Abolition and the General Intellect." *Generation Online*. http://www.generation-online.org/c/fc_rent13.htm.
Harrabin, Roger. 2013. "Fracking: RSPB Objects to Cuadrilla Plans for Two Sites." *BBC News*, August 17. http://www.bbc.com/news/science-environment-23730308.
Harrison, Conor. 2013. "The Historical-Geographical Construction of Power: Electricity in Eastern North Carolina." *Local Environment* 18 (4): 469–86.
Harrison, Robert Pogue. 1993. *Forests: The Shadow of Civilization*. Chicago: University of Chicago Press.
Harvey, David. 1993. "The Nature of Environment: The Dialectics of Social and Environmental Change." *Socialist Register* 29: 1–51.
———. 2000. *Spaces of Hope*. Berkeley: University of California Press.
———. 2001. "Globalization and the Spatial Fix." *Geographische Revue* 2: 23–30.
———. 2005. *A Brief History of Neoliberalism*. Oxford: Oxford University Press.
———. 2010. *Companion to Marx's Capital*. New York: Verso.
Harvey, Steve. 2010. "California's Legendary Oil Spill." *Los Angeles Times*, June 13. http://articles.latimes.com/2010/jun/13/local/la-me-then-20100613.
Hatherly, Owen. 2011. "Crude Awakening." *Frieze* 136. http://www.frieze.com/article/crude-awakening.
Hawkins, Gay. 2013. "Made to Be Wasted: PET and Disposability." In *Accumulation: The Material Politics of Plastic*, edited by Gay Hawkins, Mike Michael, and Jennifer Gabrys, 49–67. London: Routledge.
Hazlett, Alex. 2012. "Instagram Still Down after Storm Cuts Power during Brutal Heat Wave." *Mashable*, June 30. http://mashable.com/2012/06/30/instagram-down-power-heat-wave/.
Hecht, Gabrielle. 2012. *Being Nuclear: Africans and the Global Uranium Trade*. Cambridge, MA: MIT Press.
Heede, Richard. 2013. "Tracing Anthropogenic Carbon Dioxide and Methane Emissions to Fossil Fuel and Cement Producers, 1854–2010." *Climatic Change* 122 (1–2): 229–41.
Hegel, G. W .F. 1956. *The Philosophy of History*. Translated by J. Sibree. New York: Dover.
———. 1975. *Lectures on the Philosophy of World History: Introduction*. Translated by H. B. Nisbet. Cambridge: Cambridge University Press.
———. 1977. *The Phenomenology of Spirit*. Translated by A. V. Miller. Oxford: Oxford University Press.
———. 2015. *The Science of Logic*. Translated by George Di Giovanni. Cambridge: Cambridge University Press.
Heidegger, Martin. 1969. *Discourse on Thinking*. Translated by J. M. Anderson and E. Hans Freund. New York: Harper.
———. 1977a. "The Question Concerning Technology." *Basic Writings*. Translated by David Farrell Krell, 307–42. New York: Harper Collins.

———. 1977b. *The Question Concerning Technology and Other Essays.* Translated by William Lovitt. New York: Harper and Row.
———. 1977c. "The Question Concerning Technology." In *The Question Concerning Technology and Other Essays.* Translated by William Lovitt. New York: Harper and Row.
———. 1982. *The Basic Problems of Phenomenology.* Translated by Albert Hofstadter. Bloomington: Indiana University Press.
———. 1996. *Being and Time.* Translated by Joan Stambaugh. Albany: SUNY Press.
Hein, Carola. 2009. "Global Landscapes of Oil." In *New Geographies 2: Landscapes of Energy*, edited by Rania Ghosn, 67–74. Cambridge, MA: Harvard University Press.
Hendersen, Mathew. 2012. "The Tank." In *The Lease*, 9. Toronto: Coach House Books.
Herder, Johann Gottfried. 1994. "Gott, einige Gespräche." In *Schriften zu Philosophie, Literatur, Kunst und Altertum 1774–1787. Werke in zehn Bänden.* Vol. 4. Edited by Jürgen Brummack and Martin Bollacher, 679–794. Frankfurt: Deutscher Klassiker Verlag.
Heymann, Matthias. 1995. *Die Geschichte der Windenergienutzung: 1890–1990.* Frankfurt: Campus.
Higgins, Hannah B. 2009. *The Grid Book.* Cambridge, MA: MIT Press.
Hinchliffe, Steve, and Sarah Whatmore. 2008. "Living Cities: Toward a Politics of Conviviality." In *Environment: Critical Essays in Human Geography*, edited by Kay Anderson and Bruce Braun, 555–70. Aldershot, England: Ashgate.
Hitchcock, Peter. 2010. "Oil in an American Imaginary." *New Formations* 69: 81–97.
Hobbes, Thomas. 2009. [1651]. *Leviathan.* Project Gutenberg. October 11. http://www.gutenberg.org/files/3207/3207-h/3207-h.htm.
Hochfelder, David. 2013. *The Telegraph in America, 1832–1920.* Baltimore, MD: Johns Hopkins University Press.
Hoffmeyer, Jesper. 2008. *Biosemiotics: An Examination into the Signs of Life and the Life of Signs.* Translated by Jesper Hoffmeyer and Donald Favareau. Scranton, PA: University of Scranton Press.
Hogan, Mél. 2015. "Data Flows and Water Woes: The Utah Data Center." *Big Data & Society*, 1–12.
Holling, Crawford Stanley. 1973. "Resilience and Stability of Ecological Systems." *Annual Review of Ecology and Systematics* 4: 1–24.
Holloway, Hilton. 2013. "China Faces New Car Explosion." *Autocar*, October 24. http://www.autocar.co.uk/car-news/industry/china-faces-new-car-explosion.
Holt, Jennifer, and Patrick Vonderau. 2015. "Where the Internet Lives: Data Centers as Cloud Infrastructure." In *Signal Traffic: Critical Studies of Media Infrastructures*, edited by Lisa Parks and Nicole Starosielski, 71–93. Champaign-Urbana: University of Illinois Press.
"Holy Land Energy Vision Statement: A Movement for Clean Energy." 2014. *Loretto Community.* http://www.lorettocommunity.org/energy/.
Honarvar, Afshin, Jon Rozhon, Dinara Millington, Thorn Walden, Carlos A. Murillo, and Zoey Walden. 2011. *Economic Impacts of New Oil Sands Projects in Alberta (2010–2035).* Calgary: Canadian Energy Research Institute/University of Calgary.
Honnef, Hermann. 1932. *Windkraftwerke.* Braunschweig, Germany: F. Vieweg und Sohn.
Hopkins, Rob. 2010. *The Transition Handbook: From Oil Dependency to Local Resilience.* Creative Commons Attribution-Share Alike 3.0. http://www.cs.toronto.edu/~sme/CSC2600/transition-handbook.pdf.
Hopper, Tristin. 2013. "If a Hunger Striker Is Well Hydrated, the Human Body Can Survive for Weeks or Even Months without Food." *National Post*, January 4. http://news.nationalpost.com/news/canada/theresa-spence-hunger-strike.
Horkheimer, Max, and Theodor Adorno. 2002. *Dialectic of Enlightenment: Philosophical Fragments.* Translated by Edmund Jephcott. Stanford, CA: Stanford University Press.
Hornborg, Alf. 2013. "The Fossil Interlude: Euro-American Power and the Return of the Physiocrats." In *Cultures of Energy*, edited by S. Rupp, T. Love, and S. Strauss, 49–51. Walnut Creek, CA: Left Coast Press.
The House of World Cultures. 2013–2014. "The Anthropocene Project." *HKW.* http://www.hkw.de/en/programm/projekte/2014/anthropozaen/anthropozaen_2013_2014.php.

"How the National Grid Responds to Demand." 2008. YouTube video, 4:30, from *Britain from Above*. Posted by "Robert Woodman," November 9. http://www.youtube.com/watch?v=UTM2Ck6XWHg.

Howard, Emma. 2015. "Keep It in the Ground Campaign: Six Things We've Learned." *Guardian*, March 25. http://www.theguardian.com/environment/keep-it-in-the-ground-blog/2015/mar/25/keep-it-in-the-ground-campaign-six-things-weve-learned.

Howe, Cymene. 2014. "Anthropocenic Ecoauthority: The Winds of Oaxaca." *Anthropological Quarterly* 86 (1): 381–404.

Huawen, William, Tianyi Luo, and Tien Shiao. 2013. "China's Response to Air Pollution Poses Threat to Water." *World Resources Institute*, October 23. http://www.wri.org/blog/2013/10/china's-response-air-pollution-poses-threat-water.

Huber, Matt. 2011. "Oil, Life, and the Fetishism of Geopolitics." *Capitalism Nature Socialism* 22 (3): 32–48.

———. 2013. "What Do We Mean by 'Energy Policy?' Life, Capitalism, and the Broader Field of Energy Politics." *State of Nature*, May 4. http://www.stateofnature.org/?p=7138.

———. 2014. "Refined Politics: Petroleum Products, Neoliberalism, and the Ecology of Entrepreneurial Life." *Journal of American Studies* 46 (2): 295–312.

Hud. 1963. Directed by Martin Ritt. Paramount Pictures.

Hufeland, Christoph Wilhelm. 1798. "Mein Begriff von der Lebenskraft." *Journal der practischen Heilkunde* 6: 785–96.

———. 1979. *The Art of Prolonging Life*. New York: Amo Press.

Humboldt, Alexander von. 1797. *Versuche über die gereizte Muskel- und Nervenfaser nebst Vermuthungen über den chemischen Prozess des Lebens in der Thier- und Pflanzenwelt*. 2 Vols. Posen, Prussia: Decker.

Hussain, Yadullah. 2013a. "Fossil Fuels Reign, Even in 2050: Study." *National Post*, October 18, FP 7.

———. 2013b. "Over-fracked, Over-drilled." *National Post*, October 18: FP 1, 6.

Huxley, Julian. 1942. *Evolution: The Modern Synthesis*. London: George Allen and Unwin Ltd.

Ide III, R. William, and Joseph O. Blanco of McKenna, Long, and Aldridge, LLP. 2009. *A Report to the Board of Directors of the Tennessee Valley Authority Regarding Kingston Factual Findings*. July 21. https://jobs.tva.com/kingston/board_report/mla_kingston_report.pdf.

Ifowodo, Ogaga. 2005. *The Oil Lamp*. Trenton, NJ: Africa World Press.

Illich, Ivan. 1974. *Energy and Equity*. New York: Harper and Row.

An Inconvenient Truth. 2006. Directed by Davis Guggenheim. Lawrence Bender Productions.

"Initiative 'Unser Hamburg - unser Netz': Hamburger stimmen für Rückkauf der Energienetze." 2013. *Der Spiegel*, September 22. http://www.spiegel.de/wirtschaft/unternehmen/hamburger-stimmen-fuer-rueckkauf-der-energienetze-a-923811.html.

Innis, Harold Adams. 1970. *The Fur Trade in Canada*. Toronto: University of Toronto Press.

Inside Man. 2006. Directed by Spike Lee. Imagine Entertainment.

International Energy Agency. 2015. *World Energy Outlook 2015*. Paris: IEA.

Jackson, Chuck. 2008. "Blood for Oil: Crude Metonymies and Tobe Hooper's *Texas Chain Saw Massacre* (1974)." *Gothic Studies* 10 (1): 48–60.

Jaggar, Alison. 1989. "Love and Knowledge: Emotion in Feminist Epistemology." *Inquiry* 32 (2): 151–76.

Jakob, Michael. 2001. "Conversation with Paul Virilio." *2G* 18 (2): 4–7.

James, C. L. R. 1989. *The Black Jacobins. Toussaint L'Overture and the San Domingo Revolution*. New York: Vintage Books.

James, William. 1880. "The Feeling of Effort." In *Anniversary Memoirs of the Boston Society of Natural History*. Boston: Boston Society of Natural History.

Jameson, Fredric. 1994. *The Seeds of Time*. New York: Columbia University Press.

———. 2005. *Archaeologies of the Future: The Desire Called Utopia and Other Science Fictions*. London: Verso.

Jancovici, Jean-Marc. 2013. [2005]. "How Much of a Slave Master Am I?" *Manicore*, August. http://www.manicore.com/anglais/documentation_a/slaves.html.

Jarhead. 2005. Directed by Sam Mendes. Universal Pictures.
Javed, Noor. 2013. "Attawapiskat Chief Theresa Spence Hunger Strike: How's Her Health and How Long Can She Continue?" *Star*, January 7. https://www.thestar.com/news/canada/2013/01/07/attawapiskat_chief_theresa_spence_hunger_strike_hows_her_health_and_how_long_can_she_continue.html.
Jevons, William Stanley. 1865. *The Coal Question: An Inquiry Concerning the Progress of the Nation, and the Probable Exhaustion of Our Coal-Mines*. London: Macmillan.
Johnson, Bob. 2010. "Coal, Trauma, and the Origins of the Modern Self." *Journal of American Culture* 33 (4): 265–79.
———. 2014. *Carbon Nation: Fossil Fuels in the Making of American Culture*. Lawrence: University Press of Kansas.
Johnson, Bobbie. 2009. "How Much Energy Does the Internet Really Use?" *Guardian*, May 14. https://www.theguardian.com/technology/2009/may/14/internet-energy-savings.
Jørgensen, Dolly. 2012. "OSPAR's Exclusion of Rigs-to-Reefs in the North Sea." *Oceans and Coastal Management* 58: 57–61.
Joseph, Miranda. 2002. *Against the Romance of Community*. Minneapolis: University of Minnesota Press.
Juhasz, Antonia. 2011. *Black Tide: The Devastating Impact of the Gulf Oil Spill*. Hoboken, NJ: Wiley.
Kanigher, Robert (w), Carmine Infantino (p), and Joe Kubert (i). 1956. *Showcase #4*. September–October. DC Comics.
Kant, Immanuel. 1983. *Kritik der Urteilskraft. Werke in zehn Bänden*. Vol. 5. Darmstadt, Germany: Wissenschaftliche Buchgesellschaft.
Karl, Terry Lynn. 1997. *The Paradox of Plenty: Oil Booms and Petro-States*. Berkeley: University of California Press.
Kaufman, Scott. 2013. *Project Plowshare: The Peaceful Use of Nuclear Explosives in Cold War America*. Ithaca, NY: Cornell University Press.
Kay, Barbara. 2013. "Barbara Kay on Theresa Spence: You Call That a 'Hunger Strike?'" *National Post*, January 4. http://news.nationalpost.com/full-comment/barbara-kay-on-theresa-spence-you-call-that-a-hunger-strike.
Keil, Roger. 2007. "Sustaining Modernity, Modernizing Nature: The Environmental Crisis and the Survival of Capitalism." In *The Sustainable Development Paradox: Urban Political Economy in the United States and Europe*, edited by Rob Krueger and David Gibbs, 41–66. New York: Guilford.
Kelly, Kevin. 1994. *Out of Control: The New Biology of Machines, Social Systems, and the Economic World*. New York: Addison-Wesley.
Kenny, James L., and Andrew G. Secord. 2001. "Public Power for Industry: A Re-Examination of the New Brunswick Case, 1940–1960." *Acadiensis: Journal of the History of the Atlantic Region* 30 (2): 84–108.
———. 2010. "Engineering Modernity: Hydroelectric Development in New Brunswick, 1945–70." *Acadiensis: Journal of the History of the Atlantic Region* 39 (1): 3–26.
Kerouac, Jack. 1976. [1957]. *On the Road*. New York: Penguin.
Khodorkovsky. 2011. Directed by Cyril Tuschi. LaLa Films!
Kilian, Crawford. 2013. "What Comes after Capitalism?" *Salon*, August 7. http://www.salon.com/2013/08/07/is_there_a_viable_alternative_to_capitalism_partner/.
Kirby, Did. 2011. "Made in China: Our Toxic, Imported Air Pollution." *Discover Magazine*, March 18. http://discovermagazine.com/2011/apr/18-made-in-china-our-toxic-imported-air-pollution.
Kirsch, Scott. 2005. *Proving Grounds: Project Plowshare and the Unrealized Dream of Nuclear Earthmoving*. New Brunswick, NJ: Rutgers University Press.
Kistner, Rocky. 2013. "Health Problems Still Plague Arkansas Residents Near ExxonMobil Tar Sands Spill." *Huffington Post*, October 26. http://www.huffingtonpost.com/rocky-kistner/health-problems-still-pla_b_3818110.html.
Klare, Michael T. 2011a. "The New Thirty Years' War: Winners and Losers in the Great Global Energy Struggle to Come." *Common Dreams*, June 27. http://www.commondreams.org/

views/2011/06/27/new-thirty-years-war-winners-and-losers-great-global-energy-struggle-come.
———. 2011b. *The Race for What's Left*. New York: Metropolitan.
Klein, Naomi. 2007. *The Shock Doctrine: The Rise of Disaster Capitalism*. Toronto: Knopf.
Kolbert, Elizabeth. 2014. *The Sixth Extinction: An Unnatural History*. New York: Henry Holt.
———. 2015. "We Are the Asteroid." Carr Distinguished Interdisciplinary Lecture Series, Skidmore College, November 3. https://www.skidmore.edu/carr/lectures/fall-2015-lecture.php.
Koring, Paul. 2013. "Ottawa Pitches the Oil Sands as 'Green.'" *Globe and Mail*, March 5. http://www.theglobeandmail.com/report-on-business/industry-news/energy-and-resources/ottawa-pitches-the-oil-sands-as-green/article9306257/.
Kosek, Jake. 2006. *Understories*. Durham, NC: Duke University Press.
Koselleck, Reinhart. 2004. *Futures Past: On the Semantics of Historical Time*. New York: Columbia University Press.
Kosman, L. A. 1969. "Aristotle's Definition of Motion." *Phronesis* 14 (1): 40–62.
Kowalsky, Nathan, and Randolph Haluza-DeLay. 2013. "Homo Energeticus: Technological Rationality in the Alberta Tar Sands." In *Jacques Ellul and the Technological Society in the 21st Century*, edited by Helena M. Jerónimo, José L. Garcia, and Carl Mitcham, 159–75. Dordrecht, Netherlands: Springer.
Krige, John. 2006. "Atoms for Peace, Scientific Internationalism, and Scientific Intelligence." *Osiris* 21 (1): 161–81.
Kunzig, Robert. 2009. "The Canadian Oil Boom: Scraping Bottom." *National Geographic* 215 (3): 34–59.
Kyd, Thomas. 1793. *A Treatise on the Law of Corporations*. London: J. Butterworth.
Ladurantaye, Steve. 2011. "Canada Called Prime Real Estate for Massive Data Computers." *Globe and Mail*, June 22. http://www.theglobeandmail.com/report-on-business/canada-called-prime-real-estate-for-massive-data-computers/article5841631/.
Lafargue, Paul. 2011. *The Right to Be Lazy*. Edited by Bernard Marszalek. Oakland, CA: AK Press.
Landon, Stuart, and Constance Smith. 2010. "Energy Prices and Alberta Government Revenue Volatility." *C.D. Howe Institute Commentary*, November. https://www.cdhowe.org/public-policy-research/energy-prices-and-alberta-government-revenue-volatility.
Lanier, Jaron. 2010. *You Are Not a Gadget*. New York: Alfred A. Knopf.
L'Annunziata, Michael F. 2007. *Radioactivity: Introduction and History*. Amsterdam: Elsevier.
Laplanche, Jean, and Jean-Bertrand Pontalis. 1988. *The Language of Psychoanalysis*. London: Karnac Press.
Larkin, Brian. 2013. "The Politics and Poetics of Infrastructure." *Annual Review of Anthropology* 42: 327–43.
Latour, Bruno. 1993. *We Have Never Been Modern*. Cambridge, MA: Harvard University Press.
———. 2004. *The Politics of Nature*. Cambridge, MA: Harvard University Press.
———. 2007. *Reassembling the Social: An Introduction to Actor-Network-Theory*. Oxford: Oxford University Press.
———. 2011. "Love Your Monsters." *Breakthrough Journal* 2: 21–28. http://thebreakthrough.org/images/main_image/Breakthrough_Journal_Issue_2.pdf.
Leahy, Stephen. 2006. "Environment: Burning Energy to Produce It." *InterPress Service*, July 24. http://www.ipsnews.net/2006/07/environment-burning-energy-to-produce-it/.
Lee, Stan (w), Jack Kirby (p), and George Klein (i). 1961. *Fantastic Four #1*. November. Marvel Comics.
Lee, Stan (w), Jack Kirby (p), and Joe Sinnott (i). 1968. *Fantastic Four #72*. March. Marvel Comics.
Leffler, Melvyn P. 2005. "National Security and U.S. Foreign Policy." In *Origins of the Cold War: An International History*, edited by Melvyn P. Leffler and David S. Painter, 15–41. London: Routledge.
LeMenager, Stephanie. 2012. "The Aesthetics of Petroleum, after Oil!" *American Literary History* 24 (1): 59–86.

———. 2014. *Living Oil: Petroleum Culture in the American Century*. Oxford: Oxford University Press.
Leopold, Aldo. 1992. "Game and Wild Life Conservation." In *The River of the Mother of God: And Other Essays by Leopold*, 164–68. Madison: University of Wisconsin Press.
Lessons of Darkness. 1992. Directed by Werner Herzog. Canal+.
Levant, Ezra. 2010. *Ethical Oil: The Case for Canada's Oil Sands*. Toronto: Mclelland & Stewart.
———. 2013b. "The Scandals of Theresa Spence and Attawapiskat." *Toronto Sun*, January 3.
Levine, Marc, and Kate Sears. 2013. "Fracking for Oil Is Environmentally Risky." *Marin Independent Journal*, August 16. http://www.marinij.com/article/ZZ/20130816/NEWS/130818661.
Levi-Strauss, Claude. 1991. "Toward the Intellect." *Totemism*. Translated by Rodney Needham, 72–91. London: Merlin.
Lindee, M. Susan. 1994. *Suffering Made Real: American Science and the Survivors at Hiroshima*. Chicago: University of Chicago Press.
Lippard, Lucy R. 1998. *The Lure of the Local: Senses of Space in a Multi Centered Society*. New York: The New Press.
Lippit, Akira Mizuta. 2000. *Electric Animal: Toward a Rhetoric of Wildlife*. Minneapolis: University of Minnesota Press.
"Lithium." 2013. U.S. Geological Survey, Mineral Commodity Summaries, January. http://minerals.usgs.gov/minerals/pubs/commodity/lithium/mcs-2013-lithi.pdf.
Loadman, John. 2005. *Tears of the Tree: The Story of Rubber, a Modern Marvel*. Oxford: Oxford University Press.
Logar, Ernst. 2011. *Invisible Oil*. Vienna: Springer.
Loomis, Carol J. 2001. "The Value Machine: Warren Buffett's Berkshire Hathaway Is on a Buying Binge." *Fortune Magazine*, February 19. http://archive.fortune.com/magazines/fortune/fortune_archive/2001/02/19/296860/index.htm.
Louisiana Story. 1948. Directed by Robert J. Flaherty. Standard Oil.
Lovell, Bryan. 2010. *Challenged By Carbon: The Oil Industry and Climate Change*. Cambridge: Cambridge University Press.
Lucier, Paul. 2008. *Scientists and Swindlers: Consulting on Coal and Oil in America, 1820–1890*. Baltimore, MD: Johns Hopkins University Press.
Lukács, Georg. 1971. "The Ideology of Modernism." *Realism in Our Time*. New York: Harper Torchbooks.
Lukacs, Martin. 2013. "New Brunswick Fracking Protests Are the Frontline of a Democratic Fight." *Guardian*, October 21. http://www.theguardian.com/environment/2013/oct/21/new-brunswick-fracking-protests.
Luke, Timothy W. 2009. "Power Loss or Blackout: The Electricity Network Collapse of August 2003 in North America." In *Disrupted Cities: When Infrastructure Fails*, 55–68. London: Routledge.
Luong, Pauline Jones, and Erika Weinthal. 2010. *Oil Is Not a Curse: Ownership Structure and Institutions in Soviet Successor States*. Cambridge: Cambridge University Press.
Lynch, Michael J., Michael A. Long, Kimberly L. Barrett, and Paul B. Stretesky. 2013. "Is It a Crime to Produce Ecological Disorganization? Why Green Criminology and Political Economy Matter in the Analysis of Global Ecological Harms." *British Journal of Criminology* 53 (6): 997–1016.
M., P. 2011. *Bolo'Bolo*. New York: Autonomedia.
Maass, Peter. 2009. *Crude World*. New York: Knopf.
Mackenzie, Adrian. 2010. *Wirelessness: Radical Empiricism in Network Cultures*. Cambridge, MA: MIT Press.
Magstadt, Thomas. 2013. "The Coming Age of Coal." *Nation of Change*, November 6. http://www.nationofchange.org/coming-age-coal-1383733704.
Major, Claire, and Tracy Winters. 2013. "Community by Necessity: Security, Insecurity, and the Flattening of Class in Fort McMurray, Alberta." *Canadian Journal of Sociology* 38 (2): 141–66.
Malthus, Thomas. 1993. *An Essay on the Principle of Population*. Oxford: Oxford University Press.

Mann, Simon. 2011. *Cry Havoc*. London: John Blake.
Marcuse, Herbert. 1960. *Reason and Revolution: Hegel and the Rise of Social Theory*. New York: Beacon Press.
Marder, Jenny. 2011. "Nuclear Reactors and Nuclear Bombs: What's the Difference?" *PBS News Hour*, April 6. http://www.pbs.org/newshour/rundown/what-is-the-difference-between-the-nuclear-material-in-a-bomb-versus-a-reactor/.
Marglin, Stephen A. 2010. *The Dismal Science: How Thinking Like an Economist Undermines Community*. Cambridge, MA: Harvard University Press.
"Marie Curie and the Radioactivity: The 1903 Nobel Prize in Physics." 2014. *Nobelprize.org*. Nobel Media AB. http://www.nobelprize.org/educational/nobelprize_info/curie-edu.html.
Marriott, James, and Mika Minio-Paluello. 2012. *The Oil Road: Journeys from the Caspian Sea to the City of London*. London: Verso.
Marsden, William. 2008. *Stupid to the Last Drop: How Alberta Is Bringing Environmental Armageddon to Canada (And Doesn't Seem to Care)*. Toronto: Knopf.
Marsh, George P. 1874. *Earth as Modified by Human Action*. New York: Scribner.
Marshall, Christa. 2014. "Toxic Mercury Pollution May Rise with Arctic Meltdown." *Scientific American*, January 16. http://www.scientificamerican.com/article/toxic-mercury-pollution-may-rise-with-arctic-meltdown/.
Martinez-Alier, Joan. 2009. "Social Metabolism and Environmental Conflicts." *Socialist Register* 43. http://socialistregister.com/index.php/srv/article/view/5868#.V1XodFc3FTc.
Marvin, Simon, and Stephen Graham. 2001. *Splintering Urbanism: Networked Infrastructures, Technological Mobilities and the Urban Condition*. London: Routledge.
Marx, Karl. 1957. *Capital, Volume I*. Translated by Samuel Moore and Edward Aveling. London: Allen & Unwin.
———. 1959. *The Economic and Philosophical Manuscripts of 1844*. Moscow: Progress.
———. 1967. *Capital, Volume III*. New York: International Publishers.
———. 1973. *Grundrisse: Introduction to the Critique of Political Economy*. Translated by Martin Nicolaus. New York: Vintage Books.
———. 1977. *Capital, Volume I*. Translated by Ben Fowkes. New York: Vintage Books.
———. 1979. "Letter to Pavel Vassilyevich Annenkov (in Paris), Brussels, December 28, 1846." In *The Letters of Karl Marx*. Translated by Saul K. Padover, 44–54. Englewood Cliffs, NJ: Prentice-Hall.
———. 1990. *Capital, Volume I*. Translated by Ben Fowkes. Harmondsworth, England: Penguin.
———. 1991. *Capital: Volume III*. Translated by David Fernbach. New York: Penguin.
———. 1992. *Capital: A Critique of Political Economy*. Vol. 2. *The Process of Circulation of Capital*. New York: Penguin Classics.
———. 2007. *Economic and Philosophical Manuscripts of 1844*. Mineola, NY: Dover.
Marx, Karl, and Friedrich Engels. 1988. *The German Ideology*. New York: International Publishers.
———. 1998. *The German Ideology*. New York: Prometheus.
Marx, Leo. 1964. *The Machine in the Garden: Technology and the Pastoral Ideal in America*. New York: Oxford University Press.
Matheny, Jim. 2013. "TVA Ash Spill Measured in Billions Five Years Later." *WBIR*, December 20. http://www.wbir.com/news/local/tva-ash-spill-measured-in-billions-five-years-later/95026355.
Mattern, Shannon. 2013. "Infrastructural Tourism." *Places Journal*, July 2. https://placesjournal.org/article/infrastructural-tourism/.
Maxwell, Richard, and Toby Miller. 2012. *Greening the Media*. Oxford: Oxford University Press.
Mbembe, Achille, and Sarah Nuttall. 2008. "Introduction: Afropolis." In *Johannesburg: The Elusive Metropolis*, edited by Achille Mbembe and Sarah Nuttall, 1–36. Johannesburg: Wits University Press.
Mayer, Jane. 2010. "Covert Operations: The Billionaire Brothers Who Are Waging a War against Obama." *New Yorker*, August 30. http://www.newyorker.com/magazine/2010/08/30/covert-operations.
McCarthy, Cormac. 1992. *All the Pretty Horses*. New York: Vintage.

———. 2007. *The Road*. New York: Vintage.

McCarthy, Tom. 2008. *Auto Mania: Cars, Consumers and the Environment*. New Haven, CT: Yale University Press.

McCoy, Roger. 2006. *Ending in Ice: The Revolutionary Idea and Tragic Expedition of Alfred Wegener*. Oxford: Oxford University Press.

McDonald, David. 2009. "Introduction: The Importance of Being Electric." In *Electric Capitalism: Recolonising Africa on the Power Grid*, xv–xxiii. Cape Town: HSRC Press.

McDougall, Dan. 2009. "In Search of Lithium: The Battle for the 3rd Element." *Daily Mail*, April 5. http://www.dailymail.co.uk/home/moslive/article-1166387/In-search-Lithium-The-battle-3rd-element.html.

McGowen, Elizabeth, and Lisa Song. 2012. "The Dilbit Disaster: Inside the Biggest Oil Spill You've Never Heard Of." *Inside Climate News*, June 26. http://insideclimatenews.org/news/20120626/dilbit-diluted-bitumen-enbridge-kalamazoo-river-marshall-michigan-oil-spill-6b-pipeline-epa.

McKibben, Bill. 2010. "We're Hot as Hell and We're Not Going to Take It Anymore: Three Steps to Establish a Politics of Global Warming." *Tom Dispatch*, August 4. http://www.tomdispatch.com/post/175281/tomgram%3A_bill_mckibben,_a_wilted_senate_on_a_heating_planet/.

———. 2011. *Eaarth: Making a Life on a Tough New Planet*. New York: Vintage.

McKim, Joel. 2009. "Of Microperception and Micropolitics: An Interview with Brian Massumi." *Inflexions* 3. http://www.inflexions.org/n3_massumihtml.html.

McMurtry, Larry. 2002. *Horseman, Pass By*. New York: Scribner.

McNeill, J. R. 2000. *Something New under the Sun: An Environmental History of the Twentieth Century*. London: Allen Lane.

Meadows, Donna, Dennis Meadows, Jørgen Randers, and William Bahrens. 1972. *The Limits to Growth*. New York: Universe Books.

Medicus, Friedrich Casimir. 1774. *Vorlesung von der Lebenskraft*. Mannheim, Germany: Hof- und Akademische Buchdruckerei.

Medovoi, Leerom. 2010. "A Contribution to the Critique of Political Ecology: Sustainability as Disavowal." *New Formations* 69 (1): 129–43.

Meikle, Jeffrey. 1995. *American Plastic: A Cultural History*. New Brunswick, NJ: Rutgers University Press.

Meisel, Steven. 2010. "Water & Oil." *Vogue Italia*, August 2. http://www.vogue.it/en/fashion/cover-fashion-stories/2010/08/02/water-oil/.

Meillassoux, Quentin. 2008. *After Finitude: An Essay on the Necessity of Contingency*. Translated by Ray Brassier. London: Continuum.

Melaina, Marc. 2007. "Turn of the Century Refueling: A Review of Innovations in Early Gasoline Refueling Methods and Analogies for Hydrogen." *Energy Policy* 35 (10): 4919–34.

"Melancholy Catastrophe." 1805. *Times*, Issue 6539. October 1: 3.

Melville, Herman. 1956. *Moby Dick*. Edited by Alfred Kazin. Boston: Houghton-Mifflin-Riverside.

———. 2007. *Moby-Dick*. Edited by John Bryant and Haskell Springer. New York: Longman.

Merrill, Karen R. 2012. "Texas Metropole: Oil, the American West, and US Power in the Postwar Years." *Oil in American History* 99 (1): 197–207.

Merton, Thomas. 2003. *When Trees Say Nothing: Writings on Nature*. Edited by Kathleen Deignan. Notre Dame, IN: Sorin.

Mies, Maria. 1986. *Patriarchy and Accumulation on a World Scale*. London: Zed.

Miéville, China. 2011. "Covehithe." *Guardian: Oil Stories*, April 22. https://www.theguardian.com/books/2011/apr/22/china-mieville-covehithe-short-story.

"Mike Beard: Should You Buy Small Oil and Gas Stocks for Deep Profits?" 2014. *Kapital Wire*, March 19. http://wire.kapitall.com/investment-idea/oil-and-gas-stocks-mike-beard/.

Miller, Daniel, ed. 2001. *Car Cultures*. New York: Berg.

Miller, Perry. 1956. *Errand into the Wilderness*. New York: Harper and Row.

Miller, Rich. 2013. "Intel: Pollution in Asia Shortens Server Component Life." *Data Center Knowledge*, November 25. http://www.datacenterknowledge.com/archives/2013/11/25/intel-pollution-asia-shortens-server-component-life/.

Miller, Shawn William. 2000. *Fruitless Trees: Portuguese Conservation and Brazil's Colonial Timber*. Stanford, CA: Stanford University Press.

Milman, Oliver. 2015. "Mining Industry's New 'Coal Is Amazing' TV Ad Labelled Desperate." *Guardian*, September 6. http://www.theguardian.com/environment/2015/sep/06/mining-industrys-new-coal-is-amazing-tv-ad-slammed-as-desperate.

Minqi, Li. 2008. *The Rise of China and the Demise of the Capitalist World Economy*. New York: Monthly Review Press.

Mirzoeff, Nicholas. 2014. "Visualizing the Anthropocene." *Public Culture* 26 (2): 213–32.

Misrach, Richard. 2013. *Petrochemical America*. Photographs. Fraenkel Gallery, San Francisco.

Mitchell, Katharyne, Sallie A. Marston, and Cindi Katz. 2004. *Life's Work: Geographies of Social Reproduction*. Oxford: Antipode.

Mitchell, Timothy. 2009. "Carbon Democracy." *Economy and Society* 38 (3): 399–432.

———. 2011. *Carbon Democracy: Political Power in the Age of Oil*. London: Verso.

Mitchell, W. J. T. 1994. *The Reconfigured Eye: Visual Truth in the Post-Photographic Era*. Cambridge, MA: MIT Press.

Mitchell, W. J. T., and Mark B. N. Hansen, eds. 2010. *Critical Terms for Media Studies*. Chicago: University of Chicago Press.

Mitropoulos, Angela. 2013. *Contract and Contagion*. New York: Minor Compositions.

Monbiot, George. 2007. *Heat: How to Stop the Planet from Burning*. Toronto: Anchor.

Moore, Jason. 2003. "The Modern World-System as Environmental History? Ecology and the Rise of Capitalism." *Theory and Society* 32: 307–77.

———. 2010. "The End of the Road? Agricultural Revolutions in the Capitalist World-Ecology, 1450–2010." *Journal of Agrarian Change* 10 (3): 389–413.

———. 2011a. "Ecology, Capital, and the Nature of Our Times." *American Sociology Association* 17 (1): 108–47.

———. 2011b. "Transcending the Metabolic Rift: A Theory of Crises in the Capitalist World-Ecology." *Journal of Peasant Studies* 38 (1): 1–46.

———. 2012. "Cheap Food and Bad Money." *Review* 33 (2–3): 1–29.

———. 2013. "From Object to Oikeios: Environment-Making in the Capitalist World-Ecology." Unpublished, Fernand Braudel Center, Binghamton University. http://www.jasonwmoore.com/uploads/Moore__From_Object_to_Oikeios__for_website__May_2013.pdf.

———. 2014a. "The Capitalocene, Part I: On the Nature and Origins of Our Ecological Crisis." Unpublished, Fernand Braudel Center, Binghamton University. http://www.jasonwmoore.com/uploads/The_Capitalocene__Part_I__June_2014.pdf.

———. 2014b. "The Capitalocene, Part II: Abstract Social Nature and the Limits to Capital." Unpublished, Fernand Braudel Center, Binghamton University. http://www.jasonwmoore.com/uploads/The_Capitalocene___Part_II__June_2014.pdf.

———. 2014c. "The End of Cheap Nature." In *Structures of World Political Economy and the Future of Global Conflict and Cooperation*, edited by Christian Suter and Christopher Chase-Dunn, 285–314. Berlin: LIT Verlag.

Moore, Lisa. 2009. *February*. Toronto: Anansi.

Moores, Shaun. 2014. "Digital Orientations: 'Ways of the Hand' and Practical Knowing in Media Uses and Other Manual Activities." *Mobile Media and Communication* 2 (2): 196–208.

Moors, Kent. 2011. *The Vega Factor: Oil Volatility and the Next Global Crisis*. New York: Wiley.

Morales, Alex. 2013. "Fossil Fuels Need to Stay Unburned to Meet Climate Target." *Bloomberg*, September 27. http://www.bloomberg.com/news/articles/2013-09-27/fossil-fuels-need-to-stay-unburned-to-meet-climate-target.

Morel, Edmund Dene. 1920. *The Black Man's Burden: The White Man in Africa from the Fifteenth Century to World War I*. New York: Monthly Review Press.

Moretti, Franco. 1998. *Atlas of the European Novel, 1800–1900*. London: Verso.

———. 2007. *Graphs, Maps, Trees*. London: Verso.

Morton, Oliver. 2007. *Eating the Sun: How Plants Power the Planet*. New York: Harper Collins.

Morton, Timothy. 2009. *Ecology without Nature: Rethinking Environmental Aesthetics*. Boston: Harvard University Press.

———. 2013. *Hyperobjects: Philosophy and Ecology after the End of the World*. Minneapolis: University of Minnesota Press.

Mouhot, Jean-François. 2011. "Past Connections and Present Similarities in Slave Ownership and Fossil Fuel Usage." *Climatic Change* 105 (1): 329–55.

Muecke, Stephen. 2005. "Country." In *New Keywords: A Revised Vocabulary of Culture and Society*, edited by Tony Bennett, Lawrence Grossberg, and Meaghan Morris, 61–62. London: Blackwell.

Mulvey, Kathy, and Seth Shulman. 2015. "The Climate Deception Dossiers: Internal Fossil Fuel Industry Memos Reveal Decades of Corporate Misinformation." *Union of Concerned Scientists*. http://www.ucsusa.org/global-warming/fight-misinformation/climate-deception-dossiers-fossil-fuel-industry-memos#.VodAh1c3FTd.

Mumford, Lewis. 1964. "Authoritarian and Democratic Technics." *Technology and Culture* 5 (1): 1–8.

———. 2000. *Art and Technics*. New York: Columbia University Press.

———. 2010. *Technics and Civilization*. Chicago: University of Chicago Press.

Munif, Abdelrahman. 1987. *Cities of Salt*. New York: Vintage.

———. 1989a. *Al-munbatt (The Uprooted)*. Beirut: al-Muassasa al-Arabiyya lid-Dirasat wan-Nashr.

———. 1989b. *Badiyat al zulumat (The Desert of Darkness)*. Beirut: al-Muassasa al-Arabiyya lid-Dirasat wan-Nashr.

———. 1991. *The Trench*. New York: Vintage.

———. 1993. *Variations on Night and Day*. New York: Vintage.

Munro, Al. n.d. "Textile Geometries: A Speculation on Stretchy Space." *inter-disciplinary*, 1–4. http://www.inter-disciplinary.net/critical-issues/wp-content/uploads/2013/08/AlMunro_sp4-wpaper.pdf.

Nace, Ted. 2009. *Climate Hope: On the Front Lines of the Fight against Coal*. San Francisco: CoalSwarm.

Nan, Xu. 2013. "Forget the New Air Pollution Plan, GDP Growth Is Still King in China." *China Dialogue*, November 1. https://www.google.ca/?client=safari#q=Forget+the+New+Air+Pollution+Plan,+GDP+Growth+Is+Still+King+in+China.+china+dialogue&gfe_rd=cr.

Nardizzi, Vin. 2013. *Wooden Os: Shakespeare's Theatres and England's Trees*. Toronto: University of Toronto Press.

Nason, Riel. 2011. *The Town that Drowned*. Fredericton: Goose Lane Editions.

National Research Council. 2012. *Terrorism and the Electric Power Delivery System*. Washington, DC: National Academies Press.

"Natural Oil 'Spills': Surprising Amount Seeps into Sea." 2009. *LiveScience*, May 20. http://www.livescience.com/5422-natural-oil-spills-surprising-amount-seeps-sea.html.

Ndiaye, Pap. 2006. *Nylon and Bombs: DuPont and the March of Modern America*. Translated by Elborg Foster. Baltimore, MD: Johns Hopkins University Press.

Needham, Joseph. 1984. *Science and Civilization in China*. Vol. 6. *Biology and Biological Technology: Part II Agriculture*. Cambridge: Cambridge University Press.

Nef, J. U. 1966. *The Rise of the British Coal Industry*. London: Archon Books.

Neocleous, Mark. 2008. *Critique of Security*. Edinburgh: University of Edinburgh Press.

Neslen, Arthur. 2012. "Israel Set to Demolish EU-funded Renewables Projects." *EurActiv*, December 14. http://www.euractiv.com/section/development-policy/news/israel-set-to-demolish-eu-funded-renewables-projects/.

Neuhart, John, Marilyn Neuhart, and Ray Eames, eds. 1989. *Eames Design: The Work of Charles and Ray Eames*. New York: Abrams.

Neuman, Justin. 2012. "Petromodernism." Presentation at Petrocultures, University of Alberta, Edmonton, September 7–8.

Niblett, Michael. 2012. "World-Economy, World-Ecology, World Literature." *Green Letters* 16 (1): 15–30.

Nietzsche, Frederick. 1997. *Daybreak Thoughts on the Prejudices of Morality*. Cambridge: Cambridge University Press.

Nikiforuk, Andrew. 2010. *Tar Sands: Dirty Oil and the Future of a Continent*. Vancouver: Greystone.
———. 2012. *The Energy of Slaves: Oil and the New Servitude*. Vancouver: Greystone.
Nixon, Rob. 2011. *Slow Violence and the Environmentalism of the Poor*. Cambridge, MA: Harvard University Press.
Noah, Timothy. 2012. *The Great Divergence: America's Growing Inequality Crisis and What We Can Do about It*. New York: Bloomsbury.
Noreng, Oystein. 2002. *Crude Power: Politics and the Oil Market*. London: Taurus.
Notley, Tanya, and Anna Reading. 2013. "Rare Earths and Our Insatiable Appetite for Digital Memory." *The Conversation*, November 28. http://theconversation.com/rare-earths-and-our-insatiable-appetite-for-digital-memory-20938.
Nowatzki, Mike. 2014. "1 in 7 Private Sector Jobs Tied to Oil, Gas." *The Dickinson Press*, July 30. http://www.thedickinsonpress.com/content/1-7-private-sector-jobs-tied-oil-gas-industry-accounts-285-percent-wages-data-could-affect.
"Nuclear Boy (Subbed)." 2011. YouTube video, 4:33, posted by "Shibata Bread," March 16. http://www.youtube.com/watch?v=5sakN2hSVxA.
"Nuns Versus Pipeline." 2014. *Religion and Ethics News Weekly*, January 31. http://www.pbs.org/wnet/religionandethics/2014/01/31/january-31-2014-sisters-of-loretto-nuns-versus-pipeline/21909/.
Nwokeji, G. Ugo. 2008. "Slave Ships to Oil Tankers." In *Curse of Black Gold: 50 Years of Oil in the Niger Delta*, edited by Michael Watts, 63–65. Brooklyn: powerHouse.
Nye, David D. 1994. *American Technological Sublime*. Cambridge, MA: MIT Press.
Nye, David E. 1990. *Electrifying America: Social Meanings of a New Technology*. Cambridge, MA: MIT Press.
O'Brien, Charles, Nora Volkow, and T. K. Li. 2006. "What's in a Word? Addiction versus Dependence in DSM-V." *American Journal of Psychiatry* 163: 764–65.
O'Connor, James. 1998. *Natural Causes: Essays in Ecological Marxism*. New York: Guilford Press.
OECD. 2007. "Infrastructure to 2030: Volume 2, Mapping Policy for Electricity, Water, and Transport." *Organization for Economic Co-Operation and Development*. https://www.oecd.org/futures/infrastructureto2030/40953164.pdf.
Office for Metropolitan Architecture. 2010. *Roadmap 2050: A Practical Guide to a Prosperous, Low-Carbon Europe Volume III: Graphic Narrative*. The Hague: European Climate Foundation.
Offshore: A Feature-Length Interactive Documentary. 2013. Directed by Brenda Longfellow. Glen Richards and Helios Design Labs..
The Oil (Neft). 2003. Directed by Murad Ibragimbekov.
Oil Wells of Baku: Close View. 1896. Directed by Lumière Brothers. Kino DVD.
Ojaide, Tanure. 2010. [2006]. *The Activist*. Princeton: AMV Publishing.
Oklahoma Crude. 1973. Directed by Stanley Kramer. Stanley Kramer Productions.
Okri, Ben. 1988. "What the Tapster Saw." In *Stars of the New Curfew*, 183–94. New York: Penguin.
Olien, Diana Davids, and Roger M. Olien. 2002. *Oil in Texas: The Gusher Age 1895–1945*. Austin: University of Texas Press.
O'Malley, Pat. 2010. "Resilient Subjects: Uncertainty, Warfare and Liberalism." *Economy and Society* 39 (4): 488–509.
"OPEC Share of World Crude Oil Reserves, 2014." 2016. *Organization of the Petroleum Exporting Countries*. http://www.opec.org/opec_web/en/data_graphs/330.htm.
Osodi, George. 2011. *Delta Nigeria: The Rape of Paradise*. London: Trolley Books.
Owen, David. 2010. *Green Metropolis: Why Living Smaller, Living Closer, and Driving Less are the Keys to Sustainability*. New York: Riverhead.
Oxford English Dictionary. 2013. Oxford: Oxford University Press.
Papanin, Ivan. 1939. *Life on the Icefloe*. New York: Messner.
Parenteau, Bill. 1992. "The Woods Transformed: The Emergence of the Pulp and Paper Industry in New Brunswick." *Acadiensis: Journal of the History of the Atlantic Region* 22 (1): 5–43.
Parikka, Jussi. 2010. *Insect Media: An Archaeology of Animals and Technology*. Minneapolis: University of Minnesota Press.

———. 2012. "New Materialism as Media Theory: Medianatures and Dirty Matter." *Communication and Critical/Cultural Studies* 9 (1): 95–100.
Parisi, Luciana, and Tiziana Terranova. 2000. "Heat Death: Emergence and Control in Genetic Engineering and Artificial Life." *ctheory*, May 19. https://journals.uvic.ca/index.php/ctheory/article/view/14604/5455.
Park, Chris. 2013. *Acid Rain: Rhetoric and Reality*. New York: Routledge.
Parks, Lisa, and Nicole Starosielski. 2015. *Signal Traffic: Critical Studies of Media Infrastructures*. Champaign-Urbana: University of Illinois Press.
Parsi, Trita. 2007. *Treacherous Alliance: The Secret Dealings of Israel, Iran, and the United States*. New Haven, CT: Yale University Press.
Paster, Gail Kern. 2004. *Humoring the Body: Emotions and the Shakespearean Stage*. Chicago: University of Chicago Press.
Paterson, Matthew, and Johannes Stripple. 2010. "My Space: Governing Individuals' Carbon Emissions." *Environment and Planning D: Society and Space* 28: 341–62.
"Peace on Earth: Dream or Possibility?" 1946. *LOOK Magazine*, December 24.
Pearson, Chris. 2006. "'The Age of Wood': Fuel and Fighting in French Forests, 1940–1944." *Environmental History* 11 (4): 775–803.
Pérez Alonso, Juan. 1976. *Hundiéndonos en el exremento del Diablo*. Caracas: Editorial Lisbona.
Perlin, John. 2005. *A Forest Journey: The Story of Wood and Civilization*. Woodstock, VT: Countryman.
Pesek, William. 2013. "China Is Choking on Its Success." *Bloomberg*, November 4. https://www.bloomberg.com/view/articles/2013-11-04/china-is-choking-on-its-success.
Petroleum Economist. 2012. *Major Pipelines of the World, 4th Edition*. London: Petroleum Economist Cartographic.
Pettner, Julian, and Nigel Turner. 1984. *Automania: Man and the Motor Car*. London: Collins.
"Pharrell Williams Announces Clothing Created from Plastic Recycled from the Oceans." 2014. *Look to the Stars: The World of Celebrity Giving*, February 7. https://www.looktothestars.org/news/11490-pharrell-williams-announces-clothing-created-from-plastic-recycled-from-the-oceans.
Pietri, Arturo Úslar. 2013. [1936.] *Sembrar el Petróleo*. October 18. http://webdelprofesor.ula.ve/economia/ajhurtado/lecturasobligatorias/sembrar%20el%20petroleo.pdf.
Pinkus, Karen. 2011. "The Risks of Sustainability." In *Criticism, Crisis, and Contemporary Narrative: Textual Horizons in an Age of Global Risk*, edited by Paul Crosthwaite, 62–80. London: Routledge.
Pioneer. 2013. Directed by Eric Skjoldbjaerg. TrustNordisk.
Planet Green. 2014. "How to Go Green: Fashion and Beauty." *Planet Green*, March 9. http://planetgreen.dis%20covery.com/go-green/green-index/fashion-beauty-guides.html.
Plato. 2008. "Republic." In *Great Dialogues of Plato*. Translated by W. H. D. Rouse, 118–422. New York: Signet.
Podobnik, Bruce. 1999. "Toward a Sustainable Energy Regime: A Long-Wave Interpretation of Global Energy Shifts." *Technological Forecasting and Social Change* 62: 155–72.
———. 2006. *Global Energy Shifts: Fostering Sustainability in a Turbulent Age*. Philadelphia: Temple University Press.
Polanyi, Karl. 2001. *The Great Transformation: The Political and Economic Origins of Our Time*. [1944.] New York: Beacon.
Pollan, Michael. 2006. *The Omnivore's Dilemma: A Natural History of Four Meals*. New York: Penguin.
Pomeranz, Kenneth. 2000. *The Great Divergence: China, Europe, and the Making of the Modern World Economy*. Princeton, NJ: Princeton University Press.
Poovey, Mary. 1998. *A History of the Modern Fact: Problems of Knowledge in the Sciences of Wealth and Society*. Chicago: University of Chicago Press.
Porter, Eduardo. 2014. "Old Forecast of Famine May Yet Come True." *New York Times*, April 1. http://www.nytimes.com/2014/04/02/business/energy-environment/a-200-year-old-forecast-for-food-scarcity-may-yet-come-true.html?_r=0.

Posner, Eric, and David Weisbach. 2010. *Climate Change Justice*. Princeton, NJ: Princeton University Press.
Povinelli, Elizabeth. 2006. *The Empire of Love: Toward a Theory of Intimacy, Genealogy and Carnality*. Durham, NC: Duke University Press.
Pratt, Joseph. 1980. "The Petroleum Industry in Transition: Antitrust and the Decline of Monopoly Control in Oil." *Journal of Economic History* 40 (4): 815–37.
Press, Jordan. 2013. "Attawapiskat Audit Shows Questionable Spending Practices after Receiving More Than $100 Million in Federal Funding in Six Years." *National Post*, January 7. http://news.nationalpost.com/news/canada/canadian-politics/attawapiskat-audit-shows-questionable-spending-practices-after-receiving-more-than-100-million-in-federal-funding-in-six-years.
Prochaska, Georg. 1820. *Physiologie oder Lehre von der Natur des Menschen*. Vienna: Beck.
Provost, Claire, and Matt Kennard. 2014. "Hamburg at Forefront of Global Drive to Reverse Privatization of City Services." *Guardian*, November 12. http://www.theguardian.com/cities/2014/nov/12/hamburg-global-reverse-privatisation-city-services.
"pub AREVA - Une histoire qui n'a pas fini de s'écrire," 2011. YouTube video, 1:00, posted by "SuperMax242," January 16. http://www.youtube.com/watch?v=rUg1Cpnj8KA.
Pumzi. 2009. Directed by Wanuri Kahiu. Inspired Minority Pictures.
Quijano, Aníbal. 2003. "Colonialidad del Poder, Eurocentrismo, y América Latina." In *La Colonialidade del Saber: Eurocentrismo y Ciencias Sociales*, edited by Edgardo Lander, 201–46. Buenos Aires: CLACSO.
Quinn, Jill Sisson. 2010. *Deranged: Finding a Sense of Place in the Landscape and in the Lifespan*. Baltimore: Apprentice House.
Raban, Jonathan. 1992. *The Oxford Book of the Sea*. Oxford: Oxford University Press.
Rabinow, Paul, and Nikolas Rose. 2006. "Biopower Today." *BioSocieties* 1 (2): 195–217.
Radford, Tim. 2004. "Do Trees Pollute the Atmosphere?" *Guardian*, May 13. https://www.theguardian.com/science/2004/may/13/thisweekssciencequestions3.
Radin, Margaret Jane. 1993. *Reinterpreting Property*. Chicago: University of Chicago Press.
Radway, Janice. 1981. "The Utopian Impulse in Popular Literature: Gothic Romances and 'Feminist' Protest." *American Quarterly* 33 (2): 140–62.
Rajan, Sudhir Chella. 2006. "Automobility and the Liberal Disposition." In *Against Automobility*, edited by Steffen Böhm, Campbell Jones, Chris Land, and Matthew Paterson, 113–29. New York: Wiley.
Reddy, William. 2001. *The Navigation of Feeling: A Framework for the History of Emotions*. Cambridge: Cambridge University Press.
Reil, Johann Christian. 1910. *Von der Lebenskraft*. [1795.] Edited by Karl Sudhoff. Leipzig: Barth.
Reilly, Evelyn. 2009. *Styrofoam*. New York: Roof Books.
Renwick, D. W. S., T. Redman, and S. Maguire. 2013. "Green Human Resource Management: A Review and Research Agenda." *International Journal of Management Reviews* 15 (1): 1–14.
Revkin, Andrew. 1992. *Global Warming: Understanding the Forecast*. New York: Abbeville Press.
Richards, John. 2003. *The Unending Frontier: An Environmental History of the Early Modern World*. Berkeley: University of California Press.
Richmond, Holly. 2014. "Pharrell Is Making Jeans Out of Recycled Ocean Plastic." *Grist*, February 18. http://grist.org/living/pharrell-is-making-jeans-out-of-recycled-ocean-plastic/.
Ricoeur, Paul. 1986. *The Symbolism of Evil*. Boston: Beacon Press.
Rifkin, Jeremy. 1993. *Beyond Beef: The Rise and Fall of the Cattle Culture*. New York: Penguin.
Riggs, Mike. 2013. "Intense Smog Is Making Beijing's Massive Surveillance Network Practically Useless." *Atlantic*, November 5. http://www.citylab.com/tech/2013/11/intense-smog-making-beijings-massive-surveillance-network-practically-useless/7481/.
"Risk." *Merriam-Webster*. http://www.merriam-webster.com/dictionary/risk.
Ritter, Johann Wilhelm. 1798. *Beweis, daß ein beständiger Galvanismus den Lebensprozess im Thierreich begleite*. Weimar, Germany: Industrie-Comptoir.
Roberts, Adam. 2006. *The Wonga Coup: Guns, Thugs, and a Ruthless Determination to Create Mayhem in an Oil Rich Corner of Africa*. London: Profile Books.

Robertson, Grant. 2013. "U.S. Officials Were Probing Safety of Bakken Oil Months Before Lac-Mégantic." *Globe and Mail*, August 29. http://www.theglobeandmail.com/report-on-business/industry-news/energy-and-resources/us-officials-were-probing-safety-of-bakken-oil-route-months-before-lac-megantic/article14032762/.

Robertson, Kirsty. Forthcoming. "Oil Futures/Petro-Fabrics." In *Petrocultures: Oil, Energy, Culture*, edited by Sheena Wilson, Adam Carlson, and Imre Szeman. Montreal: McGill-Queen's University Press.

Robinson, Kim Stanley. 1993. *Red Mars*. New York: Spectra.

———. 1994. *Green Mars*. New York: Spectra.

———. 1996. *Blue Mars*. New York: Spectra.

Rogers, Douglas. 2012. *Oil without Money: The Significance of Bartered Oil in (and beyond) Russia*. Seattle: National Council for Eurasian and East European Research.

Rogoway, Mike. 2013. "Intel Finds Asian Pollution Makes Computers Sick, Too." *Oregon Live*, October 19. http://www.oregonlive.com/silicon-forest/index.ssf/2013/10/intel_finds_asian_pollution_ma.html.

Rose-Redwood, Reuben. 2008. "Genealogies of the Grid: Revisiting Stanislawski's Search for the Origin of the Grid-pattern Town." *Geographical Review* 98 (1): 42–58.

Rosenwein, Barbara. 2008. *Emotional Communities in the Early Middle Ages*. Ithaca, NY: Cornell University Press.

Ross, Deveryn. 2013. "Idle No More's Real Challenge." *Winnipeg Free Press*, January 24. http://www.winnipegfreepress.com/opinion/analysis/idle-no-mores-real-challenge-188173011.html.

Ross, Kristin. 1996. *Fast Cars, Clean Bodies: Decolonization and the Reordering of French Culture*. Cambridge, MA: MIT Press.

Ross, Michael. 2012. *The Oil Curse: How Petroleum Wealth Shapes the Development of Nations*. Princeton, NJ: Princeton University Press.

Roy, Arundhati. 2007. *The Cost of Living*. New York: Random House.

———. 2008. *The God of Small Things*. New York: Random House.

Ruddiman, William. 2003. "The Anthropogenic Greenhouse Era Began Thousands of Years Ago." *Climatic Change* 61: 261–93.

———. 2005. *Plows, Plagues, and Petroleum: How Humans Took Control of Climate*. Princeton, NJ: Princeton University Press.

———. 2013. "The Anthropocene." *Annual Review of Earth and Planetary Sciences* 41 (4): 22–38.

Ruggles, Kate. 2014. "Learning God in North Dakota." *On Second Thought*: 12–14.

Rumsey, Abby Smith. 2016. *When We Are No More: How Digital Memory Is Shaping Our Future*. London: Bloomsbury Publishing. Kindle Edition.

Runyan, Keith. 2013. "Kentucky Nuns Issue 'Energy Vision,' Urge Halt to Pipeline." *Huffington Post*, December 12. http://www.huffingtonpost.com/keith-runyon/kentucky-nuns-issue-energy_b_4427853.html.

Russell, Karl, and Keith Bradsher. 2016. "Selling More Cars in China." *New York Times*, March 28. http://www.nytimes.com/interactive/2016/03/24/business/international/china-car-sales.html?emc=eta1&_r=0.

Rutherford, Anne. 2014. "Mimetic Innervation." In *The Routledge Encyclopedia of Film Theory*, edited by Edward Branigan and Warren Buckland, 285–89. New York: Routledge.

Sabin, Paul. 2005. *Crude Politics: The California Oil Market, 1900–1940*. Berkeley: University of California Press.

Sachs, Joe. n.d. "Aristotle: Motion and Its Place in Nature." *Internet Encyclopedia of Philosophy*. http://www.iep.utm.edu/aris-mot/.

Said, Edward. 1978. *Orientalism*. New York: Vintage.

———. 1982. "Opponents, Audiences, Constituencies, and Community." *Critical Inquiry* 9 (1): 1–26.

———. 1993. *Culture and Imperialism*. New York: Vintage.

———. 2004. *Humanism and Democratic Criticism*. New York: Columbia University Press.

Sampirisi, Jenny. 2011. *Croak*. Toronto: Coach House Books.

Sampson, Anthony. 1975. *The Seven Sisters: The Great Oil Companies and the World They Shaped.* New York: Bantam Books.
Samson, Daniel, ed. 1994. Introduction to *Contested Countryside: Rural Workers and Modern Society in Atlantic Canada, 1800–1950*, 1–33. Fredericton, New Brunswick: Acadiensis Press for the Gorsebrook Research Institute for Atlantic Canada Studies.
Sanders, Scott Russell. 1985. *In Limestone Country*. Boston: Beacon Press.
Sassen, Saskia. 2001. *The Global City: New York, London, Tokyo*. Princeton, NJ: Princeton University Press.
Sawyer, Suzana. 2001. "Fictions of Sovereignty: Of Prosthetic Petro-capitalism, Neoliberal States, and Phantom-like Citizenship in Ecuador." *Journal of Latin American Anthropology* 6 (1): 156–97.
Shaftesbury, Anthony Ashley Cooper, Third Earl of. 2001. *Characteristics of Men, Manners, Opinions, Times*. Indianapolis: Liberty Fund.
Shamah, David. 2015. "Israeli Robot-cleaning System Promises Brighter Future for Solar Power." *Times of Israel*, December 13. http://www.timesofisrael.com/israeli-robot-cleaning-system-promises-brighter-future-for-solar-power/.
Scheer, Hermann. 2004. *The Solar Economy: Renewable Energy for a Sustainable Global Future*. London: Earthscan.
Scheerer, Eckart. 1989. "On the Will: A Historical Perspective." In *Volitional Action: Conation and Control*, edited by Wayne A. Hersberger, 39–60. Amsterdam: Elsevier Science Publishers.
Schewe, Philip. 2007. *The Grid: A Journey through the Heart of Our Electrified World*. Washington, DC: Joseph Henry Press.
Schiller, Naomi. 2011. "'Now That the Petroleum Is Ours': Community, Media, State Spectacle and Oil Nationalism in Venezuela." In *Crude Domination: An Anthropology of Oil*, edited by Andrea Behrends, Stephen P. Reyna, and Günther Schlee, 190–219. New York: Berghahn Books.
Schneider, Eric D., and Dorion Sagan. 2005. *Into the Cool: Energy Flow, Thermodynamics, and Life*. Chicago: University of Chicago Press.
Schneider-Mayerson, Matthew. 2015. *Peak Politics: Apocalyptic Environmentalism and Libertarian Political Culture in the United States*. Chicago: University of Chicago Press.
Schofield, Barry. 2002. "Partners in Power: Governing the Self-Sustaining Community." *Sociology* 36 (3): 663–83.
Schoolhouse Rock! 1979. "Electricity, Electricity!" Aired January 6.
Schopenhauer, Arthur. 1962. "Über den Willen in der Natur." In *Sämtliche Werke: Kleinere Schriften*. Vol. 3. Edited by Wolfgang von Löhneysen, 301–479. Darmstadt, Germany: Wissenschaftliche Buchgesellschaft.
Schumacher, E. F. 2010. *Small Is Beautiful: Economics as if People Mattered*. [1973.] New York: Perennial.
Scranton, Roy. 2013. "Learning How to Die in the Anthropocene." *New York Times*, November 10. http://opinionator.blogs.nytimes.com/2013/11/10/learning-how-to-die-in-the-anthropocene/.
Seigworth, Gregory J., and Melissa Gregg. 2010. "An Inventory of Shimmers." In *The Affect Theory Reader*, edited by Melissa Gregg and Gregory J. Seigworth, 1–28. Durham, NC: Duke University Press.
Sekula, Allan. 2003. *Black Tide/Merea Negra*. Santa Monica, CA: Christopher Grimes Gallery. June 7–July 12.
Seiler, Cotton. 2012. "Welcoming China to Modernity: US Fantasies of Chinese Automobility." *Public Culture* 24 (2): 357–84.
Serres, Michel. 1995a. *Atlas*. Paris: Flammarion.
———. 1995b. *The Natural Contract*. Translated by Elizabeth MacArthur and William Paulson. Ann Arbor: University of Michigan Press.
———. 1998. *Les Cinq Sens*. Paris: Hachette.
Sesser, Stan. 2010. "An Oil Rig's Second, Scuba-Diving Life." *Wall Street Journal*, September 18. http://www.wsj.com/articles/SB10001424052748703376504575491662467118800.

"Settlers Destroy Solar Panels in South Hebron Hills." 2014. *Ma'an News Agency*, March 27. http://www.maannews.com/Content.aspx?id=685211.

Shaffer, Brenda, and Taleh Ziyadov. 2012. *Beyond the Resource Curse*. Philadelphia: University of Pennsylvania Press.

Shakespeare, William. 1997. *The Complete Works of Shakespeare*. Edited by David Bevington. New York: Longman.

———. 1999. *The Tempest*. Edited by Virginia Mason Vaughan and Alden T. Vaughan. London: Cengage Learning.

"SHAMS 1 CSP Solar Power Plant, Abu Dhabi, Inauguration Video, 2013," 2013. Vimeo video, 4:46, posted by JBM, March 20. http://vimeo.com/62266148.

Shapiro, Stephen. 2008. "Transvaal, Transylvania: *Dracula*'s World-System and Gothic Periodicity." *Gothic Studies* 10 (1): 29–47.

Sheeler, Charles. 1930. *American Landscape*. Painting. Museum of Modern Art, New York.

———. 1940. "Power: Six Paintings by Charles Sheeler." *Fortune Magazine*, December.

Shell Energy Scenarios to 2050. 2008. Shell International BV. http://www.shell.com/content/dam/shell/static/public/downloads/brochures/corporate-pkg/scenarios/shell-energy-scenarios 2050.pdf.

Sheller, Mimi. 2004. "Automotive Emotions: Feeling the Car." *Theory, Culture & Society* 21 (4–5): 221–42.

Shelley, Mary. 2002. *Frankenstein*. New York: Penguin.

Shelley, Percy Bysshe. 1973. "A Defence of Poetry." In *The Oxford Anthology of English Literature*. Vol. 4. *Romantic Poetry and Prose*, edited by Harold Bloom and Lionel Trilling, 744–61. New York: Oxford University Press.

Shever, Elana. 2008. "Neoliberal Associations: Property, Company, and Family in the Argentine Oil Fields." *American Ethnologist* 35 (4): 701–16.

"A Shocking Catastrophe Occurred after the Race at Reading on Wednesday." *Times*, Issue 6425. August 31: 3.

Shove, Elizabeth, Mika Pantzar, and Matt Watson. 2012. *The Dynamics of Social Practice: Everyday Life and How It Changes*. London: Sage.

Shukin, Nicole. 2009. *Animal Capital*. Minneapolis: University of Minnesota Press.

Siegel, Jerry, and Joe Shuster. 1938. *Action Comics Vol. 1*. June. DC Comics.

Signals and Signposts: Shell Energy Scenarios to 2050. 2011 Shell International BV. http://www2.warwick.ac.uk/fac/soc/pais/research/researchcentres/csgr/green/foresight/energyenvironment/2011_shell_international_signals_and_signposts_-_shell_energy_scenarios_to_2050.pdf.

Silko, Leslie Marmon. 1998. "Interior and Exterior Landscapes: The Pueblo Migration Stories." In *Speaking for the Generations: Native Writers on Writing*, edited by Simon J. Ortiz, 2–24. Tucson: University of Arizona Press.

Silverstein, Jake. 2012. "The Cold Eye of Larry McMurtry." *Texas Monthly*, November. http://www.texasmonthly.com/the-culture/the-cold-eye-of-larry-mcmurtry/.

Simmel, Georg. 2011. *The Philosophy of Money*. London: Routledge Classics.

Sinclair, Upton. 1927. *Oil!* New York: Grosset & Dunlap.

Skocz, Dennis. 2010. "Husserl's Coal-Fired Phenomenology: Energy and Environment in an Age of Whole-House Heating and Air-Conditioning." *Environmental & Architectural Phenomenology Newsletter* 21 (2): 16–21.

Smil, Vaclav. 1999. "Preindustrial Societies." In *Energies: An Illustrated Guide to the Biosphere and Civilization*, 105–32. Cambridge, MA: MIT Press.

———. 2008. *Oil: A Beginner's Guide*. Oxford: Oneworld Publications.

Smith, Crosbie. 1998. *The Science of Energy: A Cultural History of Energy Physics in Victorian Britain*. Chicago: University of Chicago Press.

Smith, Melanie. 2014. "Fritos and Beans." *On Second Thought*: 16–17.

Smith, Michael V. 2011. *Progress; A Novel*. Toronto: Cormorant Books.

Smith, Neil. 1984. *Uneven Development: Nature, Capital, and the Production of Space*. New York: Blackwell.

Smith, Richard. 2013. "Capitalism and the Destruction of Life on Earth: Six Theses on Saving the Humans." *Truthout*, November 10. http://www.truth-out.org/opinion/item/19872-capitalism-and-the-destruction-of-life-on-earth-six-theses-on-saving-the-humans.

Smith, Rick, and Bruce Lourie. 2009. *Slow Death by Rubber Duck: How the Toxic Chemistry of Everyday Life Affects Our Health*. Toronto: Alfred A. Knopf Canada.

Smithson, Robert. 1966. "Entropy and the New Monuments." *Artforum* 4 (10): 26–31.

———. 1969. *Asphalt Rundown*. Photograph. James Cohan Gallery, New York.

———. 1970. *Partially Buried Woodshed*. Photograph. James Cohan Gallery, New York.

———. 1996. "Art Through the Camera's Eye." In *Robert Smithson: The Collected Writings*, edited by Jack Flam, 371–76. Berkeley: University of California Press.

Sofia, Zoe. 2000. "Container Technologies." *Hypatia* 14 (2): 181–200.

Soni, Vivasvan. 2010. *Mourning Happiness: Narrative and the Politics of Modernity*. Ithaca, NY: Cornell University Press.

Soraghan, Mike. 2011. "Baffled About Fracking? You're Not Alone." *New York Times*, May 13.

Sørensen, Bent. 2012. *A History of Energy: Northern Europe from the Stone Age to the Present Day*. Abingdon, VA: Earthscan.

Sörgel, Herman. 1931. "Das Panropa-Projekt als städtebauliche Darstellungsstudie." *Baumeister* 29 (5): 216.

Sorkin, Andrew Ross. 2015. "A New Tack in the War on Mining Mountains: PNC Joins Banks Not Financing Mountaintop Coal Removal." *New York Times*, March 9. http://www.nytimes.com/2015/03/10/business/dealbook/pnc-joins-banks-not-financing-mountaintop-coal-removal.html?_r=0.

Sörlin, Sverker. 2012. "Environmental Humanities: Why Should Biologists Interested in the Environment Take the Humanities Seriously?" *BioScience* 62 (9): 788–89.

Spahr, Juliana. 2007. *The Transformation*. Berkeley, CA: Atelos.

Spengler, Oswald. 2006. *The Decline of the West*. New York: Vintage.

Sperling, Daniel, and Deborah Gordon. 2009. *Two Billion Cars: Driving toward Sustainability*. Oxford: Oxford University Press.

Speth, James Gustave. 2008. *The Bridge at the End of the World: Capitalism, the Environment, and Crossing from Crisis to Sustainability*. New Haven, CT: Yale University Press.

Spinoza, Benedict de. 1959. *Ethics: On the Correction of Understanding*. Translated by Andrew Boyle. London: Everyman's Library.

———. 2009. [1677]. *The Ethics (Ethica Ordine Geometrico Demonstrata)*. Part 3. Translated by Robert Harvey Monro Elwes. Project Gutenberg, May 28. http://www.gutenberg.org/files/3800/3800-h/3800-h.htm.

Spittles, Brian. 1995. *Britain Since 1960: An Introduction*. Houndmills, Hampshire, UK: Macmillan.

Standish, Arthur. 1611. *The Commons Complaint*. London: William Stansby.

Starosielski, Nicole. 2015. "Fixed Flow: Undersea Cables as Media Infrastructure." In *Signal Traffic: Critical Studies of Media Infrastructures*, edited by Lisa Parks and Nicole Starosielski, 53–70. Chicago: University of Illinois Press.

Stearns, Peter. 1989. *Jealousy: The Evolution of an Emotion in American History*. New York: New York University Press.

Stedman, Edmund Clarence, ed. 1905. *The New York Stock Exchange*. New York: Stock Exchange Historical Co.

Steffen, Will, Paul J. Crutzen, and John R. McNeill. 2007. "The Anthropocene: Are Humans Overwhelming the Great Forces of Nature?" *Ambio: A Journal of the Human Environment* 36 (8): 614–21.

Steffen, Will, Jacques Grinevald, Paul J. Crutzen, and John McNeill. 2011. "The Anthropocene: Conceptual and Historical Perspectives." *Philosophical Transactions of the Royal Society A* 369: 842–67.

Steigerwald, Joan. 2014. "Treviranus' Biology: Generation, Degeneration and the Boundaries of Life." In *Gender, Race, and Reproduction: Philosophy and the Early Life Sciences in Context*, edited by Suzanne Lettow, 105–27. Albany: SUNY Press.

Steinberg, Philip E., Elizabeth Nyman, and Mauro J. Caraccioli. 2012. "Atlas Swam: Freedom, Capital, and Floating Sovereignties in the Seasteading Vision." *Antipode* 44 (4): 1532–50.

Stelzig, Christine, Eva Ursprung, and Stefan Eisenhofer, eds. 2012. *Last Rites Niger Delta: The Drama of Oil Production in Contemporary Photographs*. Munich: Staatliches Museum fur Volkerkunde Munchen.

Stendhal. 2003. [1830]. *The Red and the Black*. Translated by Roger Gard. London: Penguin Classics.

Sterman, David. 2009. "Israel's Solar Industry: Reclaiming a Legacy of Success." *Climate Institute*, July. http://climate.org/archive/topics/international-action/israel-solar.html.

Sterne, Jonathan. 2006. "Transportation and Communication, Together as You've Always Wanted Them." In *Thinking with James Carey: Communications, Transportation, History*, edited by Jeremy Packer and Craig Robertson, 117–36. New York: Peter Lang.

———. 2007. "Out with the Trash: On the Future of New Media." *Residual Media*, edited by Charles R. Acland, 16–31. Minneapolis: University of Minnesota Press.

Stewart, Jon. 2010. "An Energy-Independent Future." *The Daily Show*, Comedy Central, July 16.

Stoler, Ann Laura. 2013. *Imperial Debris: On Ruins and Ruination*. Durham, NC: Duke University Press.

Stone, Andrea. 2008. "Oil Boom Creates Millionaires and Animosity in North Dakota." *USA Today*, September 9. http://abcnews.go.com/Business/story?id=5768171&page=1&single Page=true.

Strauss, Sarah, Stephanie Rupp, and Thomas Love. 2012. *Cultures of Energy: Power, Practices, Technologies*. Walnut Creek, CA: Left Coast Press.

Stromberg, Joseph. 2013. "What Is the Anthropocene and Are We in It?" *Smithsonian Magazine*, January. http://www.smithsonianmag.com/science-nature/what-is-the-anthropocene-and-are-we-in-it-164801414/?no-ist.

Sturgis, Sue. 2014. "Alabama Oil Train Disaster Met with Official Neglect." *Institute for Southern Studies*, January 21. http://www.southernstudies.org/2014/01/alabama-oil-train-disaster-met-with-official-negle.html.

Subdhan, Abigale. 2014. "Saskatchewan School Officials Backtrack after Banning 'Got Land? Thank an Indian' Hoodie." *National Post*, January 15. http://news.nationalpost.com/news/canada/saskatchewan-school-officials-backtrack-after-banning-got-land-thank-an-indian-hoodie.

Sumner, Jennifer. 2005. *Sustainability and the Civil Commons: Rural Communities in the Age of Globalization*. Toronto: University of Toronto Press.

Superman IV: The Quest for Peace. 1987. Directed by Sidney J. Furie. Cannon Films.

Suzuki, David. 2013. *The Carbon Manifesto*. http://trialofsuzuki.ca/#manifesto.

"Swedes Find 'World's Oldest Tree.'" 2008. *BBC News*, April 17. http://news.bbc.co.uk/2/hi/europe/7353357.stm.

Swimme, Brian Thomas, and Mary Evelyn Tucker. 2011. *Journey of the Universe*. New Haven, CT: Yale University Press.

Swyngedouw, Erik. 2007. "Impossible 'Sustainability' and the Postpolitical Condition." In *The Sustainable Development Paradox: Urban Political Economy in the United States and Europe*, edited by Rob Krueger and David Gibbs. New York: Guilford.

Szeman, Imre. 2007. "System Failure: Oil, Futurity, and the Anticipation of Disaster." *South Atlantic Quarterly* 106 (4): 805–23. http://saq.dukejournals.org/content/106/4/805.full.pdf+html.

———. 2011. "Literature and Energy Futures." *PMLA* 126 (2): 323–25.

———. 2014. "Blind Faith." *Policy Options* (January–February): 16–20.

Szeman, Imre, and Maria Whiteman. 2012. "Oil Imag(e)inaries: Critical Realism and the Oil Sands." *Imaginations* 3 (2): 46–67.

Taine, John. 1946. *The Time Stream*. Buffalo: The Buffalo Book Company.

Tait, Carrie. 2012. "Ottawa Launches Alberta Counterterrorism Unit." *Globe and Mail*, June 6. http://www.theglobeandmail.com/report-on-business/industry-news/energy-and-resources/ottawa-launches-alberta-counterterrorism-unit/article4236422/.

Takach, Geo. 2010. *Will the Real Alberta Please Stand Up?* Edmonton: University of Alberta Press.
Taku River Tlingit First Nation v. British Columbia (Project Assessment Director). 2004. 3 S.C.R. 550, 2004 SCC 74.
Talbot, William Henry Fox. 1968. *The Pencil of Nature*. [1844.] New York: Da Capo.
Tann, Ken. 2010. "Imagining Communities: A Multifunctional Approach to Identity Management in Texts." *New Discourse on Language: Functional Perspectives on Multimodality, Identity, and Affiliation*, edited by Monika Bednarek and J. R. Martin, 153–94. London: Continuum.
Tarr, Joel, and Clay McShane. 2008. "The Horse as an Urban Technology." *Journal of Urban Technology* 15 (1): 5–17.
Tate, Curtis. 2014. "More Oil Spilled from Trains in 2013 than in Previous 4 Decades, Federal Data Show." *McClatchy DC*, January 20. http://www.mcclatchydc.com/news/nation-world/national/economy/article24761968.html.
Taylor, Stephen. 2013. "Theresa Spence, Living It Up." *Stephen Taylor*, January 10. http://www.stephentaylor.ca/2013/01/theresa-spence-living-it-up/.
Tennessee Valley Authority/US Environmental Protection Agency. 2012. "Kingston Ash Recovery Project Non-Time Critical Removal Action River System Baseline Human Health Risk Assessment." Prepared by Jacobs Engineering Group Inc., Document No. EPA-AO-052. July 11. http://www.tva.gov/kingston/admin_record/pdf/NTC/NTC83/App_H_BHHRA_2012-07-11.pdf.
Térranova, Charissa. 2014. *Automotive Prosthetic: Technological Mediation and the Car in Conceptual Art*. Austin: University of Texas Press.
Terranova, Tiziana. 2004. *Network Culture: Politics for the Information Age*. Ann Arbor, MI: Pluto Press.
There Will Be Blood. 2007. Directed by Paul Thomas Anderson. Paramount Vantage.
Thibodeau, Patrick. 2007. "That Sound You Hear? The Next Data Center Problem." *Computerworld*, July 31. http://www.computerworld.com/article/2542868/data-center/that-sound-you-hear—the-next-data-center-problem.html.
———. 2012. "Hurricane Sandy Leaves Wounded Servers in Its Wake." *Computerworld*, November 19. http://www.computerworld.com/article/2493139/data-center/hurricane-sandy-leaves-wounded-servers-in-its-wake.html.
Thomas, Michael. 2007. *American Policy Toward Israel: The Power and Limits of Beliefs*. London: Routledge.
Thomson, Ian, and Robert G. Boutilier. 2011. "Social License to Operate." In *SME Mining Engineering Handbook*, edited by Peter Darling, 1779–96. Littleton, CO: Society for Mining, Metallurgy and Exploration.
Tierney, John. 2008. "Are We Ready to Track Carbon Footprints?" *New York Times*, March 25. http://www.nytimes.com/2008/03/25/science/25tier.html?_r=0.
Tierney, Matthew. 2009. *The Hayflick Limit*. Toronto: Coach House.
"Timing Is Everything." 2013. *Under the Influence*. CBC Radio, January 19. http://www.cbc.ca/undertheinfluence/season-2/2013/01/19/timing-is-everything/.
Tinker Salas, Miguel. 2009. *The Enduring Legacy: Oil, Culture, and Society in Venezuela*. Durham, NC: Duke University Press.
Tipping Point: The Age of the Tar Sands. 2011. Directed by Tom Radford and Niobe Thompson. CBC Television. January 27.
Toffler, Alvin. 1970. *Future Shock*. New York: Bantam.
———. 1980. *The Third Wave*. New York: Bantam.
Tolstoy, Leo. 2002. [1873–1877]. *Anna Karenina*. Translated by Richard Pevear and Larissa Kolokhonsky. London: Penguin.
Tønnessen, Johan Nicolay, and Arne Odd Johnson. 1982. *The History of Modern Whaling*. Berkeley: University of California Press.
Topouzelis, Konstantinos. 2008. "Oil Spill Detection by SAR Images: Dark Formation Detection, Feature Extraction and Classification Algorithms." *Sensors* 8: 6642–59.

Torrance, Morag. 2009. "The Rise of a Global Infrastructure Market Through Relational Investing." *Economic Geography* 85 (1): 75–97.
Toscano, Alberto. 2011. "Logics and Opposition." *Mute* 3 (2): 30–41.
Totman, Conrad. 1989. *The Green Archipelago: Forestry in Preindustrial Japan*. Berkeley: University of California Press.
Traffic. 2000. Directed by Steven Soderbergh. Bedford Falls Company.
Treviranus, Gottfried Reinhold. 1802. *Biologie, oder Philosophie der lebenden Natur für Naturforscher und Aerzte*. Göttingen, Germany: Röwer.
Trotsky, Leon. 1991. *Literature and Revolution*. Translated by Rose Strunsky. London: RedWords.
Trouillot, Michel-Rolph. 1995. *Silencing the Past: Power and the Production of History*. Boston: Beacon Press.
Tsing, Anna. 2005. *Friction: An Ethnography of Global Connection*. Princeton, NJ: Princeton University Press.
Tully, John. 2011. *The Devil's Milk: A Social History of Rubber*. New York: Monthly Review Press.
Tulsa. 1949. Directed by Stuart Heisler. Walter Wanger Productions.
Turcotte, Heather. 2011a. "Contextualizing Petro-Sexual Politics." *Alternatives: Global, Local, Political* 36 (3): 200–20.
———. 2011b. *Petro-Sexual Politics: Global Oil, Legitimate Violence and Transnational Justice*. Charleston, SC: BiblioBazaar.
Turner, Terisa, and Leigh Brownhill. 2004. "Why Women Are at War with Chevron: Nigerian Subsistence Struggles against the International Oil Industry." *Journal of Asian and African Studies* 39 (1–2): 63–93.
Turner, Wallace. 1986. "A Clash Over Aid Effort on the First 'Skid Row.'" *New York Times*, December 2. http://www.nytimes.com/1986/12/02/us/a-clash-over-aid-effort-on-the-first-skid-row.html.
"A Twirling Toy Run by Sun: Gadget is Forerunner of Future Solar Power Machine." 1958. *Life Magazine*, March 24.
"Two Cars in Every Garage and Three Eyes on Every Fish." 1990. *The Simpsons*. Directed by Wesley Archer. 20th Century Fox Television.
United Nations, Department of Economic and Social Affairs, Population Division. 2015. *World Population Prospects: 2015 Revision*. New York: United Nations.
United Nations Economic and Social Council: Report of the Secretary General on Problems of the Human Environment, 47th Session. 1969. Agenda Item 10. May 26.
"Unobtanium." 2016. *Avatar Wiki*. Last modified February 26. http://james-camerons-avatar.wikia.com/wiki/Unobtanium.
Urry, John. 2005. "The 'System' of Automobility." 2005. In *Automobilities*, edited by Mike Featherstone, Nigel Thrift, and John Urry, 25–40. London: Sage.
———. 2006. "Inhabiting the Car." *Against Automobility*. Edited by Steffan Böhm, Campbell Jones, Chris Land, and Matthew Paterson, 17–31. Oxford: Blackwell.
———. 2007. *Mobilities*. Cambridge: Polity.
US Energy Information Administration. 2013. "Highlights: International Energy Outlook 2013." *US Department of Energy*. http://www.eia.gov/countries/cab.cfm?fips=CH.
———. 2014. "Drilling Productivity Report (Petroleum and Other Liquids)." *US Department of Energy*, August 11. http://www.eia.gov/petroleum/drilling/archive/2014/08/.
US Senate Committee on Foreign Relations. 2011. *Avoiding Water Wars: Water Scarcity and Central Asia's Growing Importance for Stability in Afghanistan and Pakistan*. A Majority Staff Report Prepared for the Use of the Committee on Foreign Relations, US Senate, 112th Congress, First Session, February 22. http://www.foreign.senate.gov/imo/media/doc/Senate%20Print%20112-10%20Avoiding%20Water%20Wars%20Water%20Scarcity%20and%20Central%20Asia%20Afgahnistan%20and%20Pakistan.pdf.
Valencia, Alexandra. 2013. "Ecuador Congress Approves Yasuni Basin Oil Drilling in Amazon." *Reuters*, October 4. http://uk.reuters.com/article/uk-ecuador-oil-idUKBRE99302820131004.

Valentine, Katie. 2013. "People Who Live Downwind from Alberta's Oil and Tar Sands Operations Are Getting Blood Cancer." *ThinkProgress*, October 28. http://thinkprogress.org/climate/2013/10/28/2845411/alberta-tar-sands-pollution-cancer/.

Vanderklippe, Nathan. 2013. "How to Fix China's Pollution Problem? It May Not Be Able to Afford It." *Globe and Mail*, November 2. http://www.theglobeandmail.com/news/world/how-to-fix-chinas-pollution-problem-it-may-not-be-able-to-afford-the-solution/article15216255/?page=all.

van Ham, Peter. 2010. *Social Power in International Politics*. London: Routledge.

Vico, Giambattista. 1999. *New Science: Principles of the New Science Concerning the Common Nature of Nations*. Translated by David Marsh. New York: Penguin.

Viney, Leslie. 2010. "Fuelling the Future." *BP Magazine*. http://www.bp.com/en/global/corporate/bp-magazine/bp-magazine-archive.html.

Virilio, Paul. 1998. "The Suicidal State." In *The Virilio Reader*, edited by James Der Derian, 29–45. Oxford: Blackwell.

Voigt, Wolfgang. 1998. *Atlantropa: Weltbauen am Mittelmeer, ein Architektentraum der Moderne*. Hamburg: Dölling und Galitz.

Voser, Peter. 2011. "Resilience and the Energy Sector." Conference Paper. *Cambridge Sustainability Leadership Programme Alumni Reunion*. June 9. http://www.shell.com/media/speeches-and-articles/2011/resilience-and-the-energy-sector.html.

Wackernagel, Mathias, and William Rees. 1996. *Our Ecological Footprint: Reducing Human Impact on the Earth*. Gabriola Island, BC: New Society Publishers.

Wainaina, Binyavanga. 2007. "Glory." *Bidoun* 10. http://bidoun.org/issues/10-technology#glory.

Wainer, David, and Anna Hirtenstein. 2015. "Israel's 300 Days of Sun No Help as Offshore Gas Eclipses Solar." *Bloomberg News*, September 9. http://www.bloomberg.com/news/articles/2015-09-10/israel-s-300-days-of-sun-no-help-as-offshore-gas-eclipses-solar.

Wainwright, David, and Michael Calnan. 2002. *Work Stress: The Making of a Modern Epidemic*. Philadelphia: Open University Press.

Wald, Matthew. 2013. "As Worries over the Power Grid Rise, a Drill Will Simulate a Knockout Blow." *New York Times*, August 16. http://www.nytimes.com/2013/08/17/us/as-worries-over-the-power-grid-rise-a-drill-will-simulate-a-knockout-blow.html.

Waldie, Paul. 2013. "English Hamlet Becomes Unlikely Hub for Global Fracking Debate." *Globe and Mail*, July 30. http://www.theglobeandmail.com/report-on-business/industry-news/energy-and-resources/fracking-debate-heats-up-in-britain/article13522403/.

Walker, Jeremy, and Melinda Cooper. 2011. "Genealogies of Resilience: From Systems Ecology to the Political Economy of Crisis Adaptation." *Security Dialogues* 42 (2): 143–61.

Walker, Martin. 1994. "Lethal Exposure." *Guardian*, June 20. Wall, Wendy. 2002. *Staging Domesticity: Household Work and English Identity in Early Modern Drama*. Cambridge: Cambridge University Press.

Wallace, David Foster. 1996. *Infinite Jest*. New York: Little, Brown and Company.

Wallace, Leslie. 2010. "The Transition Handbook: From Oil Dependency to Local Resilience." *Alternatives Journal*, June 2. http://www.alternativesjournal.ca/community/reviews/transition-handbook-oil-dependency-local-resilience.

Warde, Paul. 2006. *Ecology, Economy and State Formation in Early Modern Germany*. Cambridge: Cambridge University Press.

Watkins, Mel. 2013. "Bitumen as a Staple." *Progressive Economic Forum*, December 26. http://www.progressive-economics.ca/2013/12/26/the-staple-theory-50-mel-watkins/.

Watts, Michael. 2004. "Antinomies of Community: Some Thoughts on Geography, Resources and Empire." *Transactions of the Institute of British Geographers* 29 (2): 195–216.

———. 2011. "Blood Oil: The Anatomy of a Petro-Insurgency in the Niger Delta, Nigeria." In *Crude Domination: An Anthropology of Oil*, edited by Andrea Berhrends and Stephen P. Reyna, 49–80. New York: Berghan Books.

Watts, Michael, and Ed Kashi. 2008. *Curse of the Black Gold*. New York: powerHouse Books.

Wearden, Graeme. 2014. "Oxfam: 85 Richest People as Wealthy as Poorest Half of the World." *Guardian*, January 20. https://www.theguardian.com/business/2014/jan/20/oxfam-85-richest-people-half-of-the-world.
Webb, Walter Prescott. 1964. *The Great Frontier*. Austin: University of Texas Press.
Weeks, Kathi. 2011. *The Problem with Work: Feminism, Marxism, Antiwork Politics, and Postwork Imaginaries*. Durham, NC: Duke University Press.
Wegener, Alfred. 1966. *The Origins of Continents and Oceans*. Translated by John Biram. New York: Dover.
Weis, Tony. 2013. *The Ecological Hoofprint*. London: Zed.
Weisenthal, Joe. 2013. "Here's David Cameron Calling for Permanent Austerity in Front of All Kinds of Ridiculous Gold Things." *Business Insider*, November 12. http://www.businessinsider.com/picture-of-david-cameron-calling-for-austerity-2013-11.
"Welcome in Wildpoldsried." 2014. *Wildpoldsried: Das Energiedorf*. http://www.wildpoldsried.de/index.shtml?homepage_en.
Wells, H. G. 1895. *The Time Machine: An Invention*. New York: H. Holt.
Wenzel, Jennifer. 2006. "Petro-Magic-Realism: Toward a Political Ecology of Nigerian Literature." *Postcolonial Studies* 9 (4): 449–64.
———. 2014a. "How to Read for Oil." *Resilience: A Journal of Environmental Humanities* 1 (2): 156–61.
———. 2014b. "Petro-Magic-Realism Revisited: Un-imagining and Re-imagining the Niger Delta." In *Oil Culture*, edited by Daniel Worden and Ross Barnett, 211–25. Minneapolis: University of Minnesota Press.
Weszhalyns, Gisa. 2013. "Oil's Magic: Materiality and Contestation." In *Cultures of Energy*, edited by Sarah Strauss, Stephanie Rupp, and Thomas Love, 267–83. Walnut Creek, CA: Left Coast Press.
"What is the Earth Charter?" 2012. *The Earth Charter Initiative*. http://earthcharter.org/discover/what-is-the-earth-charter/.
White, Leslie. 1943. "Energy and the Evolution of Culture." *American Anthropologist* 45 (3): 335–56.
White, Richard. 1995. *The Organic Machine*. New York: Hill & Wang.
White, Rob. 2008. *Crimes against Nature: Environmental Criminology and Ecological Justice*. New York: Routledge.
Whitman, Walt. 1982. "I Sing the Body Electric." In *Leaves of Grass: The Complete Poetry and Selected Prose*, 77–84. New York: Library of America.
Wilber, Tom. 2012. *Under the Surface: Fracking, Fortunes, and the Fate of the Marcellus Shale*. New York: Routledge.
Wild, Christopher Paul. 2005. "Complementing the Genome with an 'Exposome': The Outstanding Challenge of Environmental Exposure Measurement in Molecular Epidemiology." *Cancer Epidemiology, Biomarkers and Prevention: A Publication of the American Association for Cancer Research, Cosponsored by the American Society of Preventive Oncology* 18: 1847–50.
Wilder, Robert. 1945. *Written on the Wind*. New York: Putnam.
Williams, Mark et al. 2011. "The Anthropocene: A New Epoch of Geological Time?" *Philosophical Transactions of the Royal Society* 369: 835–1111.
Williams, Michael. 2003. [2006 abridgment]. *Deforesting the Earth: From Prehistory to Global Crisis*. Chicago: University of Chicago Press.
Williams, Raymond. 1973. *The Country and the City*. New York: Oxford University Press.
———. 1980. *Problems in Materialism and Culture*. London: Verso.
———. 1983. *Keywords: A Vocabulary of Culture and Society*. New York: Oxford University Press.
Wilson, Alexander. 1992. *The Culture of Nature: North American Landscape from Disney to the Exxon Valdez*. London: Between the Lines.
Wilson, Sheena. 2014. "Gendering Oil: Tracing Western Petro-sexual Relations through Colonialist and Capitalist Petro-discourses." In *Oil Culture*, edited by Ross Barrett and Daniel Worden, 244–66. Minneapolis: University of Minnesota Press.

Winner, Langdon. 1986. *The Whale and the Reactor: A Search for Limits in an Age of High Technology*. Chicago: University of Chicago Press.
Winther, Tanja. 2011. *The Impact of Electricity: Development, Desires and Dilemmas*. New York: Berghahn Books.
Wolde-Rufael, Yemane. 2006. "Electricity Consumption and Economic Growth: A Time Series Experience for 17 African Countries." *Energy Policy* 34 (10): 1106–114.
Wolfe, Joel. 2010. *Autos and Progress: The Brazilian Search for Modernity*. Oxford: Oxford University Press.
Wordsworth, William. 1979. *The Prelude, 1799, 1805, 1850: Authoritative Texts, Context, and Reception*. Edited by Jonathan Wordsworth, M. H. Abrams, and Stephen Gill. New York: Norton.
Worldwatch Institute. 2013. "The State of Consumption Today." *Worldwatch Institute*. http://www.worldwatch.org/node/810.
Wong, Fayen. 2013 "China's Smog Threatens Health of Global Coal Projects." *Reuters Shanghai*, November 14. http://www.reuters.com/article/us-china-coal-idUSBRE9AD19L20131114.
Wong, Kristine. 2015. "6 Horrible Oil Spills Since Deepwater Horizon that You Probably Didn't Hear About." *takepart*, April 15. http://www.takepart.com/article/2014/04/15/6-horrible-oil-spills-have-happened-deepwater-horizon-you-probably-didnt-hear.
Woo, C., Y. Chung, D. Chun, S. Han, and D. Lee. 2014. "Impact of Green Innovation on Labor Productivity and Its Determinants: An Analysis of the Korean Manufacturing Industry." *Business Strategy and the Environment* 23 (8): 567–76.
Woods Hole Oceanographic Institution. 2015. "North Pole Drifting Stations." Woods Hole Oceanographic Institution Beaufort Gyre Exploration Project. http://www.whoi.edu/page.do?pid=66677.
World Bank. 2009. *World Development Report*. http://go.worldbank.org/FAV9CBBG80.
———. 2016. "Energy Overview." *The World Bank*, April 5. http://www.worldbank.org/en/topic/energy/overview.
"Worldwide Views on Climate and Energy." 2015. http://climateandenergy.wwviews.org.
Wright, Lawrence. 2010. "Lithium Dreams: Can Bolivia Become the Saudi Arabia of the Electric-Car Era?" *New Yorker*, March 22. http://www.newyorker.com/magazine/2010/03/22/lithium-dreams.
Written on the Wind. 1956. Directed by Douglas Sirk. Universal Studios. DVD.
Wundt, Wilhelm. 1904. "Grundzüge." In *Principles of Physiological Psychology (1886)*. Vol. 1. Translated by Edward Titchener, 1–10. New York: Macmillan.
XTC. 1986. "Season Cycle." *Skylarking*. Virgin Records.
Yaeger, Patricia. 2011. "Literature in the Ages of Wood, Tallow, Coal, Whale-Oil, Gasoline, Atomic Power and Other Energy Sources." *PMLA* 126 (2): 305–10.
Yaeger, Patricia, Laurie Shannon, Vin Nardizzi, Ken Hiltner, Saree Makdisi, Michael Ziser, and Imre Szeman. 2011. "Editor's Column: Literature in the Ages of Wood, Tallow, Coal, Whale Oil, Gasoline, Atomic Power, and Other Energy Sources." *PMLA* 126 (2): 305–25.
Yardley, Jim, and Gardiner Harris. 2012. "2nd Day of Power Failures Cripples Wide Swath of India." *New York Times*, July 31. http://www.nytimes.com/2012/08/01/world/asia/power-outages-hit-600-million-in-india.html.
Yergin, Daniel. 1991. *The Prize: The Epic Quest for Oil, Money, and Power*. New York: Simon and Schuster.
———. 2011. *The Quest: Energy, Security, and the Remaking of the Modern World*. New York: Penguin Press.
Youngs, Robert L. 1982. "Every Age, the Age of Wood." *Interdisciplinary Science Reviews* 7 (3): 211–19.
Zalasiewicz, Jan, Mark Williams, Alan Smith, Tiffany L. Barry, Angela L. Coe, Paul R. Bown, Patrick Brenchley, David Cantrill, Andrew Gale, Philip Gibbard, F. John Gregory, Mark W. Hounslow, Andrew C. Kerr, Paul Pearson, Robert Knox, John Powell, Colin Waters, John Marshall, Michael Oates, Peter Rawson, and Philip Stone. 2008. "Are We Now Living in the Anthropocene?" *GSA Today* 18 (2): 4–8.

———. 2011. "Stratigraphy of the Anthropocene." *Philosophical Transactions of the Royal Society A* 369: 1036–55.
Zalasiewicz, Jan, Paul J. Crutzen, and Will Steffen. 2012. "The Anthropocene." In *The Geologic Time Scale*. Vol. 2. Edited by Felix M. Gradstein, James Ogg, Mark Schmitz, and Gabi Ogg, 1033–40. Amsterdam: Elsevier.
Zalik, Anna. 2010. "'Oil futures': Shell Scenarios and the Social Constitution of the Oil Market." *Geoforum* 41: 553–64.
Zebrowski, Christopher. 2009. "Governing the Network Society: A Biopolitical Critique of Resilience." *Political Perspectives* 3 (1): 1–38.
Žižek, Slavoj. 2009. *First as Tragedy, Then as Farce*. New York: Verso.
———. 2012. "How Did Marx Invent the Symptom?" In *Mapping Ideology*, edited by Slavoj Žižek, 296–331. New York: Verso.
———. 2013. "Trouble in Paradise." *London Review of Books*, July 18.
Zylinska, Joanna. 2009. *Bioethics in the Age of New Media*. Cambridge, MA: MIT Press.
———. 2014. *Minimal Ethics for the Anthropocene*. London: Open Humanities Press.

CONTRIBUTORS

Alan Ackerman is professor of English at the University of Toronto.

Philip Aghoghovwia is postdoctoral research fellow in the Department of English Language and Literature at the University of Cape Town.

Daniel Gustav Anderson is a graduate student in the Cultural Studies Department at George Mason University.

Claudia Aradau teaches in the Department of War Studies at King's College London.

Lynn Badia is a Banting Postdoctoral Fellow in the Department of English and Film Studies at the University of Alberta.

Georgiana Banita is assistant professor of literature and media studies at the University of Bamberg.

Daniel Barber is assistant professor of architecture at Penn Design.

Darin Barney teaches in the Department of Art History and Communication Studies at McGill University.

Crystal Bartolovich is associate professor of English at Syracuse University.

Bart Beaty teaches in the Department of English at the University of Calgary.

Brent Ryan Bellamy is postdoctoral fellow at Memorial University Newfoundland.

Franco Berardi teaches Social History of the Media at the Accademia di Brera, Milan.

Amanda Boetzkes is assistant professor in the School of Fine Arts and Music at the University of Guelph.

Dominic Boyer is professor of anthropology at Rice University.

Ian Buchanan is Director of the Institute for Social Transformation Research at the University of Wollongong.

Frederick Buell teaches in the Department of English at the City University of New York—Queens College.

D. Graham Burnett is professor of history at Princeton University.

Gerry Canavan is assistant professor of twentieth- and twenty-first-century literature at Marquette University.

Warren Cariou is associate professor of English at the University of Manitoba.

Dipesh Chakrabarty teaches in the Department of History at the University of Chicago.

Sharad Chari teaches in the Department of Anthropology at the University of Witwatersrand.

Juan Cole is Richard P. Mitchell Collegiate Professor of History at the University of Michigan.

Claire Colebrook teaches in the English Department at Pennsylvania State University.

Patricia Corcoran teaches in the Department of Earth Sciences at the University of Western Ontario.

Ashley Dawson teaches in the English Department at the City University of New York—College of Staten Island.

Jeff Diamanti is postdoctoral fellow in Media@McGill.

Adam Dickinson teaches in the Department of English Language and Literature at Brock University.

Arif Dirlik is retired professor of history and anthropology at the University of Oregon.

Kit Dobson teaches in the Department of English at Mount Royal University.

Sara Dorow is associate professor of sociology at the University of Alberta.

Todd Dufresne teaches philosophy at Lakehead University.

Danine Farquharson teaches in the Department of English Language and Literature at Memorial University Newfoundland.

Matthew Flisfeder is assistant professor in the Department of Rhetoric, Writing, and Communications at the University of Winnipeg.

Lisa Gitelman is professor of English and media, culture, and communication at New York University.

Louise Green teaches in the Department of English at Stellenbosch University in South Africa.

Lindsey Green-Simms teaches African and post-colonial film and literature at American University.

Richard Grusin is professor of English at the University of Wisconsin—Milwaukee.

Susie Hatmaker is lecturer in social and cultural anthropology at the University of California—Berkeley.

Gay Hawkins teaches in the Centre for Critical and Cultural Studies at the University of Queensland.

Melissa Haynes is a graduate student in the Department of English and Film Studies at the University of Alberta.

Gabrielle Hecht is professor of History at the University of Michigan.

Peter Hitchcock is professor of English at the City University of New York—Baruch College.

Werner Hofer teaches in the Faculty of Science, Agriculture, and Engineering of Newcastle University.

Mél Hogan is assistant professor of environmental media at the University of Calgary.

Cymene Howe teaches in the Department of Anthropology at Rice University.

Caren Irr teaches in the Department of English at Brandeis University.

Jennifer Jacquet is an American environmental studies professor at New York University.

Kelly Jazvac is an artist and teaches in the Department of Visual Arts at the University of Western Ontario.

Bob Johnson teaches in the College of Letters and Science at National University.

Antonia Juhasz is an American oil and energy analyst, author, and investigative journalist.

Timothy Kaposy teaches English and communications at Niagara College.

Ed Kashi is an American photojournalist and member of VII Photo based in the Greater New York area.

Donald V. Kingsbury teaches in the Political Science Department at the University of Toronto.

Alice Kuzniar teaches English language and literature and Germanic and Slavic studies at the University of Waterloo.

Philipp Lehmann is a graduate student in the History Department at Harvard University.

Stephanie LeMenager teaches in the Department of English at the University of Oregon.

Ernst Logar is an artist and has studied culture and organization at the University of Vienna.

Andrew Loman teaches in the Department of English Language and Literature at Memorial University Newfoundland.

Graeme Macdonald is associate professor of English and comparative literary studies at the University of Warwick.

Brenda K. Marshall is a novelist who teaches in the English Department at the University of Michigan.

Joseph Masco teaches in the Department of Anthropology at the University of Chicago.

Leerom Medovoi teaches in the Department of English at the University of Arizona.

Toby Miller is Professor Emeritus in the Department of Media and Cultural Studies at the University of California—Riverside.

Jason W. Moore teaches sociology at the State University of New York—Binghamton.

Spencer Morrison is postdoctoral fellow in the Department of English and Film Studies at the University of Alberta.

Erin Morton teaches in the History Department at the University of New Brunswick.

Timothy Morton teaches in the Department of English at Rice University.

Vin Nardizzi teaches in the Department of English at the University of British Columbia.

Michael Niblett teaches in the Yesu Persaud Centre for Caribbean Studies at the University of Warwick.

Rob Nixon is professor of humanities and the environment at Princeton University.

Susie O'Brien is associate professor of English and cultural studies at McMaster University.

LISA PARKS is professor of film and media studies at the University of California—Santa Barbara.

DONALD PEASE is professor of English and comparative literature at Dartmouth College.

ANDREW PENDAKIS is assistant professor of theory and rhetoric at Brock University.

ALEXEI PENZIN is reader in art at the University of Wolverhampton.

KAREN PINKUS is professor of Italian and comparative literature at Cornell University.

FIONA POLACK teaches in the Department of English Language and Literature at Memorial University Newfoundland.

PEDRO REYES is an artist living and working in Mexico City.

KIRSTY ROBERTSON teaches in the Department of Visual Arts at the University of Western Ontario.

RAFICO RUIZ holds a PhD in communication studies and the history and theory of architecture from McGill University.

ROBERT RYDER teaches in the Germanic Studies Department at the University of Illinois—Chicago.

DEENA RYMHS teaches in the Department of English at the University of British Columbia.

ANNA SAJECKI is a graduate student in the Department of English and Film Studies at the University of Alberta.

GORDON SAYRE teaches in the Department of English at the University of Oregon.

MATTHEW SCHNEIDER-MAYERSON is postdoctoral fellow at Rice University's Center for Energy and Environmental Research in the Human Sciences.

LAURIE SHANNON is associate professor of English at Northwestern University.

ELIZABETH SHOVE is professor in the Sociology Department at Lancaster University.

STEVPHEN SHUKAITIS is lecturer in work and organization at the University of Essex.

LISA SIDERIS is associate professor in the Religious Studies Department at Indiana University Bloomington.

MARK SIMPSON teaches in the Department of English and Film Studies at the University of Alberta.

JOHN SOLURI teaches in the Department of History at Carnegie Mellon University.

Vivasvan Soni teaches British literature and literary theory at Northwestern University.

Janet Stewart is professor in visual culture at Durham University.

Allan Stoekl is professor of French and comparative literature at Penn State University.

Imre Szeman is professor of communications and culture at the University of Waterloo.

Geo Takach teaches in the School of Communication and Culture at Royal Roads University.

Noah Toly is associate professor of politics and international relations at Wheaton College.

Michael Truscello is associate professor in the Department of English at Mount Royal University.

Susan Turcot is an artist and has studied visual art and philosophy at Middlesex University in London.

Priscilla Wald is professor of English and women's studies at Duke University.

Gordon Walker is a professor at the Lancaster Environment Centre at Lancaster University.

Michael Watts is professor of geography at the University of California—Berkeley.

Jennifer Wenzel teaches English and African studies at Columbia University.

Sheena Wilson is assistant professor at the University of Alberta—Campus Saint-Jean.

Daniel Worden teaches in the School of Individualized Study at the Rochester Institute of Technology.

Patricia Yaeger was Henry Simmons Frieze Collegiate Professor of English and Women's Studies at the University of Michigan. She passed away in 2014.

Amy Zhang is a graduate student in the Department of Anthropology at Yale University.

Joanna Zylinska is professor of new media and communications at Goldsmiths, University of London.